Routledge Handbook of Human-Animal Studies

Human-animal studies, an academic field that has grown exponentially over the past decade, explores the whys, hows, and whats of human-animal relations: *why* animals are represented and configured in different ways in human cultures and societies around the world; *how* they are imagined, experienced, and given significance; *what* these relationships might signify about being human; and *what about* these relationships might be improved for the sake of the individuals as well as the communities concerned.

The *Routledge Handbook of Human-Animal Studies* presents a collection of original essays from artists and scholars who have established themselves internationally on the basis of specific and significant new contributions that are shaping the field. Offering a broad interpretive account of the development and present configurations of human-animal studies across many cultures, continents, and times, this volume explores such subjects as:

- wildness and tameness of in field and laboratory science
- ancient, indigenous, and industrial practices of animal husbandry
- politics and aesthetics of representing feral life
- human and animal implications of mourning, grief, death, and extinction.

This international, interdisciplinary handbook will be of interest to students and scholars of human-animal studies, offering entry points through anthropology, archaeology, art history, biology, cultural studies, education, environmental studies, ethology, fine art, gender studies, geography, history, literature, media studies, museum studies, performance studies, philosophy, psychology, religious studies, sociology, and visual studies.

Garry Marvin is a social anthropologist and Professor of Human-Animal studies at the University of Roehampton, London.

Susan McHugh is a literary theorist and Professor and Chair of English at the University of New England, USA.

"The *Routledge Handbook of Human-Animal Studies* is both a great read and a provocation to rendering each other capable of knowing entangled human and animal worldings better. The diverse essays are unsettling and enticing. Feral protagonists in contemporary South African novels, mourning Hawaiian crows and shared grief in the face of extinction, commercial wild meat and green exterminism in New Zealand, Sami reindeer and their herders, ancient dogs, crowds of Swedish cats in nonconforming homes, Arabian babblers reshaping their scientists, laboratory and other rats escaping more than one kind of maze, and many more human and non-human critters of temporally deep and spatially differentiated worlds: All of these both confront the arrogance of human exceptionalism and nurture the perceptual and analytical skills to inhabit bumptious and challenging human-animal studies. This *Handbook* is an ambitious, rich webbing of essays in an emergent, multi-dimensional niche-space of contemporary trans-disciplinary studies. As the editors put it in their seductive introduction: critters and scholars alike, we are in it together. The *Handbook* layers the kind of compost that can remix the wastes and resources of posthumanism into something more nourishing for these times of excess death and still possible resilience." – *Donna Haraway, author of* When Species Meet

"Human-animal studies is only a few decades old (and so, born yesterday in academic terms), but in that brief span it has attracted scholars from across the humanities, the social sciences, and the arts. Garry Marvin and Susan McHugh have produced a collection that demonstrates the striking disciplinary reach and methodological variety of this innovative field." – *Harriet Ritvo, Arthur J. Conner Professor of History, MIT*

"A new and necessary survey of a rapidly evolving field, this marvellous collection succeeds in being inviting as well as authoritative; taking on the challenge of reconceptualising the wild, the domesticated and the feral, these exceptional essays amply demonstrate Marvin and McHugh's conviction that the question of how we live with animals is fundamental to how we live with ourselves." – *Philip Howell, Department of Geography, University of Cambridge*

"There is a crucial phrase in this book about animals and humans coinventing the practice of knowledge together that sums up the spirit of the new Human-Animal Studies, and runs through this book as an essential seam. This superb collection of essays balances detailed case studies, subtle reworkings of the big questions of human-animal relations, and an appreciation of the centrality of animals in life and culture even when they appear to be at their most marginal. From Ecuador to New Zealand, from Iceland to prehistoric Crete, from science to aesthetics, from theatre to agriculture, from woodpeckers to reindeer, from the humble squeaky dog toy to the mighty mammoth, from life to death, from the collaboration of Arabian Babblers in their own ornithology to butterfly aesthetics, the range of this book is constantly fascinating and never superficial. Above all the contributors to this collection demonstrate that Human Animal Studies is no longer a minor field, but a major discipline in its own right." – Jonathan Burt, Series Editor, *Animal* – Reaktion Books

Routledge Handbook of Human-Animal Studies

Edited by Garry Marvin and Susan McHugh

Routledge
Taylor & Francis Group
LONDON AND NEW YORK

First published 2014 by Routledge

2 Park Square, Milton Park, Abingdon, Oxfordshire OX14 4RN
52 Vanderbilt Avenue, New York, NY 10017

Routledge is an imprint of the Taylor & Francis Group, an informa business

First issued in paperback 2019

British Library Cataloguing in Publication Data
A catalogue record for this book is available from the British Library

Library of Congress Cataloging-in-Publication Data
Routledge handbook of human-animal studies / edited by Garry Marvin and Susan McHugh.—First edition.
pages cm.
Includes bibliographical references and index.
1. Human-animal relationships. 2. Animals—Social aspects. 3. Animals and civilization. 4. Animal welfare.
I. Marvin, Garry. II. McHugh, Susan.
QL85.R68 2014
590—dc23

2013034762

ISBN: 978-0-415-52140-6 (hbk)
ISBN: 978-0-367-86671-6 (pbk)

Typeset in Bembo
by RefineCatch Limited, Bungay, Suffolk

For our fathers
Ed McHugh (1934–2011) and Leonard John Marvin (1926–2012)
who left us with another turn of the kaleidoscope

Contents

List of illustrations

Notes on contributors

Giovanni Aloi is a Lecturer in History of Art and Visual Cultures, working for Queen Mary University of London, Sotheby's Institute of Art, Christie's Education, and Tate Galleries. His interest in animals and nature in general dates back to an early age spent exploring the colourful and diverse flora and fauna of the south of his native Italy. In 1995, he obtained his first degree in Fine Art Theory and Practice, and subsequently moved to London in order to further his education. He found fertile ground in the unconventional and radical approaches to art theory provided by Goldsmiths University, and proceeded to gain a postgraduate diploma in History of Art and a Master's degree in Visual Cultures; his PhD on the subject of Taxidermy in Contemporary Art is presently in its final stages. Aloi also worked as a gallery manager and education officer at Whitechapel Art Gallery while curating film programmes at Prince Charles Cinema, and he has been a commercial and fine art photographer as well as a freelance artist. He has an extensive publishing profile, counting essays in English, French, Italian, Romanian, and Polish, in addition to the book *Art & Animals* (IB Tauris, 2011). He is founder and editor in chief of *Antennae: The Journal of Nature in Visual Culture.*

Associate Professor **Philip Armstrong** is Co-Director of the New Zealand Centre for Human-Animal Studies at the University of Canterbury. He is the author of *What Animals Mean in the Fiction of Modernity* (Routledge, 2008) and the co-editor, with Laurence Simmons, of *Knowing Animals* (Brill, 2007). His next book is *Animals in New Zealand Art, Literature and Everyday Life*, co-authored with Annie Potts and Deidre Brown, and forthcoming from Auckland University Press.

Steve Baker is Emeritus Professor of Art History at the University of Central Lancashire. His books include *Artist | Animal* (Minnesota, 2013), *The Postmodern Animal* (Reaktion, 2000), *Picturing the Beast* (Illinois, 2001), and, with the Animal Studies Group, *Killing Animals* (Illinois, 2006). Chapters from his books have been translated into French, German, Dutch, Swedish, and Italian. A central concern of his recent writing has been to articulate the distinctive contribution of contemporary artists to the understanding of the more-than-human world. Examples of his own recent roadkill imagery have been included in group exhibitions in London, New Orleans, and Melbourne, and have been reproduced and discussed in the journals *Art & Research, Antennae, Tierstudien*, and *Angelaki*.

Lynda Birke is a feminist biologist and Visiting Professor of Biological Sciences at the University of Chester. Her doctorate is in ethology and she still does some research in animal behaviour and welfare, but she is best known for her work in feminist science studies. In particular, she has focused on a range of issues to do with the human/animal relationship, but especially the controversy surrounding animal experimentation; she is currently doing research on horse welfare and cultures of humans/horses. She has published widely in all these fields, and her books include *Feminism, Animals, and Science: The Naming of the Shrew* (Open University Press, 1994), *Feminism and the Biological Body* (Edinburgh, 1999), and *The Sacrifice: How Scientific Experiments Transform Animals and People* (Purdue, 2007). She is an associate editor of *Society & Animals* and *Humanimalia.* When not writing about animals she does, of course, spend a great deal of time with them; among other things she loves to jump.

Henry Buller is Professor of Geography at the University of Exeter. His principal research interests are animal geographies, animal welfare, human/animal relations, geographies of nature and environmental politics. He is editor of the international rural social science journal *Sociologia Ruralis* and is an appointed member of the UK's Farm Animal Welfare Committee. He has written a number of journal articles and book chapters on animal geographies and is currently preparing a co-authored book on farm animals for Bloomsbury Press.

Una Chaudhuri is Collegiate Professor and Professor of English, Drama, and Environmental Studies at New York University. After completing undergraduate and master's level training at Delhi University, India, Chaudhuri did her doctoral work at Columbia University. She has lectured internationally and published extensively on modern drama, performance theory, and eco-criticism. She is the author of *No Man's Stage: A Semiotic Study of Jean Genet's Plays* (UMI, 1986) and the award-winning *Staging Place: The Geography of Modern Drama* (Michigan, 1997); editor of *Rachel's Brain and Other Storms: The Performance Scripts of Rachel Rosenthal* (Continuum, 2001); and co-editor, with Elinor Fuchs, of *Land/Scape/Theater* (Michigan, 2002). She was guest editor of a special issue of *Yale Theater Journal* on 'Ecology and Performance', and a special issue of *TDR: The Journal of Performance Studies* on 'Animals and Performance'. She chairs the panel of judges for the Callaway Prize for the Best Book on Drama or Theatre, and she has been a judge of the Obie and the Alpert Awards. She is a voting member of the American Theatre Wing, which awards Broadway's Tony Awards. Her current research and publications explore 'zooësis', the discourse and representation of species in contemporary culture and performance, which is also the subject of her forthcoming book, *Animal Acts: Performing Species Today*, co-edited with Holly Hughes.

Vinciane Despret is Maître de Conferences at the University of Liège and at the Free University of Brussels. She is a philosopher of science. Her first fieldwork was in the Negev desert in Israel, where she explored the possibility of innovating an 'ethology of ethologists'. Since then, her research has focused on animals with humans who observe them, live with them, or simply know them. In addition to publishing several books, she was the scientific curator of the exhibition Bêtes et Hommes held in 2007–08 at the Grande Halle de la Villette, Parc de La Villette, Paris. An English translation of her most recent book, *Que diraient les animaux si . . . on leur posait de bonnes questions?* (La Découverte, 2012) will be published in 2014.

Thom van Dooren lectures in Environmental Humanities at the University of New South Wales, Australia. His research focuses on ethical, philosophical, and political issues arising out of people's changing relationships with species of plants and animals at the edge of extinction.

His chapter has been adapted from his forthcoming book *Flight Ways: Life and Loss at the Edge of Extinction* (Columbia, 2014).

Leesa Fawcett lives beside the Nottawasaga River, along the spine of the Niagara Escarpment in the Hockley Valley, with a varied assortment of rescued, wild, companion, and feral creatures – sometimes, surprisingly, in the house together, but more often outside in woods, gardens, waterways, on trails, or visiting friends. When she is not at home, or in the untamed city of Toronto, playing or travelling about, she works. Currently, she is Associate Dean, Associate Professor, and the Coordinator of the Graduate Diploma in Environmental and Sustainability Education in the Faculty of Environmental Studies at York University in Toronto, Canada.

Adrian Franklin trained as an anthropologist in the UK, and has held professorial positions at the University of Bristol and the University of Oslo. He is currently Professor of Sociology at the University of Tasmania. His interests in human-animal relations and posthumanism were prompted in the 1990s by the realization that animals had been inexplicably left out of accounts of modern societies and cultures and yet were never more relevant. His books on human-animal studies include *City Life* (Sage, 2010), *Animal Nation: The True Story of Animals and Australia* (University of New South Wales, 2007), *Nature and Social Theory* (Sage, 2003), and *Animals and Modern Cultures* (Sage, 1999).

Katherine C. (Kasey) Grier is Professor in the Department of History and Director of the Museum Studies Program at the University of Delaware. She is the author of *Pets in America: A History* (North Carolina, 2006). A specialist in the history of domestic life in the nineteenth and twentieth centuries, her published work features innovative uses of material culture as a primary source for historical research. Her current research includes a study of animal 'fancies' and animal bodies as a form of material culture, and a book-length project on the changing roles of animals in one American city from the eighteenth century to the present. Kasey Grier received the Lifetime Achievement Award from the International Society of Anthrozoölogy in 2007.

Yannis Hamilakis is Professor of Archaeology at the University of Southampton. His main research interests are the archaeology of the bodily senses, social zooarchaeology, the politics of the past, memory and multi-temporality, and Aegean archaeology. He coordinates the Laboratory for Social Zooarchaeology at the University of Southampton. Among his many publications are *The Nation and Its Ruins: Antiquity, Archaeology, and National Imagination in Greece* (Oxford, 2007), and the co-edited collections *Thinking through the Body: Archaeologies of Corporeality* (Springer, 2001), and *Archaeological Ethnographies* (Maney, 2009). His *Archaeology and the Senses* is forthcoming from Cambridge University Press.

Kerry Harris is a PhD student at the University of Southampton. Her thesis examines the social role of hunting and wild animals in Late Bronze Age Crete. She has been the co-zooarchaeologist on the Ayios Konstantinos Mycenanean Sanctuary site (Greece), the Late Bronze Nopeigia feasting assemblage (Crete), the Theopetra Neolithic cave site (Greece), and the Koutroulou Magoula Neoltihic tell site (Greece). She also worked on the pilot study for the *Dama International* fallow deer project.

Elina Helander-Renvall is a Sámi scholar who works as senior scientist at the Arctic Centre, University of Lapland in Rovaniemi, Finland. She is Director of the Arctic Indigenous

Peoples and Sámi Research Office at the same institute. In addition to research on Sámi culture, subsistence activities, and customary law, she has investigated traditional ecological knowledge and biological diversity among the indigenous people of the Arctic. Moreover, she is interested in traditional stories and art. Before coming to the Arctic Center, she worked as Director for the Nordic Sámi Institute in Kautokeino, Norway.

Tora Holmberg is a sociologist and works at the Institute for Housing and Urban Research at Uppsala University in Sweden. She works in the intersection of science and technology studies, feminist theory, cultural sociology, and urban studies. Her main research interests concern animal experimentation and bio-technology and, more recently, controversies over urban animals. She is also a co-editor of *Humanimalia: A Journal of Human-Animal Interface Studies*. www.ibf.uu.se/PERSON/ tora/tora.html.

Garry Marvin is a social anthropologist at the University of Roehampton, London. In 2010 he was appointed as the first-ever Professor of Human-Animal Studies. He has conducted anthropological ethnographic fieldwork in the worlds of bullfighting, cockfighting, zoos, recreational hunting, and English foxhunting. In addition to numerous journal articles and book chapters, he has published *Bullfight* (Illinois, 1994), *Zoo Culture* (Illinois, 1999), and *Wolf* (Reaktion, 2012). He serves on the Board of Editors for *Society & Animals* and the Editorial Advisory Board for *Anthrozoös*. He is a member of the Board of Trustees of the *Powell-Cotton Museum of Natural History and Ethnography* in Kent, UK.

Susan McHugh is Professor and Chair of English at the University of New England, USA. All of her research and some of her teaching focus on literary, visual, and scientific stories of cross-species life. McHugh is the author of *Animal Stories: Narrating across Species Lines* (Minnesota, 2011), as well as *Dog* (Reaktion, 2004), along with dozens of essays and reviews. She serves as Managing Editor of the Humanities for *Society & Animals*, and she is a board member of the Animals & Society Institute, *Animal Studies Journal, Antennae, H-Animal Discussion Network, Humanimalia*, and *Environment and History*. Also a cover model for this volume, McHugh loves life on the edge of the Great North Woods in Maine with canine companion Sabine and artist **Mik Morrisey**.

Sophia Menache is a professor of medieval history at the University of Haifa. Her field of research covers different topics such as communication, propaganda, humour, the history of the Catholic Church and the papacy, and the Crusades. Her many publications – 14 books and nearly 120 articles – include *The Vox Dei: Communication in the Middle Ages* (Oxford, 1990) and *Clement V* (Cambridge, 1998). Her research on attitudes to dogs brought about the publication of many articles on the influence of hunting in pet keeping, violence toward animals in medieval society, and the approach of the monotheistic religions to pet keeping.

Associate Professor **Annie Potts** is the Co-Director of the New Zealand Centre for Human-Animal Studies at the University of Canterbury. Her most recent books are *Chicken* (Reaktion, 2011) and *Animals in New Zealand Art, Literature and Everyday Life*, co-authored with Philip Armstrong and Deidre Brown, and forthcoming from Auckland University Press.

Istvan Praet is Senior Lecturer in Social Anthropology at the University of Roehampton. He is the author of *Animism and the Question of Life* (Routledge, 2013).

Bernard E. Rollin is University Distinguished Professor of Philosophy, Biomedical Sciences, and Animal Science at Colorado State University. Rollin's scholarly interests include both traditional philosophy and applied philosophy, and he specializes in animal ethics, animal pain, and animal consciousness. He is the author of 17 books and over 500 articles, and has done over 1500 invited lectures in 28 countries. He was a principal architect of the 1985 US federal law protecting laboratory animals. Rollin is a weightlifter and motorcyclist.

Nigel Rothfels is a historian whose work focuses on changing ideas about animals. He is the author of a study of the origins of naturalistic displays in zoological gardens, *Savages and Beasts: The Birth of the Modern Zoo* (Johns Hopkins, 2002) and editor of the influential collection of essays, *Representing Animals* (Indiana, 2002). He is also co-editor, with Garry Marvin, of the book series *Animalibus: Of Animals and Cultures*, with Pennsylvania State University Press. He has held postdoctoral fellowships from Princeton University, the Australian National University, the National Endowment for the Humanities, and the University of Oslo. His current research focuses on western ideas about elephants since the eighteenth century.

Kathy Rudy received her MDiv and PhD in Ethics from Duke University's Religion Department. Her early research included reproductive ethics, religious ethics, sexuality, and feminist theory. Among her early publications are *Sex and the Church: Gender, Sexuality, and the Transformation of Christian Ethics* (Beacon, 1997) and *Beyond Pro-Life and Pro-Choice: Moral Diversity in the Abortion Debate* (Beacon, 1996). Her new work focuses on animals and ethics, nature, and food politics. She is the author of *Loving Animals: Toward a New Animal Advocacy* (Minnesota, 2011), as well as many scholarly essays in animal studies. Her new project focuses on disability studies and the work of guide, service, assistance, and therapy dogs to compensate for human disabilities. Her new book is tentatively titled *Sweetness and Devotion*, and focuses particularly on training medical alert dogs. Rudy received the Trinity College Distinguished Teaching Award in 1995 and the Alumni Distinguished Teaching Award in 2000.

Snæbjörnsdóttir/Wilson conduct their collaborative practice from bases in the north of England, Iceland, and Sweden. With a strong research grounding, their socially engaged projects explore contemporary relationships between human and nonhuman animals in the contexts of history, culture, and the environment. Their practice sets out to challenge anthropocentric systems and the thinking that sanctions loss through representations of 'the other', proposing instead alternative tropes, including 'parities in meeting'. Books based on their previous projects include *Uncertainty in the City* (Green Box, 2011), *nanoq: flat out and bluesome* (Black Dog, 2006), *(a)fly* (National Museum of Iceland, 2006), and *Big Mouth* (Tramway, 2004). Bryndís Snæbjörnsdóttir is currently Director of v.arc – Valand Artistic Research Centre – at the University of Gothenburg, Sweden. Mark Wilson is Reader in Fine Art at the University of Cumbria, UK. www.snaebjornsdottirwilson.com.

Wendy Woodward teaches southern African Literature, Animal Studies, and Creative Writing in the English Department at the University of the Western Cape, South Africa. *The Animal Gaze: Animal Subjectivities in Southern African Narratives* (Wits, 2008) was awarded the Deputy Vice Chancellor's Book Award for 2006–08. Her second volume of poetry, *Love, Hades and other Animals* (Protea, 2008) deals with intersections between family histories and colonialism, as well as nonhuman animals in Greek myth and the quotidian. In 2011, Woodward was instrumental in organizing the colloquium, Figuring the Animal in Post-Apartheid South Africa, where a decision was taken to form the Animal Studies Round Table in Africa (ASRA). She is a Research Associate of the New Zealand Centre for Human-Animal Studies and is on the academic advisory board for Minding Animals International. Her current research focuses on the urban animal photographs of Fanie Jason, and on embodiment and animality.

In it together

An introduction to human-animal studies

Garry Marvin and Susan McHugh

Who or what is human? Animal? And when these terms are coupled with a hyphen, how does that shift the grounds of studying? In human-animal studies the research and intellectual focus is on how animals figure and are configured in human worlds, but these worlds are formed through the relationships that humans share with animals. Why might studying this human-animal world be worth academic attention? Simply put, it is because animals – although not all, and not all equally – are essential in and for human societies. Human worlds are built upon animal lives and deaths, conceptually as well as physically. It is difficult to imagine how we could mark ourselves out as human without other animals, for we have become human alongside other animals. But this is actually more 'among' than 'alongside' – with its sense of humans and animals living parallel but separate lives – for, from the beginning, these were lives that have always been and remain profoundly intertwined.

Early humans, and some still today, have scavenged and hunted wild animals for their meat, skins, and other body parts. With the development of relationships of domestication, animals were not simply there to be killed for food. Borne by domesticated animals, humans could journey more widely, bringing all sorts of baggage. With animals to ride, human conflicts became more extensive and of a different order. When tended for meat, milk, fleece, and honey, animals' uses multiplied, and cross-species intimacies became more complex. Animal-drawn ploughs could dig deeper and turn over the earth more efficiently than a human with a digging stick or hoe, and from crops grown in fields fertilized with animal manure new forms of settlement became possible. Domesticated animals became animals owned by humans; as forms of property and wealth, they entered markets and other systems of exchange.

Today animals are husbanded in traditional ways in many parts of the world, while in others they are raised industrially, used as laboratory test subjects, and otherwise rendered integral to global exchange systems. All of these are material uses of animals, but animals figure in human cultures in still more ways that are less tangible but no less profound. In religious and other cosmological systems, animals are revered and sometimes feared, worshipped, and sacrificed. Animals are to be found in virtually all traditions of art and literature, including sculpture, painting, drawing, folktales, poetry, music, and philosophy, where they are used to express complex ideas about being human and being animal, and the relationships negotiated around these conditions.

Animals are enlisted in sports, perform in circuses, and are exhibited in museums and zoos. Some animals are brought into our homes and are incorporated into families as pets, while others enter on their own terms and are often treated as invaders. They 'people' our virtual worlds, whether in films, television programmes, social media, or gaming, and the lure of seeing them brings (sub)urbanites out of comfort zones to watch wildlife in situ, whether as eco-tourists, trophy hunters, or field (sometimes citizen) scientists. Animals are present, even when rendered invisible. Animal-based products have become so essential to the lubrication of heavy machinery and other industrial processes upon which so many of these practices are predicated that it is arguably impossible to pursue an animal-free human existence in the era of globalization.

Given the ubiquity of animals in human worlds, it should not be surprising that scholars are interested in the whys, hows, and whats of human-animal relations: *why* animals are represented and configured in different ways in human cultures and societies around the world; *how* they are imagined, experienced, and given significance; *what* these relationships might signify about being human; and *what about* these relationships might be improved for the sake of the individuals as well as the communities concerned. Just getting a handle on what is going on in human-animal encounters requires enormous creative, intellectual, and material commitments. Furthermore, it requires paying close attention to the lived experiences as well as the imagined nature of these relationships. The scale and difficulty of coordinating such work begins to explain why it has only been in the latter part of the twentieth century that researchers, especially in the humanities and social sciences, have brought studies of human-animal relationships from the margins of scholarly interest toward the centre of academic and intellectual attention.

With this volume, we advance an inclusive model of this sort of work under the term human-animal studies, in order to designate a field that has emerged within roughly the past three decades and now appears to be coming of age. Amid the many terms for what we do – including animal studies, anthrozoology, critical animal studies, etc. – the hyphen in 'human-animal studies', signally a linking, 'together in one', is the thing that has come to mark a precise starting point, which is to study animals with humans, and humans with animals, never forgetting that we are both animals in general, and humans in particular. From some perspectives, the reverse is not true: we are both, but they are not human; we are animals, but animals are certainly not us. For scholars in human-animal studies, it is important to remember that we cannot talk, write, or even think about animals in any sense except in the context of humans, if only because we can never get away from ourselves. We must navigate between the Scylla and Charybdis of the fantasy that we can understand animals completely and the presumption that we alone lay claim to objectivity.

To use another figure, the elements of human-animal relationships and how they can be represented or related could be likened to the pieces in a kaleidoscope where different patterns are created; depending on how you turn it, a different configuration and a different coherence is formed. The viewer sees or makes the coherence; it's not just that it's not self-evident, but rather that the task of human-animal studies is to engage with multiple convergences and conflicts of perspectives, some of which must remain ever unrecoverable for the humans involved. Animals have been unquestionably worthy of academic study in sciences such as biology, ecology, ethology, veterinary medicine, and zoology, but thinking about animals in relation to human lives seems to be what is new about human-animal studies. Further, it is thinking about animals as inseparable from human lives that constitutes a decisive break from 'animal studies' as originally defined by the natural sciences. More to the point, our collective project remains informed by mainstream animal science, but our perspective distinctly foregrounds how all animals live in human worlds that must be understood in order to know anything about animals, whether operating in ordinary or extraordinary habitats. From this vantage point, human-animal studies approaches animal worlds – along the lines of what biologist Jakob von Uexküll famously theorized as 'umwelten', or perceptual worlds, meaning the multiple, subjective, spatio-temporal, lived realities that are inhabited by different organisms even in shared environments, and that complicate any claims to pure objectivity – as vital, if unwieldy, components of our knowledge systems.

The essays gathered here demonstrate how the growth of academic and artistic interests in life between species has invigorated established disciplinary practices with the excitement of new combinations, collaborations, methods, and more. Together our contributors affirm that it is through research and creative practice that deeply and comprehensively integrate the humanities, social sciences, and natural sciences that human-animal studies becomes uniquely capable of opening perspectives on the production and authorization of knowledge itself as a more-than-human activity. Thus, this book heralds the arrival of human-animal studies as not only a multi- or trans-disciplinary but more importantly a meta-disciplinary field: a discursive environment conducive to developing positions and forms within existing academic discourses, as well as metamorphosing the status of each and all through crossings and other mutations that are peculiar to interspecies life.

The recent burst of work in human-animal studies makes this an ideal moment to take stock of the field. New explorers often arrive thinking they are entering unknown or unmapped territory. We want to make them aware that prior explorers have charted aspects for their own purposes. And we want to shed critical light on the fact that these maps all too often only circulate within territories defined by previously established disciplines. Taking a broad view, like Darwin we see evolving orders in this 'tangled bank', and through this book seek to reveal the connections that are making the field flourish.

As we elaborate below, our rudimentary structure remains purposefully simple and challenging because it arises from the range of human-animal relationships. We believe that this range can be captured through the interrelated and contested concepts of wild, domesticated, and feral, which suggest a crude, yet useful, set of coordinates. 'Wild' sets the human in the territory of the unfettered animal. 'Domesticated' marks the animal as brought into human-controlled territory. 'Feral' locates spaces between the two, and at the same time exposes the porousness of these demarcations. Involving both separations and connections, this tripartite structure configures the varied landscapes within which humans and animals encounter one another. All three are terms that define the nature of those relationships and how those relations are sensed, embodied, experienced, and communicated. In addition to being coordinates of the most provocative human-animal studies research to date, we regard them as provocations to further thought and research.

Histories

Since the 1990s, human-animal studies has begun to take shape through the efforts of people working in, for example, anthropology, archaeology, art history, biology, cultural studies, education, environmental studies, ethology, fine art, gender studies, geography, history, literature, media studies, museum studies, performance studies, philosophy, psychology, religious studies, sociology, and visual studies. Where they began is not as significant as that they chose to place animals at the centre of their research, and – this is important – as a consequence found it necessary to learn about and respond to work in other disciplines. From these beginnings, momentum keeps gathering. Every year there are new national and international conferences dedicated to the field. Academic journals in nearly all fields now regularly include articles about animals, and several specialist human-animal studies journals are now well established, with still more emerging. There are so many new books each year that it is no longer conceivable for any one person to keep up with the reading, a richness growing more abundant with several book series dedicated to human-animal studies scholarship now emerging from major international presses.

Amid this commotion, any survey of human-animal studies must confront a primary challenge, namely, becoming mired in competing and potentially irreconcilable accounts of the source of its excitements. Working together as editors of this volume has heightened our awareness of being in the curious position of experts speaking for a field in which neither of us could have trained. After all, the work of human-animal studies has involved reaching beyond the comfort zones of given knowledge structures. We realize that our individual perceptions reflect conversations we have had, and questions we have asked each other, across our respective disciplines over the past fifteen years. Consequently, our history of human-animal studies must remain at some level a snapshot of those discussions or, rather, a cyclorama or moving picture. In sketching it here, our aim is to present these scenes as unfolding maps for orientation to what we see as a new terrain staked out by this collection.

But first: where exactly are we coming from? When we started our academic careers, there were precious few conferences and certainly no specialist journals that would admit discussions of human-animal relations, let alone the degree programmes, research centres, and fellowships that are now encouraging them. Looking back to the 1980s, a group of individual, perhaps isolated texts – such as Mary Midgley's *Animals and Why They Matter* (1983), Keith Thomas's *Man and the Natural World* (1983), Margot Norris's *Beasts of the Modern Imagination* (1985), Vicki Hearne's *Adam's Task* (1986), Harriet Ritvo's *The Animal Estate* (1987), and Donna Haraway's *Primate Visions* (1989), to name a few personal favorites – can be identified as having inspired and allowed other scholars to get going. Yet giving credit where it is due in this case involves not simply drawing an intellectual genealogy, but more importantly identifying adventurers who laboured in hostile climates to break terrain to create the fertile fields of human-animal studies today. The intellectual currents that gave rise to a flow across disciplines toward this meta-disciplinary condition – at least, as we witnessed it in our respective US/UK contexts – are worth noting. Our different answers perhaps reflect our distinct disciplinary training in literature and social anthropology, as much as they mark the perniciousness of deep-seated divisions of knowledges.

Social anthropology presents an intriguing case. Although it is fundamentally concerned with exploring human cultures and societies, animals figured importantly from the very beginnings of the discipline because nonhuman organisms were highly significant creatures in many of the societies studied by anthropologists. From the start, anthropologists investigated and wrote about the place of animals from cosmologies to daily life. For example, totemism, the ways in which humans and animals are closely related in some societies, was something that immediately fascinated early anthropologists. How humans lived with and from animals through hunting, fishing, herding, pastoralism, and agriculture remains central to much anthropological work. So, rather late in his career, when Garry Marvin discovered the early work of human-animal studies, his initial thought was, 'Many generations of anthropologists have been doing this for a long time. What is new here?'

In 1977, Marvin began doctoral research on the world of the bullfight in Andalusia, southern Spain. He chose the topic because he was interested in Spanish culture, and the bullfight was a cultural event that had not been explored by anthropologists in any depth. His ethnographic challenge was to try to understand why the performing with and killing of bulls in public was such an important event in that culture and to learn directly from those involved in the events despite their understandable concerns about the interests and agendas of outsiders. For Marvin, this perspective made it exciting to discover subsequently scholars who thought through human-animal relations in terms of very different concerns. What drew them to animals? Were they simply in search of new research subjects? Or was this new interest in animals generated because of, in response to, a more significant shift in terms of who counts as an actor in the production of knowledges?

As a literary scholar trained in narrative theory, Susan McHugh cannot help but note how Marvin's story echoes those of Hearne, Haraway, and others in implicitly rejecting an obvious answer, that is, the notion that human-animal studies rode in on the coat-tails of the animal rights movement. For activists, Peter Singer's *Animal Liberation* (1975) pinpoints a watershed moment precisely because it extends a tradition of analytical philosophy that remains deeply invested in a notion of rights in turn grounded in individual (read: human) subjects. In contrast, for many academics, much of the research responsible for changing institutional and intellectual climates today pursues the wildly different potentials opened up by poststructuralist theory, which contemporaneously with Singer was gathering force to posit a farther reaching liberation of all from the foundational, hierarchical units of humanist thought.

Posthuman, nonhuman, counter-linguistic – call it what you like – a major significance of this broader intellectual turn is that it enables discussions of animals as agents who are not just humanlike subjects or thinglike objects, but actors of a different order, who can appear at breathtaking moments to be performing with people in ways that never add up to a simple sum of human and animal parts. Charting this radical intellectual work, theorists in this vein uncover alternate traditions – whether in philosophy or biology, crossing the work of von Uexküll with Henri Bergson, Alfred North Whitehead, and others – to claim legitimacy for human-animal studies research on these new terms. Quite apart from the question of how it might serve other interests, human-animal studies in this sense asserts revolutionary potential as an academic project: prioritizing the study of how people and animals actually do live together enables critical reflections that are as important to honouring our shared past as to continuing to share a future. At least, that is the view of McHugh, who entered in 1997 via doctoral research in literary and cultural studies, where and when very few people were venturing to talk about animals in such ways.

Nowadays we tend to recognize animals as having their own stories, if not always as having history, in the broadest sense. Less well understood is that the evolutionary theories and liberation movements ordinarily credited with leveraging such momentous changes would not have been possible without aesthetic developments driven by literary, visual, performative, and other creative engagements that increasingly require any sense of control to become less a matter of executing orders than of choreographing a dance, engaging others in different spaces, settings, and stages, and ultimately becoming something more than solitary selves. The focus on embodiment, surfaces, and exteriority in studies of visual representations of live animals has been arguably most influential in distinguishing animals as agents of a different order from human subjectivity. From this work it follows that we need not only to discuss but more profoundly to envision animals as actors operating in accordance with different narrative logics from that of human intentionality or psychological interiority, the characteristics that are stereotypically attributed to (and now appear to be folding back on) word-based texts. The work of many early human-animal studies practitioners may have been initially outstanding in this sense, but participation in similar, complementary processes unites all of the contributions to this volume.

Another central tenet of the emerging field of human-animal studies is that understanding the representational nature of animal agency means that it can never simply be seen as opposed to human identity, but instead reinforces the notion that animal agents are recognizable as such only when they are presented as deeply integrated with human forms or presences, and vice versa. To account for these nonhuman agency forms, we have to change our ways of doing academic work and, at the same time, work to shift notions of what constitutes creative practice as well.

Minimally, human-animal studies both requires and prompts new ways of thinking, along with heightening awareness of where and why other multidisciplinary areas have run aground. Reflecting this goal, we have assembled here the works of academic and creative people who challenge audiences to join them in developing new languages and forms that respond to the demands of lives shared across species lines, not the least of which is recognizing that the human can never be the sole agent or monopolizer of control in power relations.

Structures

This volume is designed to offer a broad interpretive account of the development and present configurations of the field of human-animal studies across (although by no means exhaustively representing) many cultures, continents, and times. The chapters are not overviews or summary reports on the state of human-animal studies from different disciplinary perspectives. Rather, they are *essays* by scholars who have responded to our invitation to participate with their own specific take on human-animal issues. These essays have been ordered to highlight questions and discussions emerging across the volume, reflecting continuities across historical and emergent developments of the field. While at one level individual essays might identify particular disciplinary concerns and issues, our hope is that reading them together brings readers to a whole new level where human-animal studies reveals itself as a creature evolved to fill an interdisciplinary niche, in which the terms wild, domestic, and feral interact with each other. Here these concepts also subtly guide the arrangement of the contents.

We see wild as focused on spaces and animals not thought of as under the direct control of humans. But we also believe that the notion of wild (and the same applies to domestic and feral) should not be regarded as a fixed entity, space, or condition. Although we imagine others to be born free forever, they (like us) remain constructed, fantasied, experienced, and interacted with conditionally, in highly contingent ways in different cultures and at different times. Our essayists explore the wild as a culturally constructed space, not so much as a given but instead as a purposefully imagined space akin to (and intersecting with) those of the domestic and feral through which humans represent for themselves the animals and people encountering each other in such spaces and the relationships that (as often as not) fail to follow.

Ideas and practices of domestication can only be understood in terms of ideas about the wild. Again, we are dealing with many histories, spaces, and conditions. The domestic suggests a process and condition of human control, of shaping, of bringing animals towards humans, or allowing them to draw near, and into a close coexistence. In addition, its usage to distinguish among human spaces (that is, the domestic in contrast to the public or commercial) indicates how animals enter into human politics and intimacies, even as the defining practices like husbandry, food production, even sterilization mark clear limits as to how we consider our species' mutual domestications of each other. Therefore, the domestic can never be simply opposed to the wild.

Feral refers to animals that are interstitial, conceived as in-between the wild and the domestic, animals that have escaped from domestication and recaptured some of the qualities of the wild. It also encompasses animals who are conceptually on the move, travelling between and never completely or comfortably fitting into categories. Feral animals are often problematic or troublesome for the people who encounter them because they inhabit or are in flight from human spaces in an uncontrolled manner, often with direct consequences for the people associated with them. Whether human, animal, or someone in between, feral creatures draw us into the spaces between our definitions of species life, that is, into the vital practices of embodying while at the same time sharing spaces that unite all life. Desire and the social are never quite so clearly in negotiation as on the feral frontlines.

Wild, domestic, and feral therefore only work in relation to each other, never as stand-alone concepts, nor as acultural, ahistorical, or otherwise essential conditions. In respect of this point, we have not strictly structured the placing of individual essays in the book according to this tripartite orientation. Our aim is to emphasize a view of the nature of human-animal relations at the blurred edges and overlaps of these zones as we are enmeshed in relations at the heart of each. Likewise, the essays do not represent the full complement of disciplines represented in human-animal studies as it is being practised around the world today. We chose our contributors to represent the most vibrant areas of research. But we take no cover in coverage.

Included here are artists and scholars who have established themselves internationally on the basis of specific and significant new contributions to human-animal studies. Rather than being assigned to step back and to review or survey some aspect of human animal studies, all were invited to offer pieces that show readers what they are engaged with now. What we were seeking were essays of human-animal studies, not about it. While deeply rooted in particular disciplines, the scholarship gathered here does not make the case for enhancing disciplinary control so much as for becoming open and responsive to different practices of and perspectives on human-animal relations.

In sum, we asked our contributors to think about these three concepts, to write with these terms in mind; however, we have arranged their work in ways which emphasize that wild, domestic, and feral only work in relation to each other. When thinking about animals, humans need wild as a condition to indicate what it is not. The domestic needs the wild to be outside in order to be at home with itself. The feral attempts returns and departures, always moving near one or the other, but never establishing either one as a starting point or destination. However different contributors use them as categories, they are presented overall as conditions in flux. Returning to our image of the kaleidoscope, we as editors have taken the glass beads of our contributors' essays and given them our particular turn. Readers are encouraged to make their own turns and find their own patterns.

Next steps

Thinking seriously about the implications of our questions presents real difficulties in how we study animals or, rather, how we in the humanities and social sciences study animals in significant ways. Even between ourselves, the use of the term 'animal studies' for our field, or any term that does not have 'human' closely attached, will remain controversial. For Marvin, it risks all too conveniently missing this point: understandings of animals only ever emerge within the context of the cultural events or perspectives that draw our attention to them. While respecting this point, McHugh is drawn to the productive confusion engendered by reappropriating 'animal studies' from science, which thereby becomes all the more clearly unsettled in its claim to the totality of knowledge: what potentials are unleashed when we begin to study animals as studying us? Where we do agree, however, is in respecting knowledges based on interactions.

Only by engaging with the people who have committed time to working with bulls, butterflies, octopi, opossums, snakes, sticklebacks, or (as is so often the case with extinct creatures) the material traces that they leave behind can we begin to grasp what it means to understand, experience, imagine, and represent those animals. Relationships between humans and animals are shared within but never strictly confined by our parameters; in other words, we represent traces of lives (our own as much as any others) in specific locations with histories of our making. Even at the limits of our relatings and representations, animals make sense to us in human proximities, as actors on stages performing with us. While a common assumption about the natural sciences is that they provide access to 'real' animals – they get out there and study actual animals – these ways of knowing verge on the destructive when they fail to account for their own entanglements in fragile social and textual webs.

Any historical perspective on animals remains filtered through representations and remains, and by extension all forms of representation insert a human filter that effectively creates distance from the very same human-animal relations that we purport to represent, let alone study. That said, the same charge can be levelled against any of the thinking through which research and creative activity gets going, let alone documented. From this perspective, the question becomes: Do natural sciences or any particular disciplines offer ways of transcending this condition? Or is it more useful to take compromise – rather, coalition – as our starting point, refusing any claims of special access to understanding animals in order to develop the particular configurations of imaginations that converge in any given cross-species encounter? At the risk of making unfair comparisons, we add that human-animal studies seems to offer the only way of accounting for what makes animal hoarders, circus trainers, slaughterhouse workers, and laboratory researchers so distinct, not to mention so distinctly problematic for those who have not made such extreme commitments to living with members of other species. For us, the openness to engaging different perspectives that inform alternative ways of knowing enables human-animal studies researchers to model meta-disciplinary perspectives.

Like all new worlds, human-animal studies has vistas of potentiality that come into sharper focus with critical self-reflection. Arguably in no other area do practitioners so consistently face the central questions of everyday existence: Who or what do we eat, wear, and love? In doing so, how do we live with ourselves? Why do we elect to live and die in the company of others who are so different from us? Rather than taking philosophical comfort in acknowledging that the (singular) animal question is being addressed, academics and artists who take this challenge seriously discover that the queries just keep coming. And we are not alone.

One effect of the approaches outlined in this volume is to open up third-way engagements with otherwise stalemated, binary deadlocks: for instance, to activate an environmentalist discussion that learns from eco-feminist critique to keep watch for the agents of actions presently flying below authorized radar systems, including animals along with plants, microbes, prion proteins, and indeed everyone involved in matters of life and death. We are in it together.

Bringing together many different voices, we have worked deliberately to forestall the formation of a hierarchy among them or even a clear sense of set objectives that connects them. A central tenet of this volume is that human-animal studies can only advance from a shared commitment to having no overarching political agenda except to advance research and foster further discussions in this area. Note the careful phrasing: an active commitment to something, rather than a lack of political commitment to something, which is a significantly different position. This formulation allows each of us to have personal views about welfare and other issues that concern rights and responsibilities regarding animals. In order to work together at our best to address effectively the enormous range and scale of problems facing humans and animals today, personal concerns must be regarded as highly contingent, possibly informing but never predetermining the outcomes of our research, and perhaps most importantly the possibilities for our research collaborations.

What we need are more multicultures and multispecies in human-animal studies, more 'multi-' in the human-animal studies community, a bringing together of our particular research and intellectual cultures to produce more crossings. Many barriers still need to be dismantled before these new and necessary ways of researching and writing about animals can take shape. Human-animal studies has not simply arrived and done good work; with concerted effort, it has a future and is going somewhere. Not knowing exactly where this will be is an exciting prospect, one that includes the possibility of leveraging some big culture shifts in the teaching, funding, and other structures of academics and the arts. And our experience so far indicates that the inevitable growing pains will all be worthwhile if they help us to continue to be in it together.

Learning from the mistaken self-assurance of colonial cartographers, we unsettle our sense of authority in order to recognize the scent marks, bark scratches, slime trails, and other nonhuman points of orientation providing access to the umwelten of other lives. Our shared project is doomed without lively dialogues which engage with a range of human traditions that are distinguished by different ways of living with animals. The deeply disturbing situations of mass killings, endangerments, and extinctions that human-animal researchers are compelled to address become more deeply disconcerting when we too capitulate to the demands of polarizing rhetorics steeped in enmity and violence that, however inadvertently, always validate victimization. To facilitate an ongoing, open-ended set of engagements with the powers that be and will become, human-animal studies must honour the lived experiences of people whose lives with, around, and among members of other species give the lie to any one-size-fits-all answers.

As researchers and artists electing to be involved with more-than-human beings across historical, continental, disciplinary, and above all species divides, the investigators brought together here remain exemplary in their accountability to the intricate particularities of the lives and deaths that their work attempts to represent. While their essays do not cover all human cultures and epochs – indeed, any attempt to claim as much might well be said to reveal the limits of anthropocentrism – they do come together at a particular time and within specific institutional conditions. Why this is happening at the turn of the twenty-first century bears further reflection, if only because the environments fostering such convergences remain, like so many of their foci, fragile and highly endangered.

1

Mammoths in the landscape

Nigel Rothfels

Thou, Mammoth, prepare for the conflict at once:
Set before thee thy terrible teeth,
And upon the Man rush, and beneath thy weight crush
The contemptible pigmy to death!

Frank Cowan, 'The Last of the Mammoths' (1879)

An unusual hunting story by a Henry Tukeman appeared in *McClure's* magazine in the fall of 1899. The account begins with a transcript of a letter addressed to 'H. Tukeman, Esq.' residing in Kent, England, from Horace P. Conradi. In the letter, Conradi – introduced as a recently deceased eccentric millionaire – releases Tukeman from his promise never to reveal how Conradi obtained an internationally famous specimen of mammoth, the taxidermized remains of which were then being exhibited at the Smithsonian. Thus begins a tale which over the last century has come to the attention most often of those interested in historical hoaxes. In many ways, in fact, the story seems to be just another one of those now predictable accounts of prehistoric megafauna surviving on uncharted islands and in hidden valleys.[1]

If, in some ways, the story might seem unremarkable, it deserves a quick retelling and some reflection because it not only gives a vivid idea of how people at the end of the nineteenth century imagined the lives of extinct mammoths, but it also shows particularly well how we actively create the 'natural histories' of animals. No modern human has ever observed living mammoths, but this apparently unintended hoax – the editors at *McClure's* felt they had been clear that the work was fiction – was somehow deeply convincing. When people later wrote to the editors and also the curators at the Smithsonian for more information about the 'Conradi mammoth', it was because the account seemed *so real*; what made it that way has largely to do with the historical moment in which it was written. In this, the story shows well at least part of what is driving the expanding interest in researching our cultural ways of understanding animals. While it is true that some quite important observations about the presence and significance of animals in human cultures stretch back to the beginnings of what have become the modern academic disciplines, it is really only in recent decades that animals have become a central and, more significantly, *persistent* focus for scholars from a wide array of fields outside the life sciences. This is not the place for a full discussion of the profound impact of this shift in focus; to oversimplify, when scholars begin to attend to the presence of animal life in their research, their perspectives on their fields seem to change quickly.

The point of this research is only partly to know more about the animals themselves. When someone devouring a hamburger says without hesitation that the thought of eating a kicking grasshopper is disgusting, the contributors to research in human-animal studies want to explore the thought not necessarily as a way to know more about grasshoppers or cattle, not necessarily as a way to make a case against the suffering of animals as sources of human food, not necessarily because they are interested in a larger story about agribusiness, but because they are curious about human cultures and how we think about animals in specifically cultural ways. Tukeman's account is important in the context of this volume because it shows how ideas about creatures so often have so little to do with what might be scientifically known about them. It shows how ideas about animals, living or extinct, are somehow always about a story we tell ourselves about the animal – that it is aggressive or retiring, noble or degenerate, mysterious or banal, beautiful or plain. These stories are not the science of animals; they are what we use to understand the importance of animals in our lives.

Tukeman's mammoth

So, the story. Tukeman writes that in 1890 he had been travelling in Alaska and decided to winter in Fort Yukon. Soon – and typically for stories like these – he found that he was spending more and more time with the local Indians and less and less with the white settlers. In particular, he writes, he enjoyed sharing stories with an older native man named Joe. One evening, the two of them were looking through a book of pictures showing African hunting scenes, when they 'came to the picture of an elephant, whereupon old Joe became very excited, and finally explained to me, with some reluctance, that he had seen one of these animals "up there", indicating the north with his hand'. Tukeman explained to the man that such creatures didn't live in the north, but he asked Joe to tell him more. Joe relates that he and his son, Soon-thai, travelled up the Porcupine River to a smaller river, which they followed to a mountain. They climbed up and found a cave, 'full of big bones', and an opening at the back which led them to the top of the mountain. From there, they could see 'a big valley, an' lakes an' trees, an' far away, on the other side of the valley, we see the mountains, an' beyond them, very far off, high mountains, with the snow on them which never goes away' (1899: 506).

After entering the valley and spending two days hunting and fishing on and around the many lakes, which were teaming with fish and small game, Joe and his son were hunting early one morning when Soon-thai found a huge footprint. Soon-thai wanted to see the mysterious animal that made the print and, despite Joe's fear, they followed the tracks until they finally came upon the creature that Joe called Tee-Kai-Koa. Joe continues:

> Presently we hear a splashing in a lake which is beyond some willows; an' there are no trees there; but we creep in very softly, an' we come to the reeds, an' wade through them to the edge, up to our knees in the water. He is there, the Tee-Kai-Koa, standing on the other side of the lake.
>
> *(1899: 506)*

According to Tukeman,

> The old man rose, and pointed before him. A strange glitter was in his eye, and the beads of perspiration stood out on his forehead. I could not doubt for a moment that he was describing what he had really seen.
>
> *(506)*

Joe continues:

> He is throwing water over himself with his long nose, an' his two teeth stand out before his head for ten gun-lengths, turned up, an' shining like a swan's wing in the sunlight. His hair is black an' long, an' hangs down his sides like driftweed from the tree branches after the floods.
>
> *(506–7)*

Before Joe could stop him, his son had lifted his gun and fired. Joe concludes, 'Ah, the noise! It is a cry like a thousand thousand geese, only shriller an' louder, an' it fills the valley till it reaches to the mountains, an' all the world seems to have nothing in it but that angry cry' (507).

Convinced by Joe's story, Tukeman relates that he decided to hunt the apparently last surviving mammoth with another native man, Paul, and the two embarked on their quest the following summer. After describing the long journey, the gigantic bones in the cave that marked the entrance to the trail up the mountain, and the spectacular vista of what Tukeman names the Tee-Kai-Koa valley, the hunters finally encountered the creature on 29 August. Tukeman writes that the mammoth was 'tearing up great masses of lichenous moss and feeding as an elephant feeds' while the hunters stood in 'awe inspired by the sight of this stupendous beast' (510). Tukeman decided that the mammoth must be fearless – 'What living thing could inspire it with fear?' (510) – and would instinctively attack smoke and stomp out fires, perceiving both as threats. He and Paul, therefore, performed a test to see if the animal would follow a trail of burning logs. Lighting logs and running to stands in trees from which they hoped to eventually shoot the animal, Tukeman describes the sound of the giant creature coming after the smoke:

> We were scarcely ensconced among the branches when a cry resounded over the valley which made the chills run down my back. I have heard the scream of an angry bull elephant, the roar of an African lion, the strange, half-human cry of the great gorilla; but none of these compare with the awe-inspiring cry of a mammoth. Perhaps the Indian's description of 'a thousand thousand geese' approaches it most nearly [. . .] but the very immensity of the volume of sound as the brute approached us confused any comparison I tried to make.
>
> *(512)*

When, finally, the day of the hunt came, the hunters lit a 25-foot-tall structure of green logs and, when the mammoth stormed on the smoky scene, they opened up on it with their guns. Tukeman writes that 'with the crack of our rifles came the most appalling scream of rage I have ever heard, and the vast brute, apparently unaffected by our shots, attacked the wood-pile with incredible fury'. After firing shot after shot at the animal, it began to weaken. Tukeman writes: 'the great brute was bleeding profusely from the mouth and ears, and staggered uncertainly back and forth.' At that moment, he notes:

> A feeling of pity and shame crept over me as I watched the failing strength of this mighty prehistoric monarch whom I had outwitted and despoiled of a thousand peaceful years of harmless existence. It was as though I were robbing nature and old Mother Earth herself of a child born to her younger days, in the dawn of time.
>
> *(512)*

After the arduous task of cutting up the creature and storing the skin and skeleton, Tukeman and his companion made their way to San Francisco, where they eventually sold the one-of-a-kind specimen to Conradi. The men returned to Alaska to retrieve the remains and Conradi arranged shipment to the Smithsonian, refusing to provide information on the creature's provenance. Tukeman concludes his account, noting that the general theory seems to have been that Conradi found the specimen frozen in an iceberg and notes that the measurements of the mammoth which Tukeman had given to Conradi were provided to the Smithsonian for publication and had thereafter been accepted as the work of Conradi.

Believability

Thus ended the original story, but the tale continued because of the interest of the readers. After apparently weeks of mail to the magazine and the museum about the spectacular animal that never was, the magazine agreed to publish a second article, this time written by Frederic A. Lucas, a member of the scientific staff at the Smithsonian, which also published the article in the museum's *Proceedings*. The new and more sedate article attempted to set the facts straight about mammoths, to make clear that they had been extinct for thousands of years, and to emphasize that Tukeman's story was both written and presented by the magazine as a work of fiction. According to the editors, the reason for the popular misunderstanding was to be found simply in Tukeman's skill. As they put it in an editorial comment before Lucas' article, 'We doubt if any writer of realistic fiction ever had a more general and convincing proof of success' (1900: 353). They are undoubtedly correct that it was Tukeman's skill which made the account so believable. His use of specific dates, the fictitious letter from Conradi – a name which sounds just enough like Carnegie – that somehow establishes the author's credibility, the names of actual locations, and the allusions to widespread reports in scientific and popular journals all help to lessen the impact of the central premise – that somehow a last representative of a species, long thought extinct, could be found in a hidden valley, shielded from man for thousands of years.[2]

However, Tukeman's story also worked because he fashioned a believable *mammoth*. But what can that possibly mean? Since he had clearly not had the opportunity to observe mammoths, Tukeman essentially had to make up his creature and put it into a scene that made sense. The idea of an ancient, giant beast standing in clear water in a valley teaming with game; a creature terribly alone, a last survivor again 'tearing up great masses of lichenous moss and feeding as an elephant feeds'; the effort to describe the animal's scream by comparing it to the roars of lions and elephants and the 'half-human cry' of a gorilla; the assertion that the animal would likely fear nothing and would (unlike any other creature) instantly attack fire – these are all features of an animal dreamt up in a very specific context, a context which included such different kinds of storytelling about amazing animals and landscapes as Paul Du Chaillu's 1861 *Explorations and Adventures in Equatorial Africa*, Rider Haggard's 1885 *King Solomon's Mines*, and even Rudyard Kipling's 1894 'mountainous mammoth, hairy, abhorrent, alone' from his 'The Story of Ung'.

For Tukeman and his contemporaries, living elephants naturally provided the ready model for extinct mammoths. This is not too surprising; today, too, most people would probably respond to a query about the lives of mammoths by saying something about living elephants, noting only that mammoths lived in colder places. With that said, the idea of 'elephant' is far from stable. First of all, if we are to make our mammoths simply by putting woolly coats on living elephants, we might want to consider 'which elephants?' The truth is that the genetics, appearance, and behaviours of even the living species and subspecies of elephants vary so widely that to talk about 'the lives of elephants' begins a series of increasingly problematic generalizations.[3] Even if one could meaningfully talk about characteristic elephant lives and behaviours, it would take an extraordinary leap to think that the many species of prehistoric *proboscidea* led lives that we would all somehow recognize as elephant-like. This is an order, after all, that includes by general reckoning over 150 distinct species and subspecies, which began appearing in the fossil record some 60 million years ago.

Beyond the question about which living species of elephant might serve as a model for the behaviours of extinct mammoths, however, we are also confronted with a second and in some ways more vexing problem. We have to recognize that our ideas about elephants, themselves, are changing all the time. In fact, among current popular western ideas about elephants – including that they are exceptionally intelligent, that they have extraordinarily long memories, that they have particularly rich emotional lives, that they have a complex understanding of death, that they can communicate over immense distances, and even that they can emote through artistic expression – are several that are only a few decades old. Indeed, it is clear that the elephants Tukeman was thinking about when he wrote his story were in many ways quite different from the elephants most people in the West think about today. This may seem a rather strange statement: surely an elephant is an elephant no matter when it lives in human history. It is also true, though, that what *we think* about elephants, how we talk about them and understand their lives and behaviours, has changed dramatically over even the past few decades.

For a simple example, most western natural history books at the beginning of the twentieth century presented elephants as living in family groups led by a large male and otherwise consisting of one or perhaps a few adult females and several young of various ages and sexes. After decades of successfully popularized field research on African bush elephants led by such figures as Iain Douglas-Hamilton, Cynthia Moss, and Joyce Poole, this image is being gradually displaced by dynamic images of matriarchal herds connected through complex extended family relationships. That we are still gaining new understanding of the lives of East-African elephants is not that surprising, but this fact signals the constant change in our ideas, our imaginings of these (and all) animals. We should not, then, look for the sources of Tukeman's mammoth and other mammoths of the late nineteenth century in our contemporary ideas about living elephants. Indeed, the conceptualization of a mammoth as a (1) giant, violent, and ultimately tragic creature, which (2) lived at the extreme edge of its habitat, just beyond human reach, is very much of the late nineteenth century.

The rogue

Although the use of the word 'rogue' to describe vagabonds, cheaters, rascals, and servants stretches back to the sixteenth century, its use to represent what the *Oxford English Dictionary* describes as 'an elephant driven away, or living apart, from the herd, and of a savage or destructive disposition' emerged in the middle of the nineteenth century in relation to the Sinhala expression *hora aliyā*. European works at the time claimed that 'rogue elephant' was a direct translation of the Sinhala, but it may be that the translation moved in the other direction. In either case, while human-elephant conflict has existed since the beginning of contact between the two species, the introduction in European thought of the idea of the rogue elephant in the nineteenth century was a new explanation for the conflict.[4] The expression seems to have first begun to appear in the 1840s, but one figure probably did more than any other to popularize the concept: the hunter and explorer Sir Samuel White Baker, who authored, among other works, *The Rifle and the Hound in Ceylon* of 1854. It is with Baker that key qualities in the behaviour of Tukeman's mammoth began to appear.

Baker begins his description of elephants – the most, he claimed, misunderstood animal – by describing them in *The Rifle and the Hound* not as noble or majestic creatures, but as 'naturally savage, wary, and revengeful, displaying as great courage when in their wild state as any animal known' (1854: 9).[5] Drawing a comparison between the animal most people had only seen in captivity and its wild cousin, Baker writes:

> A person who has never seen a wild elephant can form no idea of his real character, either mentally or physically. The unwieldy and sleepy looking beast, who, penned up in his cage at a menagerie, receives a sixpence in his trunk, and turns round with difficulty to deposit it in a box; whose mental powers seem to be concentrated in the idea of receiving buns tossed into a gaping mouth by children's hands; this very beast may have come from a warlike stock. His sire may have been the terror of a district, a pitiless highwayman, whose soul thirsted for blood; who, lying in wait in some thick bush, would rush upon the unwary passer-by, and knew no pleasure greater than the act of crushing his victim to a shapeless mass beneath his feet. How little does his tame sleepy son resemble him! Instead of browsing on the rank vegetation of wild pasturage, he devours plum-buns; instead of bathing his giant form in the deep rivers and lakes of his native land, he steps into a stone-lined basin to bathe before the eyes of a pleased multitude, the whole of whom form their opinion of elephants in general from the broken spirited monster that they see before them.
>
> *(8–9)*

Echoing an argument he often made about the soft existence and habits of the people of European cities, people who could have little sympathy for the wild adventures of hunting in the colonies, Baker asserts that the wild elephant of the forest and jungle is a daring and intelligent adversary, an animal whose 'great natural sagacity' made it the most dangerous game. And, not surprisingly, the most dangerous of these dangerous creatures, in Baker's distinctively hypertrophic and in today's terms 'carno-phallogocentric' (Derrida 1991) worldview, were naturally the males. While noting, then, that herds were often made up entirely of females, his interest inevitably turns to the bull elephant who is 'much larger than the female, and is generally more savage'. According to Baker, the wild bull elephant often seems to prefer 'solitude to a gregarious life', and when he does,

> he then becomes doubly vicious. He seldom strays many miles from one locality, which he haunts for many years. He becomes what is termed a 'rogue.' He then waylays the natives, and in fact becomes a scourge to the neighborhood, attacking the inoffensive without the slightest provocation, carrying destructions into the natives' paddy fields, and perfectly regardless of night fires or the usual precautions for scaring wild beasts.
>
> *(1854: 12)*

The 'rogues', according to Baker, are killers. He writes: 'The daring pluck of these "rogues" is only equalled by their extreme cunning.' Baker claims that a rogue will walk downwind so that he will smell any hunter attempting to follow up his tracks. Once he senses a hunter, the elephant will stand

> perfectly motionless [. . .] like a statue in ebony, the very essence of attention, every nerve of scent and hearing stretched to its cracking point; not a muscle moves, not a sound of a rustling branch against his rough sides; he is a mute figure of wild and fierce eagerness.
>
> *(12–13)*

So much for Baker's general introduction to this beast. When it comes to actual hunting adventures, the creatures do not disappoint; indeed, they seem uncannily bent on human destruction. Out hunting, for example, with his brother after two rogues who had teamed up – and we should note that living together actually conflicts with his own definition of rogues – the hunters found themselves led by the elephants into a small muddy clearing surrounded by impenetrable jungle. Of course, it was a carefully laid trap. At the moment that Baker's brother found himself stuck in mud, the author writes that he 'suddenly heard a deep guttural sound in the thick rattan within four feet of me; in the same instant the whole tangled fabric bent over me, and bursting asunder showed the furious head of an elephant with uplifted trunk in full charge upon me' (1854: 108). Baker continues:

> I had barely time to cock my rifle, and the barrel almost touched him as I fired. I knew it was in vain as his trunk was raised. B. fired his right-hand barrel at the same moment without effect from the same cause. I jumped on one side and attempted to spring through the deep mud: it was no use, the long grass entangled my feet, and in another instant I lay sprawling in the enraged elephant's path within a *foot* of him. In that moment of suspense, I expected to hear the crack of my own bones as his massive foot would be upon me. It was an atom of time. I heard the crack of a gun; it was B.'s last barrel. I felt a spongy weight strike my heel, and, turning quickly heels over head, I rolled a few paces and regained my feet. That last shot had floored him just as he was upon me; the end of his trunk had fallen upon my heel. Still he was not dead, but he struck at me with his trunk as I passed round his head to give him a finisher with the four-ounce rifle, which I had snatched from our solitary gun-bearer.
>
> *(108–9)*

In the true nightmarish mode typical of Baker's hunts, as if all of this were not enough, at the moment Baker was firing at the elephant on the ground, the other elephant attacked with a 'savage scream'. Baker writes,

> I saw the ponderous fore-leg cleave its way through the jungle directly upon me. I threw my whole weight back against the thick rattans to avoid him, and the next moment his foot was planted within an inch of mine. His lofty head was passing over me in full charge at B., who was unloaded, when, holding the four-ounce rifle perpendicularly, I fired exactly under his throat. I thought he would fall upon me and crush me, but this shot was the only chance, as B. was perfectly helpless.
>
> *(109–10)*

Baker, of course, had jumped away 'the instant after firing'. The elephant stopped in its charge against Baker's brother, escaped with its 'death wound', and was found dead the next day (110).

It is important to realize that as much as writings like Baker's became very popular among the period's adventure hunters, not everyone found this sort of over-the-top prose very convincing. In his *Sketches of the Natural History of Ceylon* of 1861, for example, the naturalist and British colonial secretary in Ceylon from 1845 to 1850, James Emerson Tennent, argued that while it might be true that actual rogues were dangerous, they were also so rare that few hunters would likely ever encounter one. Indeed, Baker's notion that elephants would be 'thirsting for blood' while 'lying in wait in the jungle' was, for Tennent, simply ridiculous. If anything, he insists,

> the cruelties practised by the hunters have no doubt taught these sagacious creatures to be cautious and alert, but their precautions are simply defensive; and beyond the alarm and apprehension which they evince on the approach of man, they exhibit no indication of hostility or thirst for blood.
>
> *(1861: 147–8)*

In the end, it seems that the enthusiastic din arising from Baker's popular writings drowned out those critics, like Tennent, who questioned the veracity of the celebrated hunter's descriptions. By the end of the nineteenth century, after decades of writing from a variety of perspectives on 'rogues', most people appear to have accepted the idea that a male elephant, separated or perhaps driven from the society of other elephants, somehow turned, contradictorily, both mad with rage and remarkably focused on killing humans. The terrifying, crazed, morose, and vicious lone male elephant entered into the western conceptualization of elephants. There was little natural historical study behind this argument, of course, but it became an almost obsessional claim, especially of hunters. Tennent's own text, ironically, even provides part of the *OED*'s etymology of the term 'rogue elephant'. In the West, before Baker and his contemporaries, a violent elephant was typically seen as only seeking a justifiable and measured revenge against a personal or perhaps familial injustice. Beginning with Baker, the wild adult male elephant came to be seen as essentially *always*, and pathologically, a 'rogue', even in the company of other elephants. As the century wore on, the bull elephant became the ultimate quarry, a beast both cunning *and* spectacularly dangerous. Indeed, when Teddy Roosevelt went on his safari to Africa, he went before anything else after Baker's elephant; he sought a confrontation with a mighty and necessarily existential foe.

Reading accounts like these, we must remember that a hunting story is always only just that. Hunting stories are what is told, what is shared with those who were not there. They are written with a specific audience in mind and the way they are told has everything to do with the expectations of that audience and very little to do with what actually happened in the field. When Baker and the many others who wrote in this style told their stories of hunting big game, the stories were meant to be exciting, they were meant to take people to other places to experience extraordinary adventures. The truly mundane aspects of these expeditions – the days of walking, the missed shots, the poorly hit game that staggered around pathetically before dying in the bush, never to be claimed by the hunter – just never really show up. In the end, it was in the accounts of Baker and his followers, though, that the mad, rogue elephant emerged. Again, I am not claiming that violent elephants have never existed or were just dreamt up by these people. I am arguing that the *interpretation* of elephant-human conflict changes over time; the crazed, morose elephant – the elephant that would become the model for Tukeman – did not exist in western conceptualizations of elephants until the middle of the nineteenth century. In the late eighteenth-century pages of a naturalist like Buffon, for example, the elephant can be violent, but that violence is understood as measured and just and in protection of the elephant's family, not because the creature desired blood, nor because it feared nothing and attacked without provocation.

The mammoth and the landscape

If the model for the behaviour of Tukeman's mammoth came from the claim that lone male elephants were violent and aggressive quarry, creatures that quickly switched from being the hunted to the crazed hunter, it was the discovery of frozen mammoths in Siberia that completed the vision behind Tukeman's story. In a 1930 account of mammoths published in the popular science magazine, *Natural History*, Henry Fairfield Osborn of the American Museum of Natural History described what he called 'The Romance of the Woolly Mammoth'. The mammoth, Osborn declared, is 'the classic of palæontology; [. . .] the first extinct mammal to be found by man; [. . .] the first to be used as proof of a universal deluge; [. . .] the first to be used as proof of the existence of a long extinct world of mammalian life antecedent to the deluge; [. . .] the first to receive a scientific description in the Latin language; [. . .] the first to receive a scientific name – *Elephas primigenius* or "the first of all, or original, elephant" ' (1930: 227). Mammoths became a scientific 'classic', but more than that, too. Over the course of the nineteenth century, mammoths became part of a popular consciousness. Indeed, the gigantic recreation of a mammoth at the Pittsburgh Exposition in 1878 only hints at the popular resonance of this creature; this model, it should be noted, was 16 feet high and 29 feet long, but nevertheless based, it was claimed, on accurate measurements of mammoth bones in Germany.[6] Economically, too, the mammoth had an impact in the period as the ivory of literally tens of thousands of mammoths reached markets over the course of the century.

The bones, tusks, and occasionally skin and more of mammoths presented fascinating riddles for scientists and the public alike; and, in fact, many aspects of the story of mammoths still remain unsettled. Until well into the twentieth century, though, the central question came down to this: How come we have never found a living mammoth? In the eighteenth century, there was an answer reported in western accounts in an amused tone that claimed, according to folk traditions, that the creatures lived underground and that they died upon coming into contact with the air or cold; and then there was the more 'scientific' argument that the mammoths had likely moved out of the north and merged with or become one of the known modern types of elephant. These views were replaced in the nineteenth century after the French anatomist Georges Cuvier declared that the remains of mammoths came from an extinct species. Essentially two opposing groups coalesced. On one side were the catastrophists, who traced their arguments to Cuvier and who were convinced that the disappearance of the mammoth could only be explained by an environmental cataclysm, a classic work here being Henry Howarth's *The Mammoth and the Flood* of 1887. On the other side were the uniformitarians, who traced their lineage to the English geologist Charles Lyell and who argued both that the world was a lot older and that changes were more gradual than the catastrophists believed. Complicating matters for everyone came the realization that humans had lived beside mammoths, that they had hunted the beasts, and perhaps been the cause (or at least part of the cause) of their extinction.

Against this background, one of the most remarkable things in the history of paleontology occurred: the unearthing of frozen, *intact* mammoths in Siberia. Even though there were several significant finds of this type in the nineteenth century, it turns out that the two most famous perfectly bracket the century: the Adams Mammoth and the Berezovka Mammoth. The short versions of their histories go something like this. The Adams Mammoth was first glimpsed in 1799 by a hunter named Ossip Shumakhov near the Lena Delta, an area described by Bassett Digby in 1926 as 'The End of the Earth' (opposite 144). Over the course of several years, Shumakhov watched the animal emerge from a thawing and weathering bank, and, in 1804, he finally sawed off and sold the tusks to a merchant in Yakutsk, who then sketched an illustration of the beast as it had been described to him. In 1806, Michael Adams, who was affiliated with the Academy of Sciences in St Petersburg, was travelling to China with a Russian ambassador, when they stopped in Yakutsk. Hearing the story from the merchant, Adams quickly set out and eventually recovered the badly mutilated body of the mammoth and sent the skeleton and parts of the skin and hair by sledge and train to St Petersburg, where the reconstructed skeleton became the first articulated mammoth skeleton ever exhibited.

One hundred years later, in April 1901, a telegram was received in St Petersburg relaying news, again from Yakutsk, that another mammoth had been discovered the fall before in well-preserved condition along the Berezovka River, north of the Arctic Circle. Otto Herz, a zoologist with the Academy of Sciences; Eugen Pfizenmayer, a taxidermist; and a young geologist named Dimitrii Sevastianov left the following month. Herz and Pfizenmayer, both of whom published substantial accounts of the expedition, reached the site near the end of September, had constructed a cabin in which they could work on the mammoth's remains, and, by the end of October, prepared it so that the animal was ready to be shipped overland, again, by sledge and rail to St Petersburg.

In many ways, the two finds are remarkably similar. Despite the parallels between them, though, the Berezovka Mammoth caught the imagination of the public like no other specimen before. Indeed, by 1926, when Eugen Pfizenmayer recalled his feeling of standing speechless before the grave of a 'witness to prehistory', a 'diluvial monster' that had 'filled the primitive peoples of the forest and tundra with superstitious terror' (126), the idea of the mammoth had become intensely vivid for the public. Perhaps not surprisingly, then, when the Berezovka Mammoth was displayed in St Petersburg in 1903, the museum did more than just mount its skeleton, as had been done with the Adams Mammoth; the museum reconstructed in taxidermy the moment of the animal's death, as it lay helpless, unable to stand with its broken pelvis and rear leg bones. Pfizenmayer's 'diluvial monster' is a clear relative of Baker's elephant; but there is also something about this reconstruction that is very different, something that resonates with Tukeman's description of the tragic 'cry of a thousand thousand geese' and his 'feeling of pity and shame' as he 'watched the failing strength of this mighty prehistoric monarch.'

In his *The Last Dinosaur Book* of 1998, W.J.T. Mitchell argues that the dinosaur, as a modern mythical creature, 'expresses the political unconscious [. . .] of modern life, manifesting collective anxieties about disaster and extinction, epitomizing our own ambivalence toward our collective condition.' The dinosaur, he concludes, 'accompanies the disaster theme with a narrative of our own possible extinction [. . .] and function[s] as a species totem for the human race' (1998: 261–2). But if the bipedal dinosaur – the *Tyrannosaurus rex* and more recently the *Velociraptor* – came to represent, as Mitchell contends, the shocking violence, the rapaciousness of a humanity bent on destruction above all else, then the slow, ponderous, vegetarian, and ultimately family-centred mammoth became a profound lesson in what human violence has wrought; it became, in fact, a warning about what continued human gluttony might bring. The idea that the mammoth could stand as a warning about unrestrained slaughter was, in fact, part of the world of Tukeman and was echoed by many observers in the period. The German hunter and photographer Carl Georg Schillings, in a chapter on 'A Dying Race of Giants' in his 1907 *In Wildest Africa*, laments, for example, that

> by a singular coincidence, the mammoth remains preserved in the ice have been found just at a time when the craze for slaughtering their African relations has reached its climax, and when by means of arms that deal out death at great, and therefore safe distances, the work of annihilation is all too rapidly progressing.
>
> *(516)*

It is not that the Berezovka Mammoth was seen as a direct victim of human violence; like most mammoths in the period, though, its pathetic death became the central index of its existence. The story of the Berezovka Mammoth became more than the story of an individual; it became the story of a species. In the late nineteenth and early twentieth centuries, western ideas about mammoths did not turn on the literally millions of years during which this species successfully propagated across the northern hemisphere; ideas quickly settled around the meaning of a species passing into extinction. This is the tone, for example, of an arresting mural of a group of mammoths at the Field Museum in Chicago (Figure 1.1). The work was executed in the late 1920s by Charles Knight, an artist who stands even today as one of the most important illustrators of prehistoric life, an artist who, it happens, was commissioned very early on in his career by *McClure's* to illustrate Lucas' article and who discussed the influence of the Berezovka Mammoth in his popular *Before the Dawn of History*.

The mural is one of a series of some thirty works by Knight at the Field depicting the evolution of life on the planet, beginning in the Precambrian and moving through the Paleozoic, Mesozoic, and Cenozoic eras. In the work, Knight has painted four mammoths moving with determination across a snow-covered, hilly landscape with little to eat beyond a few wisps of grass in the foreground and groupings of pine in protected valleys and beside a lake in the distance. One of two woolly rhinos seen in the lower right of the image has apparently lifted his head to watch the procession of a large male leading a female and two young of different ages, the older one of which seems to be asking whether they are there yet. But where are these mammoths going? Pushed to the 'end of the world', plodding across a landscape with nothing like the forage necessary for these creatures, the mammoths in this image are portrayed as the last survivors of a species. In a description that could have been for this illustration, Knight conjured the final days of the mammoths for his book *Prehistoric Man: The Great Adventurer*:

Figure 1.1 © The Field Museum, #CK30T. Ron Testa

> One can imagine how in the depths of some terrible winter, with bushes and grasses covered deep in snow, the destruction of the great herds was finally accomplished. Gales and bitter cold perhaps first decimated the younger and weaker members of the groups, and the fate of even the strongest individuals became at length an assured fact under the pressure of months of starvation. Each morning saw many of the beaten creatures lying dead or exhausted in the huge drifts until only the patriarchs of the once great clans remained alive. With waving trunks and feeble trumpetings they greeted the coming day, but at length even their huge frames began to weaken. In the wan sunlight of early spring great mounds of white slowly melted above the prostrate forms, the pale gleam from ivory tusks like smooth curved headstones marking the resting place of this superb race of mammalian giants now forever passed into the limbo of vanished things.
>
> *(1949: 211)*

Knight's mammoths, along with the Berezovka and Tukeman mammoths that preceded them, suggest the importance of Akira Lippit's observation in his *Electric Animal* that, in modernity, animals seem to 'exist in a state of *perpetual vanishing*' (2000: 1, original emphasis). Joined by a pathetic menagerie of dodos, great auks, passenger pigeons, thylacines, and so many others, these animals were among the first to push a broader public to ponder the meaning of species extermination. Portrayed as prehistoric 'elephants', moreover, these mammoths heralded the beginning of a broader conceptualization of nature itself as a victim of human violence – a conceptualization that has stayed with us now for over a century, even while species after species disappears.

Notes

1 Among the more notable examples of these sorts of stories are Jules Verne's 1864 *Voyage au centre de la terre* [*Journey to the Centre of the Earth*] and Arthur Conan Doyle's 1912 *Lost World.*

2 These are, of course, the techniques used throughout this genre and are as present in *King Kong* as they are in ongoing fascinations with the Loch Ness and Lake Champlain monsters, sasquatch, or even the 2009 hoax rediscovery of the Carolina Parakeet.

3 Just attending to relatively free-ranging populations of elephants (as opposed to a great variety of more and less captive situations), one would have to recognize that the lives of the bush elephants (*Loxodonta africana*) in Kenya contrast in many ways quite sharply with those of the forest elephants (*Loxodonta cyclotis*) in Cameroon, and the lives of both of these often strikingly contrast with those of elephant populations in Sri Lanka (*Elephas maximus maximus*), India (*E. m. indicus*), and Sumatra (*E. m. sumatranus*). For a clear discussion of elephant evolution and speciation, see Sukumar 2003: 3–54.

4 There have been a number of accounts in recent years, including the 2006 article by Charles Siebert in the *New York Times Magazine*, that have argued that elephants and humans lived side by side in peace until relatively recently. There is simply no historical basis for this claim and there are literally thousands of years of records that suggest the opposite.

5 In my essay, 'Elephants, Ethics, and History', I discuss an earlier shift in ideas about elephants in Buffon's *Natural History* and R.G. Cummings' (1850) *Five Years of a Hunter's Life in the Far Interior of South Africa.*

6 For more on this reconstructed mammoth, see N.A. (1879), *The Mammoth.*

References

Baker, S.W. (1854) *The Rifle and the Hound in Ceylon.* London: Longman.

Buffon, G.L. (1812) *Natural History: General and Particular,* W. Smellie (trans.), London: Cadell and Davies.

Cowan, F. (1879) 'The Last of the Mammoths', in N.A. *The Mammoth: Being Articles Concerning This Celebrated Specimen which Have Appeared in Various American and European Newspapers*, Pittsburgh: Stevenson, Foster & Co. pp. 34–9.

Cumming, R.G. (1850) *Five Years of a Hunter's Life in the Far Interior of South Africa. With Notices on the Native Tribes, and Anecdotes of the Chase, of the Lion, Elephant, Hippopotamus, Giraffe, Rhinoceros, &c.*, New York: Harper Brothers.

Derrida, J. (1991) 'Eating Well, or The Calculation of the Subject: An Interview with Jacques Derrida', in E. Cadava, P. Connor, and J-L. Nancy (eds), *Who Comes after the Subject?*, P. Connor and A. Ronell (trans.), New York: Routledge, pp. 96–119.

Digby, B. (1926) *The Mammoth and Mammoth Hunting in North-East Siberia*, New York: Appleton.

Herz, O.F. (1902) *Berichte des Leiters der von der Kaiserlichen Akademie der Wissenschaften zur Ausgrabung eines Mammuthkadavers an die Kolyma-Beresowka Ausgesandten Expedition*, St Petersburg: Kaiserlichen Akademie.

Howarth, H.H. (1887) *The Mammoth and the Flood: An Attempt to Confront the Theory of Uniformity with the Facts of Recent Geology*, London: Sampson.

Knight, C. (1935) *Before the Dawn of History*, New York: McGraw-Hill.

—— (1949) *Prehistoric Man: The Great Adventurer*, New York: Appleton-Century-Crofts.

Lippit, A.M. (2000) *Electric Animal: Toward a Rhetoric of Wildlife*, Minneapolis: University of Minnesota Press.

Lucas, F.A. (1900) 'The Truth about the Mammoth', *McClure's Magazine* 14.4 (February): 349–55. Reprinted in *The Annual Report of the Board of Regents of the Smithsonian Institution*, Washington: Government Printing Office, 1901, pp. 353–9.

Mitchell, W.J.T. (1998) *The Last Dinosaur Book: The Life and Times of a Cultural Icon*, Chicago: University of Chicago Press.

N.A. (1879) *The Mammoth: Being Articles Concerning This Celebrated Specimen which Have Appeared in Various American and European Newspapers*, Pittsburgh: Stevenson, Foster & Co.

Osborn, H.F. (1930) 'The Romance of the Woolly Mammoth', *Natural History* 30.3: 226–41.

Pfizenmayer, E.W. (1926) *Mammutliechen und Urwaldmenschen in Nordost-Sibirien*, Leipzig: Brockhaus.

Rothfels, N. (2008) 'Elephants, Ethics, and History', in C. Wemmer and C.A. Christen (eds), *Elephants and Ethics: Toward a Morality of Coexistence*, Baltimore, MD: Johns Hopkins University Press, pp. 100–19.

Schillings, C.G. (1907) *In Wildest Africa*, London: Hutchinson.

Siebert, C. (2006) 'An Elephant Crackup', *New York Times Magazine*, 8 October, available online at www.nytimes.com/2006/10/08/magazine/08elephant.html, accessed 17 May 2013.

Sukumar, R. (2003) *The Living Elephants: Evolutionary Ecology, Behavior, and Conservation*, New York: Oxford University Press.

Tennent, J.E. (1861) *Sketches of the Natural History of Ceylon*, London: Longman.

Tukeman, H. (1899) 'The Killing of the Mammoth', *McClure's Magazine* 13.6 (October): 505–15.

2

Domesticating practices

The case of Arabian babblers

Vinciane Despret

Honorary primates

Primatologists Shirley Strum and Linda Fedigan note, in their introductory chapter to the book *Primate Encounters*, the considerable change in scholars' interpretation of primate behavior:

> We have moved from a general vision that primate society revolves around males and is based on aggression, domination, and hierarchy to a more complex array of options based on phylogeny, ecology, demography, social history and chance events. The current image of primate society [. . .] would be a strong counterpoint to the earlier view. It would highlight the importance of females within society, emphasize tactics other than aggression (particularly those that rely on social finesse and the management of relationships), and argue that hierarchy may or may not have a place in primate society, but that males and females are equally capable of competition and rank ordering.
>
> *(Strum and Fedigan 2000: 5)*

Throughout the past twenty years, primatologists and scientific scholars have done substantial work to identify the reasons for these changes (Haraway 1992; Strum and Fedigan 2000). Numerous hypotheses have been suggested, some of them garnering controversial feedback. Scholars on one side have tended to overlook the actual changes of the animals themselves, instead focusing on human activities such as the socio-historical context of the researchers, theories, paradigms, practices, etc.[1] Scholars on the other side have often failed to ask the converse and correlative question: Why didn't most other animals, or at least not until recently, experience the same changes?

A striking exception can be found in the work of the primatologist, and now 'sheepologist', Thelma Rowell. Concerning our first issue, she not only insists that scientists have learned to ask primates other questions, but she also urges one to consider the possibility that primates actually did change. In other words, the behavior of primates has changed not only as scientists have learned to question them differently, but also as the animals themselves have changed. She suggests that these changes may have occurred partly due to the fact that the animals are under constant observation (Rowell 2000a).[2]

Rowell also addresses our second question: Why didn't most of the non-primate animals go through the same process of change? She recalls a provocative question that has begun to annoy her: 'Why are primates so much more intelligent and socially skilled than other animals?' Her answer to this question was to investigate the question itself. 'But are they?' (Rowell 1999; 2000a). How can we be so sure that primates have a more complex social life? Or, more precisely, how did we build the comparisons? The way primates have been studied is radically different from the way classical ethology has been carried out for other animals. For primates, long-term and individually based studies have been completed, with a great deal of attention paid to relationships and ways of communication. For other animals, short-term research has primarily approached questions about food. For example, literature about sheep, 'the epitome of the silly animal,'[3] is all focused on what they eat and how they choose food in a pasture, or, in short, how they convert grass into gigots.[4]

The question Rowell raises could be asked about a number of animals. Are ravens, as we consider them today, more intelligent or socially sophisticated than other birds? How about dolphins, magpies (Prior, Schwarz, and Güntürkün 2008), hyenas, and other charismatic animals that Rowell would call 'honorary primates'? If they are as intelligent, how did they get the 'chance' to become so 'well equipped'? (Latour 2000: 368).

This very question can be raised any time an animal unexpectedly 'crosses the Rubicon', as some scientists would describe it (Prior et al. 2008). Simply put, what happened to this animal? Who gave him/her a 'chance' and how?

Flying primates

These are some of the questions that led me to the field station of Hatzeva in Israel's Negev Desert. I wanted to see a bird, the Arabian babbler, and to meet its Israeli ornithologist, Amotz Zahavi. These babblers offered a sharp contrast to typical descriptions in the scientific literature of altruism in birds, the subject of the Master's thesis in psychology I had recently received when the project took form (Despret 1991). They did not behave like birds usually do; they didn't follow what could legitimately be considered the typical bird's script. In addition, their ethologist did not describe them as ethologists usually do, outside of those who produce popular writings like Konrad Lorenz famously did.[5] Not only were these birds described as *dancing* together in the morning sunrise, not only were they eager to offer presents to one another, not only would they *take pride* in caring for each other's nestlings or in defending an endangered comrade, but also, according to Zahavi's depiction, their relations relied on trust. These relations were enacted through constant claims for one's liability and the liability of other members of the group. In short, babblers were relentlessly *testing their bonds* (Zahavi 1976). I had the feeling that these birds, oddly enough, were actually depicted like primatologists portrayed their primates. However, there is no doubt that they belong to a species of birds. They lay eggs (or at least some of them do), they have feathers, and they fly, though they seem to spend much of their time on the ground. Something was wrong with the situation, and so I anticipated that it would provide a good case for empirical philosophy.

It is worth mentioning that my questions, up until that time, were not exactly the same as they would turn out to be. My actual purpose was, as I mentioned in the files I completed to secure the grant for my research, to identify in that 'odd situation' (odd birds and an ethologist oddly accounting for them) what could account for such oddity. Was the bird really peculiar per se, or, as I confess I did suspect, could these peculiarities be ascribed to some eccentricity of the observer that, for whatever reason, I had the ambition to disclose? Going to the field was, from that perspective, a simple (and enjoyable) delocalization of the questions philosophers might usually select from their Cartesian cabinet. Observing the ethologist observing his babblers, focusing on what he would say, how he would select what constituted a fact, and how he would construe what was to be seen would allow me to provide clear and *empirical judgment*. This empirical judgment, let us say in passing, is not far from that of the proverbial fool who, when the savant points out the moon, looks at the savant's finger.

Ultimately, I got the grant, left home, and went to the field to meet the babblers. And I saw them dancing. The field 'happened to me'.

Babblers who talk, babblers who babble

The Arabian babbler (*Turdoides squamiceps*) inhabits extreme deserts and is the only bird species in Israel that lives in groups year round. These groups are territorial, with numbers of birds per group generally being between three and five individuals, although they can range between two and twenty-two. Each group usually contains one breeding pair; young birds do not disperse for one to three years, during which time they act as helpers. Babblers are long-lived and may reach the age of twelve or fourteen – all the more since Zahavi has long worked to protect them.

Babblers seem to be very peculiar birds: not only do they help at the nest, all members of the group taking care of a single nest, but they also cooperate to protect their territory against neighboring groups of babblers and non-territorial individuals ('refugees' as Zahavi calls the latter). The birds' dealings with each other involve providing a great deal of assistance to others. They offer presents to feed each other – seeds, insects, flowers – before they are full themselves: when offered a crumb of bread (by the human observer) right after they have fed their fellows with a similar crumb, they will eat with relish – something, Zahavi explains, that satiated babblers do not do. They endanger themselves by mobbing raptors and snakes. They imperil themselves by coming to the rescue of group members who get caught in a net or by a predator or by enemy babblers during a fight. They play together. They also dance together in the morning, sometimes after bathing. In short, through their whole life, they cooperate. Of course, they may fight, but they fight seriously with a member of their group only once in their whole life – if ever. And if they sometimes compete with each other, and they actually do, they also compete to act as sentinel of the territory or to feed their comrades.

What babblers actually do, according to Zahavi, is quest for social 'status' – status being different from hierarchy since it is gained through altruistic actions (Zahavi 1990) or, as Zahavi formulated in his later writings, they are very interested in the issue of prestige (Zahavi and Zahavi 1997) – a hypothesis that has also been suggested for ravens (Heinrich 1991; 2000). When one babbler feeds another, he/she draws attention to him/herself: he/she emits a special trill and lifts his/her beak above the beak of the one he/she is feeding. But generally, a babbler tries to avoid being fed by another bird of about the same age (i.e. the same rank in the hierarchy): he/she might try to escape or, in some cases, close his/her beak tightly, despite being hungry, and, Zahavi says, he/she will eagerly accept one of the observers' crumbs of bread immediately after refusing a juicy insect from another babbler. If a lower ranked individual attempts to feed a higher-ranking one, he/she might be beaten up. The babblers (the one who is concerned, but also all the ones who attend to the scene) will construe the proposal as a declaration of open revolt.

First, it should be mentioned that each bird observed (nowadays among 250) is individually identified: they are all tagged with a combination of three colored bands and one numbered band. They are all habituated (i.e. 'used to our presence'), and observers may walk among them without scaring them (Zahavi and Zahavi 1997). Observers working with Zahavi usually give the birds tiny amounts of breadcrumbs. Most often, when going to the field and trying to find a particular group, Zahavi calls them with a whistle, and they come – which is, he explains, rather easier than trying to find them or run after them. Babblers seem to prefer to live on the ground and are slow flyers, which makes the observations still easier. As briefly mentioned earlier, Zahavi's description of the birds is rather unusual in ethological discourse. Zahavi explains:

> Because they are habituated to our presence we can hear the soft, widely varied calls they use to communicate, calls that don't carry more than a few yards. Some individuals are talkative, others taciturn, and still others grumble all day long. When a higher-ranking individual approaches a lower-ranking one, the latter often makes a soft sound to acknowledge the other, as if saying 'Yes, sir.'
>
> *(Zahavi and Zahavi 1997: 126)*

The theory itself is still stranger: babblers do help, do assist, do protect, do feed others in order to gain prestige. For example, Zahavi explains, a babbler who finds food may not swallow it right away but instead may hold it in his/her beak and look around to see whom he/she can feed. If he/she sees only individuals whose rank is higher than his/her own, he/she may waver briefly, then swallow the food.

> When the beta male, the number two male of a group, goes up to the top of a tree to stand up as sentinel, we often see the alpha male, his superior, busily looking for food; he then gives it to the sentinel in full view of the other members of the group and replaces him on guard duty. In many cases, we can tell when the sentinel notices the alpha male's preparation to feed him from the direction of the sentinel's gaze, and often because the sentinel abandons his lofty post before the alpha male arrives to displace him.
>
> *(Zahavi and Zahavi 1997: 126)*

In the field, I also met Jon, an ethologist who had studied at Oxford. Jon was young but he was already an experienced ethologist. He was also (and still is) a sociobiologist. He was there while I was visiting and he was also working with babblers. Jon definitively does not agree with Zahavi's theory. According to Jon, if babblers help, it is because they want to maximize their inclusive fitness (or because altruism has been selected for). In helping close kin, each individual may expect to pass on part of his genetic pool.

Following and listening to either Jon or Zahavi gave me the feeling that we were not dealing with the same birds. The first difference between Jon and Zahavi was obvious: they totally disagreed about what could be said or more precisely what could count as a true hypothesis. For Zahavi, a hypothesis about babblers is true in some way because babblers provide evidence of it. Or, to put it another way, humans may assume a particular hypothesis about babblers because babblers never act without good reason. For Jon, a hypothesis is true only if you can prove it. And to prove is to experiment.

In other words, Zahavi trusts what the babblers show – or what the babbler says in showing; Jon trusts the experiment that will make the babblers concede to behave in such a way that it confirms a hypothesis. Zahavi follows, moves inside the group, listens to the babblers, and explains: for example, 'we move easily among the babblers; we see what they are looking at and become aware of their intentions' (Zahavi and Zahavi 1997: 126). He has learned to pay attention to each tiny difference of vocalization: why does a babbler shout so loud when it threatens another one? Of course, *we* shout when we are angry. But in the case of the babblers, the shouting is not aimed at the purported listener but rather at other babblers that are farther away: the shouting makes them witnesses.

Let us note in passing that Thelma Rowell proposes a similar hypothesis to account for fighting in sheep:

> I think that one of the functions of the fighting is coordination, or [more precisely] the result of fighting is coordination. The females take a keen interest to the fight. When you are doing this [hitting hands loudly] it makes a lovely loud noise. And if somebody is fighting here, the ewes will come running to watch. And so I said, all right, let us say that it is not a fight, this is a display. How could a sheep make a loud noise by himself? If it is a sound [of] one hand clapping, you can't do it, but with two of them, you make a spectacular noise and the visual display. And it is very exciting to the ewes, and they all come and eat together, and then, it's all over.[6]

Only confident babblers can afford to shout their threats in front of the crowd. 'This principle,' Zahavi adds, 'is well-known in politics: a publicly declared intention is likelier to be carried out than one agreed upon in secret' (Zahavi and Zahavi 1997: 75). Why do babblers sometimes spring their wings upward when approaching a snake, making themselves more visible? They must have good reasons to act in such a dangerous way: babblers want to show off their courage to other members of their group; that is why they call attention to themselves in that situation.

What babblers show is nothing less than what they mean to show. I would suggest that Zahavi may be very confident because his theory claims that signals are always reliable, or honest. 'What you see is what it's worth' could define the way babblers behave, according to Zahavi, since signals for prestige are costly. Feeding others or acting as sentinel involves time and resources: if a babbler cannot afford it, he/she will be prevented from doing it. This is called the 'Handicap Principle theory' (Zahavi 1975). To make it clearer, we may take, for example, the relationship between prey and predator. Paradoxically at first glance, you may find the rule of honesty in the encounter between a gazelle and a wolf, when the gazelle reveals him/herself to the predator that might not spot him/her. Why does the gazelle waste time and energy jumping up and down (stotting) instead of running away as fast as it can? Why does he/she take such a risk? It has been usually interpreted as an act of altruism (as paradoxical as it can be), performed to warn the congeners. No, claims Zahavi, it is not a matter of 'sacrifice':

> The gazelle is signaling to the predator that it has seen it; by 'wasting' time and by jumping high in the air rather than bounding away, it demonstrates in a reliable way that it is able to outrun the wolf. The wolf, upon learning that it has lost its chance to surprise its prey, and that this gazelle is in tip-top physical shape [. . .] may decide to look for more promising prey.
>
> *(Zahavi and Zahavi 1997: xvi)*

In the same vein, by showing off how much they can invest in standing guard, in feeding others, in taking serious risks, babblers show off their ability to win in a fight and their desirability as group-mates. In other words, 'the altruist's investment in the altruistic act offers a reliable, concrete index of that ability' (Zahavi and Zahavi 1997: 142).

To experience or to experiment?

From Jon's point of view, one cannot simply claim whatever one wants, feels, thinks, expects when explaining the birds' behavior, unless that person can design an experiment to demonstrate these hypotheses. For example, in order to prove that the behaviors of helping at the nest and feeding chicks are not governed by motives of prestige, one would need to show that these behaviors are determined by other rules, in that precise case by sociobiological rules of kinship selection. Jon aims to show that helpers and parents provision for nestlings at similar rates, irrespective of sex or rank within the group. If, for example, the nestlings were very hungry, parents would enhance their effort; on the other hand, helpers-at-the-nest, if this were solely a matter of prestige, wouldn't be sensitive to the change of conditions. Jon put the hypothesis to the test: some playbacks of tapes showing chick-begging enhancing the feeding from both the parents and the helpers-at-the-nest showed evidence that the helpers have the same provisioning rule as the parents and therefore do not use nestling feeding as an altruistic signal to gain social prestige.

When I asked Zahavi about experimentation, he told me, 'of course we do experiments with babblers. We ask them to come, and we tell them what we want them to do and the question we are dealing with. And they do it.' I will never be sure Zahavi wasn't serious.

Was he teasing the candid philosopher? Was he being literal? Perhaps he was both. Whatever the answer to these questions, it appears that what he meant by the words, 'to make an experiment *with* babblers', exposes the porousness of the demarcation between wild and domesticated.

Zahavi offers the babblers something that is not far from the process of domestication itself: a process that transforms the beings involved in it, both humans and animals, and changes their involvement in the relationship. They *do* something, and they do it *together*. The experiment is no longer *about* the babblers or focused *upon* the babblers; it is led *with* them. The experiment, in other words, domesticates both Zahavi's practices and the babblers' *usage*. They are 'becoming with' (Despret 2004; Haraway 2008). Zahavi's babblers, therefore, would continuously pass through different states as well as different habits. They are sometimes wild. Sometimes, after becoming actively involved in a process of domestication and then going back to the wild to live amongst other wild babblers, the birds hold a new identity: they are feral. As long as this semiotic demarcation can be blurred among the unstable and living tropes, however, we should probably redefine the babblers in a way that corresponds to the Leibnizian 'quasi-causes': these creatures are successively and recursively quasi-wild, quasi-domesticated, and thus quasi-feral.

Interestingly enough, especially in the field situation, the expected often unfolds in an unexpected way. Most of the time, the way situations actually occur turns out to be much more worthwhile than previous expectations would have suggested (Latour 1999). What initially appeared to be clear-cut differences and sharp contrasts turned out to be more complex, problematic, and confusing. For example, we can assume that Jon wouldn't have thought of the birds as acting like humans or being led by anthropomorphic motives. Things are sometimes more complicated. At one point, both Jon and I were observing a group of babblers feeding at the nest. At one particular moment, I noticed that one of the birds made a vocal signal that, according to Zahavi's theory, calls the attention of their nearby comrades to the effort they are making. I was almost sure, however, that this babbler was not offering any food. When I asked Jon, he told me I was right. Therefore, I wondered why the bird had falsely vocalized the signal. If I could see that the babbler was deceiving, surely the other babblers could see this too. 'Very easy to understand,' said Jon, 'this bird is carrying out a real experiment. He just wants to assess the real state of hunger of the chicks, so he just modifies one of the variables.' I was intrigued. Clearly Jon was assigning his own scientist's *ethos* to the babblers.

Of course, this story demonstrates that what scientists call anthropomorphism is always 'someone's anthropomorphism,' 'common sense' anthropomorphism, or, more precisely, the lay-person's anthropomorphism. In other words, if the birds are acting and thinking as a *scientist* would think and act, this is not anthropomorphism; it is nature's rules. *Naturomorphism* is not a sin.

But this also shows us the real contrast between two ways of thinking: Jon and Zahavi inhabit the two extremes of a continuum between 'scientist' and 'naturalist' ways of construing behavior. You may find their position on that continuum by taking into account the role of models, the question of methodology, or the issue of anthropomorphism. However, the issue of anthropomorphism, which is generally what gives most of the characteristics to this continuum, has slightly shifted from its usual conception. Jon does not reproach Zahavi's babbler for thinking like an 'anthropos,' but he reproaches the bird for thinking like a naturalist: Zahavi's babblers neither follow regular rules nor seek them; they trust what they see; they live 'anecdotally' – all naturalists' ways of acting and thinking. By contrast, Jon's babbler, as a matter of fact, does not think like an 'anthropos'; he/she thinks like someone who obeys cognitive rules of nature – he/she thinks and acts like a 'good scientist'.

More generally, it was clear that Zahavi's ways of observing and thinking were very close to the naturalist's style, whereas Jon was working within the classical ethological tradition. Zahavi talks about his babblers as anthropologists talk about their human subjects: they have agency, intentions, hopes, and desires; and he may adopt their perspective. Jon's descriptions are highly abstract and technical: 'Arabian babblers provide an interesting focus for discussions regarding unselected helping, because this system involves apparent costs and benefits of helping that are very similar in scale to those of parental provisioning' (Wright 1999: 438). Moreover, Jon's babblers are determined by simple and rigid rules: according to kin selection theory, the variation in chick-feeding efforts by helpers reflects their evolutionary interest in provisioning the young. Helpers increase their effort 'in accordance with the net fitness returns from "investment" in the brood. Hence, more related chicks should be fed at higher rates, and helping should be adjusted according to both its energetic cost and current brood demand for food' (Wright 1997: 1440).

Eileen Crist remarks that, despite their intellectual continuity, there is a great disparity between ethologists and naturalists with respect to their use of language:

> In contrast to the naturalists' language of the life-world, ethologists use a technical vocabulary, in part constructed by themselves and in part appropriated from behaviourist psychology. The linguistic and argumentative edifice created by the pioneer ethologists led to the representation of animals as natural objects. Yet it is quite certain that neither Tinbergen nor Lorenz wanted to 'desubjectify' animals.[7] In using a technical, highly theoretical language, they aimed to establish the study of animal behavior as a rigorous science; they presupposed a specific idea of 'science', on the model of natural science as well as of comparative psychology [. . .] The inexorable if unwitting consequence of applying a technical language was the epistemological objectification of animals and ultimately their mechanomorphic portrayal.
>
> *(Crist 1999: 89)*

Sociobiology has driven this tendency to the extreme: models are the main goal of field studies.[8]

Variations and varieties

Zahavi, by contrast, is more interested in individual differences than in models. And he assumes that he ought to be interested in individual differences because that is of the utmost importance for the babblers themselves. They endlessly seek and show off differences between themselves. Indeed, Zahavi's way of thinking and construing behavior is entangled in a model, a sort of 'common sense' model – which, as a matter of fact, seems very close to anthropomorphism. But Zahavi's strong interest in individual differences leads him to refute theoretical models, be they from classical ethology or from sociobiology.[9] We find a clear example of that when he criticizes the use of 'species-specific patterns' – this terminology holding the recognizable features of classical ethology – to identify members of a group, since 'those patterns have evolved to show off differences among individuals *within* the group' (Zahavi and Zahavi 1997: 72).

This last sentence may help us to speculate about how Jon and Zahavi may observe such differences between birds. Jon is seeking a model, and he is therefore paying attention to the behaviors that can either be integrated into or that will fulfill or confirm the model. He probably actively selects. Far be it from me to suggest that Zahavi is not selecting: he does select, but he selects different events. Jon is selecting 'variations' in patterns; Zahavi is selecting 'varieties'. Jon's approach is close to the experimental way of constructing data. I like to think that Zahavi is a true successor to Darwin: he collects facts that do not fit with classical theories, he looks for varieties – in anecdotes, in little stories, in individual bird biographies. He compares the experiences of humans and birds, and, as we shall see, for example, those of dogs and birds, assessing the 'good reasons' they have to act in some odd ways. He fills their lives with motives, intentions, singular aims. Jon asks the babblers to be experimental objects, and the experimental device mirrors the work of a nature that controls events and beings: birds are being controlled and constrained by the same external rules; Zahavi reconstructs their experience, and he safeguards authorship and meanings. He works and thinks like a naturalist: 'Naturalists' attention to the detailed nuances', Eileen Crist writes (what she calls episodic descriptions), 'and variations of actions is connected with their focus on episodes of animal life. In documenting animal life, they choose to narrate concrete behavioral events, episodes actually witnessed' (Crist 1999: 73).[10]

In focusing on different events, Jon and Zahavi construct different narratives in which still more different events matter and may be observed. Jon, for example, concedes that he might not have paid attention to some subtle events while he was observing, and that (the fact of not taking these events into account) might have (but in reality has not) favored one hypothesis (babblers' help at the nest is kin-oriented) over another (babblers compete for prestige and for feeding nestlings) (Wright 1997). However, I would suggest the problem is not merely a lack of attention to events and details; it is a practical problem. The manner in which each approaches the babblers is not the same. The questions of control, place, and relationship are crucial. The distance between the observer and the observed (both concrete and subjective), the involvement of their presences and their bodies, the ways they address the birds, and the manners in which they present themselves are all different. All of these differences certainly modify the attention paid to the babblers' behavior and their own selective attention.

In suggesting the possibility that their selective attention might be bonded to the very practice itself, as well as to the embodied presence of the observer in the field, I raise a question that was studied by another scientist observing babblers at the site, Roni Osztreiher. Roni investigates whether or not the close presence of the observer may influence the way the babblers behave. He relates, for example, how a dominant bird felt forced to act as sentinel because of the observer's presence and he therefore could not feed the chicks. In order to assess this possibility, Jon – as a good experimenter – compared observations at different distances and found no differences.[11] But the question of the observer's influence, if we put it that way – as I have suggested, as an experimental question – makes another difference less visible. The difference is not only between observers; it rests between the ways observers take into account the differences between groups and how their questions will enhance or decrease these differences.

For Jon, all groups behave in approximately the same way, *with regard to his hypothesis*. This means that the hypothesis may easily discard as innocuous parasites the differences between groups, and that these differences will not be perceived. For Zahavi, by contrast, each group is unique. Of course, each group faces the same problems and they find similar solutions, but each group has its own style of facing the problem or of achieving this solution. For example, Zahavi explains that the relationship between the 'first' and the 'second' males – 'first' and 'second' denoting the place in the hierarchy's rank – is not the same in all groups. It depends a great deal on the make-up of the particular group. Even when the two males are brothers, this relationship varies from group to group. The prestige of one vis-à-vis another may be radically different from one group to another, which dramatically changes the way they behave. Since prestige means, for babblers, 'the respect accorded an individual by others' (Zahavi and Zahavi 1997: 143), it also changes the behavior of the whole group.

Groups have their own style, depending partly on the level of respect accorded in the group, and babblers have, according to Zahavi, their own personalities or temperaments. We also found evidence, however, that Zahavi adopts the perspectives of the animals he observes and, moreover, embodies those perspectives. Reading some excerpts from his work, one can see this very embodiment in his methodology, acting as a kind of bodily identification. It includes following the gaze of the sentinel and knowing exactly what it should do next as well as what the observer should look at next. This embodiment entails thinking *with* the bird, looking *with* it rather than looking at it, and knowing its intentions. Babblers and humans share common traits, built by the work and the narratives of the ethologist. These traits include body posture, the shifting of glances, common anticipations (what happens next?), and, last but not least, common stories. These common stories are not just narratives that would create (or rest upon) mere hermeneutical analogies. These common stories are enacted in, as well as produced by, a common story and a common history. Both humans and babblers create narratives, rather than just telling them. They create/disclose new scripts.

When Zahavi tells us, for example, that babblers compete for rank or in order to lay eggs, stories proliferate. Both humans and babblers seem to enjoy the narrative this competition produces, and they enjoy just as much the sharing of the same questions about these narratives: 'Such contests are complex and fascinating', Zahavi writes. 'Each is a unique story, reminiscent of historical epics, Shakespearean dramas, and biblical tableaux. In one case three males were copulating with four females, all of whom laid eggs. It was impossible for anyone – human or babbler – to tell which of the fledglings in the common nest were whose offspring; it would have taken DNA analysis for us – the babblers themselves had no idea' (Zahavi and Zahavi 1997: 130).

Although babblers may fiercely compete, an adult fights seriously with another adult only once, if ever – since, as already mentioned, the quest for prestige has replaced fighting. The 'real' fight almost always occurs between the most dominant male or female and the one next in rank: 'there is no point in risking everything for anything less than the top position'. Such a fight is savage. It starts all of a sudden, and, Zahavi explains, most observers fail to see any sign that it is imminent. 'It is as though all grievances and suppressed antagonisms between two brothers or between a father and son who have lived together peacefully for years burst out in this one life-and-death fight' (Zahavi and Zahavi 1997: 131). Each story is a scene full of actors, inhabiting an existential world, living adventures that give them a history, a bibliography, a personality, and a full repertoire of will, intention, and agency:

> In one case [. . .] an eight-month-old female tried to feed her ten-month-old sister; the latter stood up, snatched the food out of her younger sister's beak, forced her to crouch like one begging for food, and then stuffed the food down her throat. Once the younger bird had swallowed the food, the dominant sister pecked her until she fled. The frustrated younger sister then took another crumb and went to feed her subordinate younger brother, who was hunting quietly for food some thirty feet away [. . .] Babblers behave as though it was the act of giving, rather than the benefit given, that matters.
>
> *(Zahavi and Zahavi 1997: 138)*

This conclusion is worthy of our attention. For Zahavi, the right question to ask his animals – the question that produces *interest* – is: what matters to babblers?

A final babbler eccentricity will clearly illustrate this. As I mentioned in the beginning of this chapter, babblers dance. Why does it matter?

Why do babblers dance?

The most impressive social activity of the babblers is the morning dance. It happens only once every several days and it takes place almost exclusively at the first light of day. One of the babblers suddenly stops at a 'suitable dance floor' and starts preening nervously or sprawling with his/her throat touching the ground. Sometimes there is no response, and there is no dance. But a second babbler may come and join together with the first, preening his/her own body. This is an invitation for others. The dancers form a line (sometimes a circle) and they press against each other, squeezing under and over and between their partners.

This dance has raised another controversy. This time it involved another scientist we have already mentioned: Roni Osztreiher. According to Roni, babblers dance mostly in the spring and before the mating season. For him, it is clear that the dancers try to get at the center of the line formed by jumping over each other – the babblers' comrades try to prevent him/her from doing this by squeezing firmly with their bodies. According to Roni, the dance is used by the babblers to attest and assess their respective strengths: this is a ritual that will determine each babbler's rank in the hierarchy, and this is why it is done when coming out of the nest and just before mating time.

Zahavi, on the other hand, emphasizes the particular moment and the particular place chosen for the ritual: babblers dance in the open, near a bush, even though they could dance more safely under trees, and they do it before sunrise. This is the most dangerous place and the most dangerous time of day: the danger from raptors is greatest, since they can exploit the low light and surprise the babblers with relative ease. This is also the best time for babblers to feed, since many night creatures are still active – termites, insects, etc. According to Zahavi, this explains why babblers dance: they do it at the most inconvenient time and place because to undertake this is the most reliable way of showing their reliable commitment to their group:

> The dancing of babblers reminded us of stories our parents' generation told about the dancing of members of pioneering kibbutzim in Israel in the earliest, most difficult and exhausting years: For months we had nothing to eat, but we danced all night.
>
> *(Zahavi and Zahavi 1997: 117)*

Dancing is a way to test the quality of the relationship and to reassess it. Even dogs do this with their human partners. Large dogs like German shepherds, Zahavi explains, use their weight to test the bond with humans:

> They approach and stand next to a visitor in a friendly manner, leaning against the visitor's legs, gradually transferring more and more of their weight, until they are pushed off. How much of this treatment a visitor will take before pushing the dog off enables the animal to assess the visitor's attitude toward it.
>
> *(Zahavi and Zahavi 1997: 112)*

Once again, Zahavi's style looks like that of a naturalist: socially sophisticated babblers are compared to bright dogs well equipped with agency, to explain how social animals invent creative ways to assess, to undertake, and to reinforce the links they are entwined in.

The importance of a partnership to its members and the relationships between partners depend on many factors that can easily change, which makes frequent testing of the social bond essential (Zahavi 1976: 1). The testing not only assesses the bond and its reliability; it creates it. Rituals perform social links.

Mixing together in the same story babblers, human beings, and dogs is not solely a practice similar to the amateur's. It is an embodied practice that, I would suggest following Donna Haraway's analysis, recreates similarities between scientific and mundane practices, an embodied practice that redefines work with animals as companionship – not far from the process of domestication. Primatologist Barbara Smuts explains that she has been struck by the frequent enactments of brief greeting rituals, be it between the baboons she observed for years or between herself and her dog Bahati. Among baboons, both friends and non-friends greet one another all the time, 'and', Haraway comments, 'who they are is in constant becoming in these rituals' (2008: 26). The way Smuts construes ritual greetings – or the way her baboons enact them – is very similar to the way Zahavi and his babblers understand them. Haraway writes:

> [A]n embodied communication is more like a dance than a word. The flow of entangled meaningful bodies in time – whether jerky and nervous or flaming and flowing, whether both partners move in harmony or painfully out of synch or something else altogether – is communication about relationship, the relationship itself, and the means of reshaping relationship and so its enacters.
>
> *(2008: 26)*

Therefore, in dancing, babblers have literally embodied the meaning of their rituals. And they appear closer and closer to the baboons who taught Smuts how to dance – as we will see – when we read, in Haraway, Smuts' words: 'With language, it is possible to lie and say we like someone when we don't. However, [. . .] closely interacting bodies tend to tell the truth' (2008: 26). 'This is', Haraway continues, 'a very interesting definition of truth' (2008: 26). She is not referring to 'the tired philosophical and linguistic arguments about whether dogs can lie, and if so, lie about lying. The truth or honesty of nonlinguistic embodied communication depends on looking back and greeting significant others, again and again [. . .] Rather, this truth telling is about [. . .] holding in esteem, and regard open to those who look back reciprocally' (2008: 27).

The dance of the babblers leads us to a last hypothesis, one raised by Roni when he noticed that babblers tended to dance the day after he recorded a group with his rather heavy camera in a close distance. Do babblers react, or do they respond?

Domesticating practices

Might animals not only be *seen seeing* by the ethologists who observe them – as Jacques Derrida put it – but also be asked about their *response* to the asking? In other words, might ethologists think animals respond to them, might they take a position with regard to the situation that one proposes to them, and might they give an opinion about what is being asked? Eileen Crist is right when she says – anticipating Derrida's criticism of René Descartes – that the difference between response and reaction is not simply a matter of language (Crist 1997): it structures the way we see animals as passive reacting beings – driven by instincts, motivations, evolutionary rules, or genes – or as active beings who invent their own life, create, give meanings to events, anticipate, and, I would add, co-invent the practice of knowledge about themselves.

If I put it in another way, my question becomes more concrete: which kinds of practices, theories, or ethologists could make that shift possible? Theories do not only affect what scientists think they should see, and therefore what they actually see: theories affect what they do, who they are, and therefore the way animals see their scientists seeing them, the way animals respond to their scientists. Which kinds of theories or practices could afford to take that into account?

Not surprisingly, we may find some answers to this question in primatology. Hans Kummer, for example, explained that he had succeeded in habituating his Hamadryas baboons in Ethiopia – after months and months of unsuccessful approaches – only when he was able to ask, and to answer, this very question: how do these baboons judge us? How do they see us? Similarly, Thelma Rowell considers that habituation is not due to the work of the scientist, but rather to the way animals perceive the very practical role of their observer. She told me that when she was working with blue monkeys, she noticed that the eagles – which usually eat blue monkeys – were around overhead, 'but', she added,

> if they look down, of course they have very good eyesight, they see a face looking up [Rowell shows the gesture of looking up with binoculars], which is very off-putting if you are thinking of catching a monkey. And then, they went and caught other monkeys, somewhere else. I think actually the point is whether a reason why a monkey eventually got tamed was because in some sense they realized that, by allowing us to be with them, they were being protected from the eagles.

And then she concluded: 'There is much more awareness than the people watching them are aware of.'

However, the clearest example in primatology is given by the analysis Haraway offers on the work of Smuts – not so surprisingly, if one remembers Haraway's response to Derrida's lack of acknowledgment of the few scientists who have been able to meet the gaze of animals. Smuts tells us that when she began her first fieldwork with baboons, she did exactly what she had been taught to do in order to habituate her animals to her presence: according to the conventions of objective knowledge, she was advised to be as neutral as possible, to be 'like a rock, to be unavailable, so that eventually the baboons would go on about their business in nature as if data-collecting humankind were not present' (quoted in Haraway 2008: 24). Haraway remarks that, in these kinds of situations, scientists 'could query but not be queried':

> [P]eople could ask if baboons are or are not social subjects, or ask anything else for that matter, without any ontological risk either to themselves, except maybe being bitten by

> an angry baboon or contracting a dire parasitic infection, or to their culture's dominant epistemologies about what are named nature and culture.
>
> *(2008: 24)*

Progress in habituation was painfully slow: the baboons frequently looked at Smuts, and the more she ignored their looks, the less satisfied they seemed. Ignoring social cues is far from neutral social behavior: 'I imagine the baboons as seeing somebody off-category, not something', Haraway writes, 'and asking if that being were or were not educable to the standard of a polite guest' (Haraway 2008: 24).

Actively 'allowing the presence', 'getting tamed', or being a 'polite guest' are all practices that enrich the process of habituation and create another shift away from the mere continuum between domesticated and wild. Animals (observer/observed) and humans (observed/observer) are becoming attuned to each other. They create reciprocal but different interests and they learn to attune them by creating trust.

Both had to learn to be polite, in the ethical, political, and epistemological senses of the word. Smuts learned to respond, to acknowledge, to look back, perhaps to greet. She learned to respect and to make this respect the core of her practice, the condition of learning. And, as she tells us, her own being was transformed:

> I [. . .] in the process of gaining their trust, changed almost everything about me, including the way I walked and sat, the way I held my body[,] and the way I used my eyes and voice. I was learning a whole new way of being in the world – the way of baboons.
>
> *(quoted in Haraway 2008: 24)*

The process of quasi-domestication is a bi-directional process. She explains that, having learned the way baboons express their emotions or intentions, she could respond to them and be understood. Of course, this is primatology: a practice that deals with animals who can resist to a certain extent and who may have good reasons to resist this will to be 'no-one' that would prevent any interaction.

Might we, however, imagine that if babblers have become so interesting (so 'odd' as I previously thought), this might be partly due to the fact that Zahavi takes into account the opinion babblers may have about the questions scientists address to them? In that case, in some way, Zahavi and the babblers may be said to be responding to each other and to be acting according to the questioning and the response of the other. They would be making each other more responsive: Zahavi's anthropological-like practice – meeting, feeding, 'biographing', and calling his babblers – could, for example, explain the fact that babblers act so interestingly. They do more things. In other words, Zahavi's practice with 'his' babblers, if we draw the contrast with Jon's babblers, gives those birds a greater chance to be interesting.

I do not however have any clear evidence of that, and it might look like generous but candid speculation. Among the hypotheses, we could first consider the peculiarity of Zahavi's practice. First, as I have already suggested, Zahavi is undertaking with his babblers a hermeneutic approach grounded in common rationality: a kind of 'if I were him/her' way of thinking. This of course might also be suspected of heavy anthropomorphism: instead of putting himself in the babblers' shoes, Zahavi would actually be asking the birds to wear human shoes. The perspective would not be the babblers', but would only reflect a human-situated standpoint.

However, we can identify another characteristic of the practice, a characteristic I mentioned previously in passing. When arriving in a group territory that may be relatively wild, Zahavi calls the babblers. He 'invites them' and offers them presents, typically some piece of bread that is generally accepted eagerly by the birds. I could suggest that each encounter in the field actually looks like a request: 'Come and show me what you can do.' It seemed as though the babblers were being actively requested to cooperate. When Zahavi says that babblers help the experimenter, isn't he, himself, suggesting this hypothesis? However, these two hypotheses miss something that Smuts did not miss: she was able to acknowledge the hard work the baboons did to obtain from her a response to them. With my two hypotheses, babblers are not seriously involved in the change. How can we formulate that from the babblers' perspective? A consideration of the theoretical framework could help us to find a more polite way of assessing both the perspective of the babblers and their active interest in the research.

The hypothesis could be formulated in two ways. First, let us take the babblers' perspective. Because babblers are so fond of exhibition, and because scientists, thanks to Zahavi's signal theory, are looking for babblers to exhibit, babblers are 'requested' to do what they do 'well' and what they feel at ease in doing. Babblers, then, might be more sensitive to being seen and to seeing who is seeing them, since scientists are responsive to what they do.[12] That is why the birds come so easily when Zahavi calls them. It is as if he is saying to them, 'Come, babblers, and show me.' Of course, there are crumbs of bread. That is exactly what a behaviorist would say. Behaviorists made us forget that to give *is* a relationship per se, and I think that, sadly, they succeeded in making starving animals (artificially starving animals) the norm – as if animals were only interested in food. Might the babblers be construing Zahavi's offer as a greeting? Do they greet the scientist?

I do not have the answers to these questions, but I strongly affirm that Zahavi is clearly greeting his babblers. And he does that with a present, with his body, when he sits among the babblers, when they approach closer and closer, when they come on his arms, when he turns his hand; he greets them with his gaze, with his voice. And babblers do understand that the question of prestige is not at stake between them – if one remembers that they accept food from their scientists after having resisted a congener's offer.

Let us now take the second perspective, that of Zahavi. I might suggest that because Zahavi is so interested in display, because his theory is almost entirely grounded in exhibition, he has had many more chances to assume that the babblers were responding to the scientist who was interacting with them. I have noticed, in one of my research projects with farm breeders, that animals taken to participate in contests or exhibitions are very responsive, and that breeders are as acutely sensitive to the way animals may be self-conscious as they are prone to credit their animals with a clear consciousness of what is at stake: 'When cows go to competitions, they know they are beautiful,' they 'pose for the camera and know exactly that they are seen,' 'they are proud' (Despret and Porcher 2007: 80). The same may be said about trainers and their dogs. And the same happiness transpires.

This hypothesis delights me not so much because it talks about happiness, but rather because it gives a new meaning, a new constructivist and non-relativist translation, to that famous and rather tired assertion – the very one I actually shared a long time ago when I was completing the files to get a grant: 'The behavior of the animal is the product of the observer's vision.' The new translation that I propose slightly changes some of the words in this assertion. I would suggest that 'the ways animals act are the consequences of the observer's gaze'. Yes, indeed, their actions are the *responsible* consequences (from *consequi*, 'following closely') of a no less responsible gaze.

Notes

1 This claim is however far too general: in the context of scientific controversy, scientists may tend to focus upon the actual changes demonstrated by the animals (or more often 'suffered' by the other observer's animals) to explain the discrepancy between rival accounts. See, for example, Amanda Rees' analysis of the 'infanticide controversy' (2009).
2 According to Rowell, the rigid hierarchy which was unanimously observed among primates in the early accounts may be explained (among other causes) as an effect of the observation; she suggests, for example, that before the practice of habituation, the covert method often used by primatologists to achieve some (relative) proximity with their animals (in a very short time) was provisioning. This may also explain, according to her (and to Margaret Power [1991] to whom she is referring), the historical change in perspective about chimpanzee society observed by Jane Goodall's team: provisioning had progressively accentuated competition among chimpanzees and produced social disruption. But see below for other (and less problematic) cases in which animals may take an active role and benefit from the encounter.
3 Thelma Rowell, interviewed by the author, June 2003. This interview was carried out during the making of a documentary, Vinciane Despret and Didier Demorcy's *Non Sheepish Sheep* (2005), for the exhibit 'Making Things Public. Atmospheres of Democracy' at ZKM in Karlsruhe, Germany in spring 2005. See also Vinciane Despret, 'Sheep Do Have Opinions' in the catalog of this exhibition (2005).
4 Other scholars (Crist 1999; Mitman 2005; Glickman 2000) have suggested that classical (or Lorenzian) ethology could have precluded the recognition of sophisticated social or cognitive competences. On how classical ethology precluded the acknowledgment of reconciliation in sheep, see also Rowell (2000b).
5 Interestingly, Zahavi's style, and the stories he tells, are exactly the same in his scientific papers (from 1973) and in his popular book (Zahavi and Zahavi 1997).
6 Thelma Rowell, interviewed by the author, June 2003.
7 But see Marion Thomas (2003) for a more radical criticism.
8 See Crist, who talks about a descriptive style characterized as 'frequency-laden' (1999: 148).
9 Of course, I would not deny that Zahavi's model is sometimes as despotic – it aims to explain all behaviors or all strange behaviors – as others are. But this model seems to work case by case, each time taking other paths, other motivations, other experiences, in sum, other 'details'.
10 Crist (1999) points out the importance of the common and possibly shared (or shareable) experience in naturalists' narratives and modes of understanding.
11 If babblers are actually as subtle as some human observers suspect, the no-differences-between-groups hypothesis might deserve a full anthology of explanations.
12 I would not go as far as saying that goats are fond of exhibition – with goats, you can never tell – but I find convincing evidence in the work of Michel Meuret (interviewed by the author in June 2009) that they are very sensitive to being observed. According to Meuret, observing them modifies the way some of them act. For some of the goats, the interest of the human observer enhances their will to compete and they enact much more 'supplanting' in foraging activities, while, for others, this interest really bothers them – and they make it clear for the human observer, sometimes pushing him aside. Still others, having noticed that another goat is the focus of the interest of the scientist, aggressively react toward their congener. See also Agreil and Meuret (2004).

References

Agreil, C. and Meuret, M. (2004) 'An Improved Method for Quantifying Intake Rate and Ingestive Behaviour of Ruminants in Diverse and Variable Habitats Using Direct Observation', *Small Ruminant Research* 54: 99–113.
Crist, E. (1997) 'From Questions to Stimuli, from Answers to Reactions: The Case of Clever Hans', *Semiotica* 113.1–2: 1–42.
—— (1999) *Images of Animals*, Philadelphia: Temple University Press.
Despret, V. (1991) 'Éthique et Éthologie: Une histoire naturelle de l'altruisme', *Cahiers d'éthologie*, numéro spécial, 11.2: 141–266.
—— (2004) 'The Body We Care for. Figures of Anthropo-Zoo-Genesis', in *Bodies on Trial*, M. Berg and M. Akrich (eds), *Body & Society* 10.2–3: 111–34.

—— (2005) 'Sheep Do Have Opinions', in B. Latour and P. Weibel (eds), *Making Things Public*, Cambridge, MA, MIT Press, pp. 360–368.
Despret, V. and Porcher, J. (2007) *Etre bête*, Arles: Actes Sud.
Glickman, S.E. (2000) 'Culture, Disciplinary Tradition, and the Study of Behavior: Sex, Rats, and Spotted Hyenas', in S. Strum and L. Fedigan (eds), *Primate Encounters: Models of Science, Gender and Society*, Chicago: University of Chicago Press, pp. 275–95.
Haraway, D. (1992) *Primate Visions: Gender, Race, and Nature in the World of Modern Science*, London: Verso.
—— (2008) *When Species Meet*, Minneapolis: University of Minnesota Press.
Heinrich, B. (1991) *Ravens in Winter*, New York: Vintage Books.
—— (2000) *Mind of the Raven*, New York: Harper Collins.
Latour, B. (1999) *Pandora's Hope*, Cambridge, MA: Harvard University Press.
—— (2000) 'A Well Articulated Primatology: Reflections of a Fellow Traveler', in S. Strum and L. Fedigan (eds), *Primate Encounters: Models of Science, Gender and Society*, Chicago: University of Chicago Press, pp. 358–82.
Mitman, G. (2005) 'Pachyderm Personalities: The Media of Science, Politics and Conservation', in L. Daston and G. Mitman (eds), *Thinking with Animals*, New York: Columbia University Press, pp. 173–95.
Power, M. (1991) *The Egalitarian: Human and Chimpanzee*, Cambridge: Cambridge University Press.
Prior, H., Schwarz, A., and Güntürkün, O. (2008) 'Mirror-Induced Behavior in the Magpie (*Pica pica*): Evidence of Self-Recognition', *PloS Biology* 6, available online at www.plosbiology.org/article/info:doi/10.1371/journal.pbio.0060202, accessed 15 June 2013.
Rees, A. (2009) *The Infanticide Controversy: Primatology and the Art of Field Science*, Chicago: Chicago University Press.
Rowell, T. (1999) 'The Myth of Peculiar Primates', in H. Box and K. Gibson (eds), *Mammalian Social Learning: Comparative and Ecological Perspectives*, Cambridge: Cambridge University Press, pp. 6–16.
—— (2000a) 'A Few Peculiar Primates', in S. Strum and L. Fedigan (eds), *Primate Encounters: Models of Science, Gender and Society*, Chicago: University of Chicago Press, pp. 57–71.
—— (2000b) 'The Ethological Approach Precluded Recognition and Reconciliation', in F. Aureli and F. de Waal (eds), *Natural Conflict Resolution*, Berkeley: University of California Press, pp. 227–9.
Strum, S. and Fedigan, L. (2000) 'Changing Views of Primate Society: A Situated North American View', in S. Strum and L. Fedigan (eds), *Primate Encounters: Models of Science, Gender and Society*, Chicago: University of Chicago Press, pp. 3–49.
Thomas, M. (2003) 'Rethinking the History of Ethology: French Animal Behaviour Studies in the Third Republic (1870–1940)', unpublished PhD thesis, University of Manchester, Center for the History of Science, Technology, and Medicine.
Wright, J. (1997) 'Helping-at-the-Nest in Arabian Babblers: Signaling Social Status or Sensible Investment in Chicks?', *Animal Behavior* 54: 1439–48.
—— (1999) 'Adaptive versus Non-adaptive Helping in Cooperative Breeders', *Behavioral Ecology and Sociobiology* 46: 437–8.
Zahavi, A. (1973) 'Communal Nesting by the Arabian Babbler. A Case of Individual Selection', *Ibis* 116: 84–7.
—— (1975) 'Mate Selection a Selection for a Handicap', *Journal of Theoretical Biology* 53: 205–14.
—— (1976) 'The Testing of a Bond', *Animal Behavior* 25: 1.
—— (1990) 'Arabian Babblers: The Quest for Social Status in a Cooperative Breeder', in P. Stacey and W. Koenig (eds), *Cooperative Breeding in Birds*, Cambridge: Cambridge University Press, pp. 105–32.
Zahavi, A. and Zahavi, A. (1997) *The Handicap Principle: A Missing Piece of Darwin's Puzzle*, Oxford: Oxford University Press.

3

Escaping the maze

Wildness and tameness in studying animal behaviour

Lynda Birke

> [T]he program of scientific experimentation that leads you to conclude that animals are imbeciles is profoundly anthropocentric. It values being able to find your way out of a sterile maze, ignoring the fact that if the researcher who designed the maze were to be parachuted into the jungles of Borneo, he or she would be dead of starvation in a week[.]
>
> Elizabeth Costello, in J.M. Coetzee, *The Lives of Animals*

The growth of human-animal studies (HAS) represents a concern to 'bring animals in' – to mainstream sociology, or to the humanities, for example. But there are, of course, areas of scholarship in which nonhuman animals have always been present. Disciplinary divisions, separating social from natural sciences, have maintained a distinction that allots animals and their behaviour largely to biology, while human sociality is usually studied elsewhere. To be sure, there is overlap – in psychology, for instance – and the distinction is increasingly challenged by posthumanism, which seeks to overturn humanist beliefs in us versus the rest of nature (Wolfe 2003; Haraway 2008; Taylor 2012). Even so, the heritage of that separation persists: humans and nonhumans still usually appear as research subjects in different academic journals.

Nevertheless, the study of animal behaviour contributes crucially to scholarship in HAS. One source, for instance, is the work of cognitive ethologists (e.g. Bekoff 2002), whose focus on sentience and cognitive abilities has helped scientists re-evaluate species they study and to realize that whether we perceive animal abilities depends on what questions we ask – as Coetzee clearly recognized.

In this chapter, I focus on biological science; I do so in part because I am a biologist with a background in animal behaviour, but also because biology contributes fundamentally to our ideas of what counts as an 'animal' or a 'species'. I draw, then, on my experiences of working with behavioural ideas, and use this to consider the interconnecting categories of wild/feral/tame that are a theme of this book. Biology studies living organisms, such as animals; some investigations focus on those living in natural habitats – in wild nature. Others may be concerned with animals kept in constrained conditions. This includes studies of domestic or zoo animals, as well as animals kept for laboratory use. Some of the latter may be used in investigations of that species' biological processes, which can contribute to knowledge about such processes in general or to veterinary science in particular. Many more, however, are used as models of primarily human diseases, serving as proxies for us in medical research.

Put another way, science relates to nonhuman animals partly by defining them and partly by using them as objects of research. Here, I consider three case studies. The first centres on ideas of 'wild' in thinking about species, particularly in ethology; the second moves to laboratory animals, considering them as examples of domestication (taming) in the service of science; the third focuses on feral animals who were studied precisely because of their ambiguous status between wild and tame. I will use these as examples to explore ideas of wildness/tameness/ferality in relation to science, focusing particularly on understanding of animal behaviour.

Biological science defines animals in the sense of arbitrating boundaries between species, and researches their characteristics: their physiology, ecology, behaviour. The concept of species – what is a species? how is it distinguished from another one? – may seem straightforward in colloquial use, but is notoriously troublesome and has bothered biologists seeking definitions for a long time (de Queiroz 2005). The simplest definition refers to populations that can interbreed. Since domesticated animals could in principle interbreed with wild animals of the same species, the idea of 'species' cuts across wild/tame distinctions. Still, what I want to stress is that biology first and foremost focuses on *species*, particularly how they are thought to have evolved, with implications for how animal behaviour is studied.

Domesticating behaviour

> Like man [sic], the rat could be said to exist on the boundary between the natural and the unnatural. [. . .] [T]he rat is neither domesticated nor entirely wild; rather it is an unwelcome but perennial cohabitant of the built environment.
>
> E. Ramsden and J. Adams, 'Escaping the Laboratory: The Rodent Experiments of John B. Calhoun and Their Cultural Influence' (2009)

Biologists are trained to think primarily in evolutionary terms: how did that species, or that characteristic, evolve? How are organisms adapted to their way of life? Inevitably, then, what species do 'in the wild' is the default setting for biology; domesticated animals are largely seen relative to their wild counterparts (discussions of dog behaviour typically refer to behaviour of wolves, for example). Trained in biology, I often do this too; even when thinking about laboratory-bred rodents, I would ask myself what they would do 'in the wild'. Perhaps their free-living ancestors did behave similarly to these highly domesticated strains, but that is an assumption. And, as the quotation above indicates, what 'wild' means depends on species: are rats, living close to our habitations, truly as 'wild' as, say, a snow leopard in the Himalayas? Wild, as other chapters explore, is a slippery concept in biology, as elsewhere.

'Tame' is no easier to pin down. It generally implies behavioural propensity in relation to humans; that is, animals accept human handling. In this chapter, I mostly use the (equally tricky) term 'domesticated' – the word more commonly in use in biology – referring to the species' history in relation to humans, a long process of change and adaptation. To become 'domesticated', species undergo many changes in physiology, anatomy, and behaviour. Some animals, of course, may be domesticated, but not really tame (the bull used for bullfighting, for example). While some species probably played active parts by living close to humans, the process of domestication also entails humans guiding selection of particular traits (Budiansky 1992; Clutton-Brock 1987). In particular, domestication involved selecting for general docility.

Biologists also identify a 'domestication syndrome', describing changes not artificially selected but occurring alongside selection for docility, such as colour, shape, teeth, and brain function.[1] Indeed, some scientists believe that species such as dogs self-domesticated first, in the sense that less aggressive animals were more likely to survive close to humans (Coppinger and Coppinger 2001). Later, human communities presumably began artificial selection, seeking further reduction in aggression. Lest this seems relatively straightforward, biologists also identify self-domestication in free-living bonobos, on the grounds that they have evolved many characteristics of the domestication syndrome (Hare *et al.* 2012). That is, behavioural and physiological 'domestication' can occur in the wild, without human intervention.

Feral animals fall in between. To biologists, feral animals imply two things. They may represent ecological disruption if they are 'non-native'; in that case, biologists might be interested in how to manage their populations in ways less disruptive to native fauna, or with a view to culling them.[2] Alternatively, they might study feral animals with reference to wild counterparts (horses or dogs, for example), on the assumption that, even if derived from escaped domesticates, they still exhibit 'wild' behaviour patterns. What interests biologists here is how much feral and wild populations match or diverge, and how ferals have adapted their behaviour to different circumstances. Sometimes, these species provide surprises: they are at times notably more flexible than studies of wild populations would suggest. Feral cats, for example, seem to show greater diversity of social structures than do their wild ancestors (Leyhausen 1988), while feral dogs may live in wolf-like packs, or in fox-like small, territorial groups (Macdonald and Carr 1995). Feral, in short, may not be completely equivalent to wild, any more than domesticated necessarily means tamed.

Turning to my three cases, I consider, first, how much ethological theory and practice hinges on prioritizing 'the wild'. Ethologists do, of course, study behaviour of domesticated animals (though this usually falls within 'applied ethology', a specialized subdiscipline), but they typically do so through the lens of species-in-the-wild. My second case enters labs, to consider a special case of domestication, namely the transformation of previously wild animals into tools of the trade: lab animals, whose behaviour is manipulated to fit experimental needs. Here, their behaviour can be seen as tamed, though only up to a point. My third case takes one particular example, in which domesticated rats went feral; they were studied partly to understand ratty social behaviour in free-living groups, but also with reference to human behaviour. While I exploit the wild/feral/tame distinctions here, the point of examining these cases is precisely to illustrate the slipperiness of such categories.

Studying behaviour: prioritizing the wild

Broadly, scientific study of animal behaviour emerged in two areas: ethology and comparative psychology. Ethology, from the Greek for character, is a subdiscipline of biology, focusing on behaviour 'in the natural environment'; psychology has historically prioritized examining how animals learn, often testing them in constrained environments. That is obviously a simplification which many scientists now would challenge. But these disciplines do have slightly different histories and outlooks. An introduction to ethology sums them up thus: 'A traditional view of the distinction between ethology and psychology was that the psychologist put his animal in a small enclosure and peered in to see what it was doing, while the ethologist put himself in the box and looked out at what the animals round about were up to' (Slater 1985: 7).

Because of ethology's emphasis on behaviour 'as it occurs in nature', much of it involves field studies (sometimes, indeed, using a hide). That is, researchers go out into nature, spending hours in detailed observation of the species/behaviour of interest. Although observational, it is today usually systematized or experimental; researchers use field studies to test specific hypotheses – for example, theories of sexual selection might predict that females behave in particular ways, which can then be investigated in the field.

Small populations can also be studied in captivity, enabling scientists to control conditions in which they make observations; many studies of bird species, such as the European Great Tit, used captive flocks in outdoor aviaries to facilitate observation. And sometimes, ethologists study animals in laboratories in order to test specific hypotheses about, say, what stimuli trigger particular behaviours. What characterizes ethological thinking, however, is the emphasis on wild animals and species' evolutionary history. Even if researchers study behaviour in laboratory-bred rats (as I have done), then they focus on behaviour assumed to characterize the species 'in nature'.

Niko Tinbergen – one of the 'founding fathers'[3] of ethology – argued that it dealt with the 'four whys' of behaviour. These are: causal (why did the animal do this at this moment? hormonal influence would be one example); developmental (why did it happen at that stage of life or development?); function (what function does the behaviour serve in terms of adaptedness to living conditions?); and evolution (why/how did the behaviour evolve?). The last two, however, became priorities for several decades, to the detriment of questions of cause and development. After the mid-1970s in particular, ethology came to focus overwhelmingly on questions of evolution and adaption, with the rise of behavioural ecology and sociobiology, shifts that helped to cement the place of quantification in behavioural studies (Crist 1999; Alcock 2003; Ord *et al.* 2005).

The drive toward quantification and experiment is inevitable in any emerging branch of science, but, in ethology, it was influenced also by the firmly scientific footing of comparative psychology. Ethology, emphasizing instinct and 'natural' behaviour, had arisen primarily in Europe, while psychology, focusing more on experimental investigation of learning, developed more strongly in North America. For some decades in the twentieth century, these two remained apart (Wilson 2002), with considerable implications for how animals – and behaviour scholarship – were seen. In *Images of Animals*, Eileen Crist (1999) describes two different writing styles in the history of scientific reporting of animal behaviour. Descriptive styles have generally typified natural history, in contrast to the more detached styles of scientific reports.[4] Beginning with Darwin, she explores how his descriptive narrative seemed to acknowledge animal subjectivity and to display a belief in the animal mind. However, by the middle of the twentieth century, much writing about behaviour expressed scepticism about animal minds and put emphasis on experimental method and quantification. Though many early ethologists were great observers of animals, retaining the descriptive detail of earlier naturalists, they also sought to put their work onto solid scientific foundations, doing experiments and explaining behaviour in terms of mechanisms of stimuli and responses.

In similar vein, Dewsbury (1997) compared writing for popular audiences by Tinbergen with that of Frank Beach (an American psychologist who focused on effects of hormones, particularly on sexual behaviour). The contrast illustrates further the divergent styles noted by Crist and the different emphases on description and quantification. Tinbergen made much of his joy in doing research and was happy to attribute mental states to animals: he wrote of wasps 'happily digging' or 'flying leisurely', and to him one bird can 'obviously understand' another (quoted in Dewsbury 1997: 376).

Beach, by contrast, used more sombre terms to describe animals: he referred, for instance, to 'beasts' or to 'lowly rats' or to alligators showing 'stubborn resistance'; he was reluctant to impute mental states (Dewsbury 1997: 376–7); and he emphasized the centrality of experiment. The illustrations of these articles followed suit. While Tinbergen's were full of idyllic rural panoramas showing animals in situ, Beach's articles portrayed animals without context; in some, the animal appeared with all background trimmed away. Dewsbury comments, 'It is as if the animal can be shown in the abstract, independent of the environment in which it [sic] exists' (378).[5]

The scientific approach to animal behaviour brought many things. On one hand, careful, systematic analysis of different factors influencing behaviour yields understanding of how, for example, species respond to specific stimuli or have become adapted to their environments. This in turn generates predictability, which is important in understanding species survival, or how animals might be managed. On the other hand, it loses individual variations, which easily disappear in scientific generalizations. As Crist notes:

> In a discourse that is fundamentally statistical in character, individual variability, as such, is minimally significant; it does not appear within the descriptive apparatus. [. . .] Indeed, statistical operations precisely sink individual variation and cancel out idiosyncratic expressions, for the purposes of discerning and mapping out general patterns and trends.
> *(1999: 147)*

Tinbergen, like Darwin, brought individuals into his popular writings, but it was the *scientific* style, stressing objectivity and statistical verification, which predominated. Animal behaviour came to focus on *generic* animals, which then stood in for species.

Reliance on quantification and the prevalence of generic animals as prototypes of the species remain predominant themes of behavioural science, although, over time, many differences between scientists have diminished (Bateson 2003). But, in emphasizing objectivity, the individual animal's subjectivity and lifeworld tend to become obscured. So, while wildness is emphasized in the sense of the adaptations of species to their ways of life, there is less understanding of what it is like to *experience* that life.

Yet even though biologists study behaviour of species 'in nature', they face the problem that the very presence of human observers (or remote tracking equipment) can alter the behaviour they want to record. However much scientists seek distancing, animals may not necessarily be fooled by their tactics, such as the use of a hide (Martin and Bateson 1986). Mere human presence can influence behaviour in a wide range of species, from lab rats (Dewsbury 1995) to agricultural or zoo animals (Hemsworth 2003; Hosey 2005) to wild damselfly larvae (Baker and McGuffin 2007) to free-living or captive primates (Jack *et al.* 2008; Iredale *et al.* 2010). Animals, in turn, may perceive humans in many ways: as predator or as prey; as socially insignificant; as conspecific; or as symbiotic, as would be the case if the researcher were also a caretaker (Estep and Hett 1995).[6]

Even if scientists use automated equipment in the lab (such as devices to measure animals' activity over 24 hours), effects of handling or human proximity still occur. Indeed, many animals do not easily habituate to human presence and continue to alter their behaviour when people are about or in a hide (Almeida *et al.* 2006). Even techniques for remote recording like radiotracking devices can disturb behaviour, with possible implications for long-term welfare (Peniche *et al.* 2011).

To be sure, some species studied in the field do indeed seem to become habituated, so that human observers appear to make no difference (Crofoot *et al.* 2010). Field biologists often use habituation, or familiarizing the animals with the presence of observers, in order to minimize disturbance. The idea is that by repeated, gradual exposure animals will become so used to observers that they simply ignore them, and scientists can maintain the fiction of objective distance.

This is not easy, however, and is time-consuming. For example, meerkat colonies have long been studied, particularly in investigations of cooperative breeding, and of social/cultural learning (Thornton and Clutton-Brock 2011). Initial familiarization took over a year: after that, any pups born into habituated groups were themselves habituated, suggesting that getting used to people is socially learned. That is, the high level of social integration characterizing meerkat colonies is precisely the means whereby scientists can get the animals to stay put rather than flee at the sight of humans.

Indeed, so familiar have people become to meerkats that they do indeed seem to ignore humans completely, apparently carrying on their business regardless. At other times, they do not ignore humans but rather consider them as part of their world: researchers have famously had to contend with animals sitting on their heads or equipment, producing an iconic image made familiar through televised series about meerkat behaviour.

Such acts clearly muddy the waters between wild and tame; yet, as Candea points out in his study of scientists studying meerkats, researchers resist calling these animals domesticated, insisting on referring to them as habituated – a narrative which presumes a distancing, a non-engagement of researchers with animals. Habituation supposedly allows researchers to blend into the surroundings, to maintain the story that these are truly wild animals. Candea notes, too, how interactions with meerkats during routine weighings are not recorded: they cannot form part of the dataset, as they do not fit the naturalist frame of the study (2010: 250).

'Wildness', then, is carefully conserved in ethological narratives, even when talking about animals who know perfectly well that humans lurk around. 'Wild' animals form the subject matter of biological research, and research is often conducted and written about in ways that reinforce this narrative. Now I do not want to suggest that biologists should abandon studying behaviour just because of human influence. On the contrary, they need to be aware of how they might affect animals, and how animals might affect their knowledge. What I want to emphasize here is that 'wild', as characteristic behaviour of a species being observed by people, is not necessarily straightforward. The baseline of 'animals in the wild' not only pervades biological thinking, but also remains a largely unchallenged category. Moreover, our understanding of nature is produced by the way we do the studies of and write about animals. Human observers cannot necessarily assume that the animals are simply doing their own thing, unaware of the act of being observed, and unable to influence the outcome.

Domesticating in the service of science

> Under precisely controlled laboratory conditions, an animal does as he [sic] damn well pleases.
>
> The 'First Harvard Law of Animal Behavior'[7]

My second theme removes us from wild nature and toward lab animals. While animals representing the wild also enter the laboratory, they have become domesticated in the process, both in the way they respond to handling and in their representation. Lab animals' behaviour may be controlled, it may be studied for its own sake or perhaps as a model for human behaviour or mental states. Or, it may be something which otherwise could interrupt experiments, as the quotation above indicates – an unwanted intrusion in which the animal asserts something of his or her wild nature.

While some animals are kept in labs for purposes of studying something about that species, many are not, but are used instead as surrogates for illnesses of another species – almost always ourselves. Most of them are purpose-bred, and the majority are rodents. Lab animals are not, of course, 'in nature'; on the contrary, nature is largely excluded from laboratory spaces. In that sense, lab animals encapsulate a specific form of domestication. These are animals whose behaviour, through generations of artificial selection, renders them relatively domesticated, even 'tame'. This, then, is far removed from studying the wild: by implication, tamed animals make more suitable surrogates for us.

Indeed, lab animals are usually produced, through breeding or genetic manipulation, to meet specific experimental needs. 'Nature' is transformed in labs. Both apparatuses and animals in experimental labs are preconstructed, purpose-bred, removed from their context; raw materials are carefully selected and 'nature' is largely excluded (Knorr-Cetina 1983: 119). Institutions such as labs and zoos endeavour to keep wild nature out. This means not only keeping out any diseases that wild counterparts might bring in, but also preventing larger animals from entering. Labs may be full of mice, but woe betide the mouse who foolishly tries to enter from outside: her fate will be poison or a trap. If animals escape, they must be re-caught and replaced in cages. If the animal was one kept apart from potential infection, behind barriers, then escape will mean sudden exposure to pathogens: whether by this means, or by human hand, the outcome for the animal almost certainly means death.

There are many reasons why lab escapees might not be left to wander about. Both the animals themselves, and the people in the lab, are potentially at risk, whether of transmitting modified genes or disease, or of more direct injury. These are animals out of place. Stock animals may be returned to their cages, but if they had already entered experimental protocols, then this may not be possible. Just as animals who up and die by themselves should not be counted as part of an experiment (see Birke and Smith 1995), so neither do animals who have upped and made their way toward the wild life – after all, the scientists will not know what has happened to them in the meantime. Tameness and wildness are contested territories in laboratory spaces.

Historically, to become lab animals entailed a gradual transformation. A century ago, while some animals were bred for purposes of experiments, others were simply taken from their usual habitat – even, in one case, rats from local rubbish tips (Foster 1980). Increasingly, however, controlled breeding ensured specific traits and produced 'standardized models' of many species, especially rodents (Rader 2004). Animals long seen as vermin, carriers of disease, were thus turned into symbolic saviours. Domestication here is two-pronged: first, animals themselves are manipulated genetically to produce desired characteristics; then, they are symbolically 'de-wilded', as they move from sewer to salvation.

At the same time, they remain animals, who may or may not cooperate with experimenters. In doing experiments, scientists must transform animals, both metaphorically and literally, from behaving beings to becoming data. Lynch (1985), in a study of neuroscientists using rats, noted contrasting everyday ideas of 'naturalistic animals' and how the scientists spoke about the 'analytic animal' – that is, after it became data, after its brain has been sectioned, sliced, and turned into material on a slide. When they said, 'that was a good animal', they meant good results from a well-prepared specimen. This is quite different from our common sense observation of other species, he argues. Referring to 'good animals' was not likely to mean the scientists' recognition of individuals' particular qualities in life. On the contrary, what was left of them was seen as fitting certain standards – whether of presenting brain tissue or because the slide fitted accepted hypotheses about brain anatomy. However good the slide preparation, if the material did not fit the hypothesis, then I suspect the scientist would have been less keen on praising the animal from whom it came.

In lab studies, the drive toward quantification is clear. Observing laboratory practices, sociologist of science Bruno Latour wryly noted that these have to do with 'the transformation of rats and chemicals into paper' (Latour 1990: 39). Scientists, he argues, are obsessed with graphs and diagrams, but seldom see the animal who supplied the data: 'Bleeding and screaming rats are quickly dispatched. What is extracted from them is a tiny set of figures', he suggests (39).

The distancing described here is also evident in the way experimental reports are usually written; consistent use of the passive voice and reduction to numbers ensure that human involvement appears minimal. Even if animals had previously had names, they seem so often to appear as numbers in written reports. Ironically, this happens even if the animal in question in lab studies is a chimpanzee, who typically might be named by caretakers, a detail omitted in subsequent papers (Wieder 1980).[8] The 'domestication' of nature in labs thus not only entails selective breeding, but also taming nature's recalcitrance through carefully managed experimental practices and writing techniques. It involves, too, denying the animality of animals, something brought forcefully home to me by a lab technician who informed me that the scientists she worked with preferred opaque cages so they could not see the animals watching them.

Yet animal recalcitrance returns: animals may indeed watch us or be seen as 'doing what they damn well please' however much we try to control variables. Many factors, from different housing conditions, strains of animal, labs, or human handlers, can all affect – and change – experimental outcomes (Birke *et al.* 2007: 45; Balcombe 2006). Indeed, such effects can alter outcomes in, say, physiology or toxicology, and with implications for the interpretation of results in relation to human medicine (Sherwin 2004). This does not necessarily invalidate findings – which may still be extrapolable to other situations – but it does mean that interpreting data drawn from controlled lab studies requires caution.

Certainly, making inferences about animal behaviour on the basis of lab studies can be tricky. As Coetzee recognized in the quotation at the start of this chapter, there are profound problems in concluding that particular species are imbecilic if animals fail to work out our anthropocentric questions. Testing intelligence in laboratory conditions has a long history, though it tells you only that the animals did or did not work out that particular task in that particular setting. But such impoverished settings do not approximate how animals live their lives if they are *not* in captivity (Rogers 1997; Segerdahl 2012) and probably tell you more about the people setting the task than they do about the animals doing them.

Controlled testing conditions can thus constrain how scientists interpret animals' behaviour or abilities. Another example comes from rats in lab-based studies of mammalian sexual responses. While 'rat sex' can be investigated in the wild, it is also part of studies into, say, effects of hormones on developmental sequences of behaviour. For many years, scientists studying rat sexual activities focused on lordosis – the posture permitting coitus – while failing to see the active role females play in soliciting sexual activity. This oversight, suggested McClintock and Adler (1978), resulted from using tiny testing cages, which severely limited females' reactions. In such restricted spaces, these animals certainly cannot do what they please.[9]

Experimental laboratory science thus involves rigorous policing of the boundaries of 'wild'. 'Wild' is excluded from lab practices through separation of animals designated as part of lab work; through specialized breeding; through converting animals into data; through controlling variables and writing styles. Meanwhile, strenuous efforts are made to ensure that potentially recalcitrant animals are domesticated and constrained. And these animals must become habituated to human handling and presence, just like their wild counterparts in field studies.

If behaviour is under investigation in labs, it is not the free flow of living that makes behaviour meaningful to the animal, but a small fraction of specific sets of responses that can be investigated experimentally. This tight control is in many ways, of course, critical to the success of science: by doing carefully structured experiments, scientists can tease out multiple factors which can influence how biology works, thus achieving some sort of predictability. What I want to emphasize here, however, is that it also facilitates a domestication – not only of the animal per se, but also of how we come to understand nature as 'wild'.

Science in labs may partly be about studying 'wild nature', but it is also about domesticating her. There is an irony here, since those who handle animals regularly (animal technicians, more often than research scientists) know perfectly well that these are more than tools of the trade, they are 'animals' – they need feeding and cleaning, and they bite. In Tiptree's 1978 short story about a psychologist having qualms about what his lab did to rats, the protagonist muses about their role in research, concluding that emotionalism in rodents is two things: defecation and biting psychologists. His qualms, in short, come from seeing the rats as shifting back from domesticated data to 'wild', to being animals, who sometimes do as they 'damn well please'. In this guise, they both represent prototypical wild species, or stand as exemplars of (say) mammalian physiology, and also represent slippage away from wildness.

Following ferality

> I think I could turn and live with animals, they are so placid and self-contain'd;
> I stand and look at them long and long.
>
> Walt Whitman, 'Song of Myself'

How feral animals behave is less commonly investigated, not least because of ethology's focus on the wild; yet these populations are potentially interesting, as they can tell us much about adaptability, in that transition between 'tame' and 'wild'.[10] Even when they are studied, however, the framework remains with wild species – how they use territory, for example – rather than how they engage with humans or with our social spaces. For that, we need analysis from other disciplines, such as geography.

What behaviour science does less often is to study ferality-in-the-making. This particular process of transition asks many questions of how a species adapts to different situations. Here, I will consider one particular example studying the transition from domesticated to wild: J.B. Calhoun's observations in the mid-twentieth century of colonies of rodents moved into open spaces. I use it here because it follows into the wild some of the animals we have met in the lab, tracing their behavioural changes, and because, like many lab studies, it was seen as providing parallels to human behaviour.

This was not an investigation of existing communities, and the animals were arguably not truly feral, in the sense that they occupied large spaces constructed by scientists to produce what Calhoun described as 'rat or mouse utopias' (Ramsden and Adams 2009). But it was research concerned with following what happened once animals moved from relative domestication (many had been living in labs) to greater freedom; in that sense, it was a study of ferality-in-the-making. And what it showed was animals very far from Whitman's buccolic image.

As populations grew, these utopias quickly descended into hell, according to Calhoun's colleagues. Animals lost social cohesion, even after population numbers restabilized: they seemed no longer to behave like rats or mice (Ramsden and Adams 2009). Calhoun had begun his career studying free-living rat populations as part of a project of rodent control, but then became interested in questions of crowding: these studies became highly influential, as his rodent colonies provided ample illustration of social pathology and were readily extrapolated to commentaries about human urban existence.

Part of the consequent backlash against the popularization of this work was the suggestion that, unlike rats, humans could tolerate greater proximity as a consequence of culture. Rodents, that is, could go wild in the sense of turning savage, but humans could transcend that kind of ferality. 'Going feral' thus seems to be something implying breakdown in human behaviour, a lapse into animality that challenges our humanity. Referring to anthropological work that, Calhoun thought, implied social breakdown with increasing population density, he said, 'I have put mice to the same test and they failed to remain mice' (quoted in Ramsden and Adams 2009: 30).

Yet other, more recent studies of lab rats allowed to become feral have come to different conclusions, reporting that they started to adapt to their new environment within days. Reporting on *The Laboratory Rat: A Natural History*, a film based on one such experiment, Peplow (2004) notes that while animal behaviourists might not be surprised, many biomedical researchers are shocked that animals inbred for many generations are capable of fending for themselves. The scientist who produced the film, Manuel Berdoy, commented, 'This shows that while we can take the animal from the wild, we have not taken the wild out of the animal' (quoted in Peplow 2004).[11]

'Wild', then, is something held to be inherent in a species, a potential which they can quickly recover if, that is, they are not under stress. The implication here is that, despite generations of lab breeding, this 'wild' resurrects itself, and 'tamed' animals revert to states unmediated by human intervention. The rats in Berdoy's film are, moreover, framed in such a way that they fascinate; we observe them being social, making burrows, eating, and raising families. These are not the rats of the sewer, nor are they the rats of Calhoun's dystopia. They are, on the contrary, more reminiscent of Whitman's 'placid, self-contained' animals, in nature. Ferality can be constructed in many ways.

Most ethological thinking, I have argued, relies on concepts of the 'animal in the wild', representative of the species. Even when long-domesticated species are studied, their behaviour is often referenced back to putative wild ancestors. While similar discourses enter the laboratory, the use of animals as models in biomedicine entails particular forms of domestication, a taming of the animal through lab practices, and ways of asking animals questions. These 'tools of the trade', however, can occasionally revert to wildness/ferality, either through escape into the pipework of the lab building or when placed experimentally outside. How their 'wildness' is then interpreted, however, depends partly on how they adapt to their new situation, and partly on human framing of what the animals do or represent.

Thinking about animals 'in the wild' lies at the heart of biological practice, part of our belief in the centrality of evolution and natural selection, so that 'wild' is in effect an unremarked category. While I have framed this chapter with reference to these categories of wild, the categories nevertheless remain slippery, as Calhoun's rat colonies indicate. Wild animals become domesticated through our scientific narratives, while domesticated animals in labs and homes can slip back into, or demonstrate flashes of, wildness.

Putting oneself in the animal's place

> A person who studies behaviour and never treats the animal as though it were human is liable to miss some of the richness and complexity of what it does. Many experienced ethologists have found how much they are helped if they put themselves in the animal's place and consider how they would deal with the situation.
>
> P. Bateson, 'Ethology'

As an undergraduate, I had long wrestled with my conscience about studying biology. Fascinating though physiology was, I was extremely uncomfortable doing anything invasive with animals. Yet fascination for them won out, and I turned to ethology, where at least I could study whole animals, without having to dismember them. Nonetheless, you were supposed to keep a distance: emotionally and epistemologically, but also materially, by keeping as hidden from the animals as much as possible. It was, moreover, a training that prioritized testing hypotheses, producing graphs, and making generalizations about the species, as well as an abhorrence of 'anthropomorphism'.

Paradoxically, most of the ethologists I knew liked animals and had long-standing interests in natural history; indeed, my passion for animals and wanting to understand why they do what they do is precisely what drew me to ethology in the first place. Colleagues would, moreover, often tell anecdotes about specific individuals among the species they studied. Anthropomorphism seemed all right in the coffee room, even if banned from the written paper.

To some extent, that tension – between personal experience and pressure to conform to scientific standards – persists.[12] In ethology, detailed descriptions of individuals (and reference to mental states) can sit alongside narratives driven by hypotheses and distancing, often in the same textbooks or journals. In primatology particularly, descriptive modes of writing are common; field studies of great ape societies are clear examples. Similarly, while there is a long history of abjuring anthropomorphism, biologists working in ethology are also sometimes willing to exploit it, as Bateson explains in the quotation above.

What are the implications of these shifts in scientific thinking about behaviour for wild/feral/tame (domesticated) distinctions? Reading Charles Darwin, I am struck by how both 'domesticated' and 'wild' animals interweave through the text. In, for example, his work on the expression of emotions, tame cats feature as much as wild canids (1872). Yet as ethology gained strength, the emphasis on generic cases grew. As these had to stand to illustrate specific *species*, so the wild animal became more prominent; only the wild animal could truly exemplify species, while domesticated animals were often seen as departures from these idealized norms. The result was that, for many years, the study of domesticated species was relegated to the sidelines.

Seeing animals as exemplars of species was further underlined by ethology's preference for explanations in terms of function and evolution, especially in the surge of interest in behavioural ecology. Once animals come to stand in for species, then individuality drops away. 'Wild' thus separates out from 'tame' in the very provenance of biological practice and thought; the 'wild' remains the gold standard, against which the behaviour of 'domesticated' or 'feral' animals can be compared.

Ethologists – although undoubtedly aware of how the presence of human observers has an impact on wild animals – seem almost to be looking out at pristine nature, in which human presence is played down. In this sense, behavioural science can be said to be 'performing the wild' (Whatmore 2002). The very act of making observations – and being observed in return – produces a different situation, a relationship at a distance, rather than seeing wildness through a neutral lens. It is, in that sense, its own kind of domestication; the relationship with the animal, however remote, transforms wildness at the point of observation.[13]

Scientists, furthermore, must draw not only on their much-valued detachment but also on personal experiences. As Candea noted in his study of behavioural research into meerkat behaviour, scientists had to

> draw on both distant and personal modes of knowledge while remembering to keep them separate. [. . .] [T]he data that would later feed into analyses of meerkat behavioral

> biology began its life in the fine-grained knowledge of particular animals by particular people.
>
> (2010: 250)

Only thus can they begin to 'put themselves in the animal's place'.

Despite my immersion in the distancing and domestication of nature in labs, I love ethology. To be sure, there are some who insist on being 'objective scientists' (whatever that means), but there are many ethologists just as willing to meet animal others as selves, with minds of their own. There is still, in some ethological writing, a sense of wonder about the natural world, a willingness to put oneself in the animal's place and ponder who s/he is. Some are not afraid, either, to name animals they study or swap anecdotes about them. Marc Bekoff underlined the importance of telling stories in understanding animals, noting that

> Anecdotes are central to the study of behavior, as they are to much of science. As we accumulate more and more stories about behavior, we develop a solid data base that can be used to stimulate further empirical research, and yes, additional stories. The plural of anecdote is data.
>
> (2002: 47)

It is these stories that often come closest to how many people experience relationships with other animals, be that the dog by the fireside or the wild animals watching us watching them. They tap into an emotional connection that cuts across wildness or tameness: these are animals in all their wonderful individuality. They are both representatives of species and simultaneously just themselves.

At the moment, scientific study has not prioritized interspecies relationships.[14] Ethologists mostly study animals 'out there', rather than how they bond (or not) with us. Insofar as they study relationships, that usually means relationships within a species, within social groups. But observing humans are, in many ways, part of the milieu in which those observed animals find themselves, and we need to understand better how that interspecies sociality is experienced by them as well as by us. We need, I would argue, to figure out better ways in which *relationships* between humans and nonhumans, as processes of communication and interconnecting, might be studied.

Emotional engagements of human and nonhuman are too often written out of science, even while they may be part and parcel of how lab or field studies are run. What happens to the interpretation of results if we acknowledge the animality of lab animals and their responses to us? If we pause to wonder who they are? What happens to our understandings of wild, of species, of specific animals, when we focus on connections between us, the interspecies inter-minglings (Haraway 2008)? What happens to concepts of wild if we were to think about nonhumans as active participants in the processes of knowledge production? The dispassionate gaze of the scientist, calmly observing 'wild' nature, begins to make less sense. Rather, we would have to locate ourselves alongside the creatures we study, to understand what connects us; if we focus on relationships, rather than on us observing them, then 'understanding behaviour' takes on a different hue. And, for all that ethology is firmly rooted in scientific traditions that distance us, I believe it is ethologists who are among those most willing to break the chains.

Acknowledgements

This chapter is dedicated to the memory of Quin, my beloved friend, and horse of my heart, who died aged 21 in February 2012. His 'wildness' in competition arenas was legendary, but we still frequently won, as long as he got his way. He set forth his views of humans and their failures to understand animal intelligence (via an amanuensis, namely me) in Birke (2004). I am also very grateful to Jo Hockenhull for helpful suggestions and comments on an earlier draft. And last, but not least, I want to thank, profoundly, the many and varied nonhumans who have watched me watching them over the years. I have never doubted their intelligence.

Notes

1 In particular, domestication has involved modifications of the hypothalamo-pituitary-adrenal axis (HPA) involved in 'flight or fight' responses. That is, stress responses are damped down in tamed animals.
2 Garry Marvin notes that these ambiguous animals become closer to wild at the point of their deaths (2006: 18).
3 Wilson (2011) pointed out that there were several women active in ethology in its early years. As is so usual, their work has largely been overlooked, and the story of 'founding fathers' as sole progenitors persists in ethology textbooks.
4 The contrast between descriptive, naturalistic accounts and experimental ones is discussed also by Vinciane Despret in this volume.
5 Abstraction is part of many scientific representations, such as diagrams of the human body (Birke 1999). While it permits clarity, it also decontextualizes, with implications for how we perceive the part of the body/animal or whatever.
6 These authors used a categorization earlier drawn up by Swiss zoologist Hediger (1964).
7 A common laboratory joke, this line is believed to have originated with Harvard biologist George Wald (Dubos 1968).
8 Here, I refer primarily to lab use of apes. Studies in the wild, such as Jane Goodall's work with chimpanzees, typically refer to them as individuals.
9 Anne Fausto-Sterling (2000) also noted continuing use of circular arenas, to stop females backing into corners and making themselves unavailable.
10 Though, on feral cats, see Kerby and McDonald (1988); and Boitani *et al.* (1995) on feral dogs.
11 For further details of the project, see www.ratlife.org.
12 See discussion in Birke (1994).
13 The apocryphal story of Schrödinger's cat seems relevant here.
14 Though see the essays in Davis and Balfour (1995).

References

Alcock, J. (2003) 'A Textbook History of Animal Behaviour', *Animal Behaviour* 65: 3–10.
Almeida, M.V., Garrido de Paula, H.M., and Tavora, R.S. (2006) 'Observer Effects on the Behavior of Non-habituated Wild Living Marmosets (*Callithrix jacchus*)', *Revista de Etologia* 8: 81–7.
Baker, R.L. and McGuffin, M.A. (2007) 'Technique and Observer Presence Affect Reporting of Behavior of Damselfly Larvae', *Journal of the North American Benthological Society* 26: 145–51.
Balcombe, J.P. (2006) 'Laboratory Environments and Rodents' Behavioural Needs: A Review', *Laboratory Animals* 40: 217–35.
Bateson, P. (1987) 'Ethology', *Oxford Companion to the Mind*, Oxford: Oxford University Press.
—— (2003) 'The Promise of Behavioural Biology', *Animal Behaviour* 65: 11–17.
Bekoff, M. (2002) *Minding Animals: Awareness, Emotions and Heart*, New York: Oxford University Press.
Berdoy, M. (2003) *The Laboratory Rat: A Natural History*, available at www.ratlife.org, accessed 1 Februrary 2012.
Birke, L. (1994) *Women, Feminism and Animals: The Naming of the Shrew*, Buckingham: Open University Press.
—— (1999) *Feminism and the Biological Body*, Edinburgh: Edinburgh University Press.
—— (2004) 'Dreaming of Pegasus – or, Quin's Story', in J. Fook, R. Klein, and S. Hartshorn (eds), *Horse Dreams: The Meaning of Horses in Women's Lives*, Melbourne: Spinifex Press, pp. 206–9.
Birke, L. and Smith, J. (1995) 'Animals in Experimental Reports: The Rhetoric of Science', *Society and Animals* 3: 23–42.

Birke, L., Arluke, A., and Michael, M. (2007) *The Sacrifice: How Scientific Experiments Transform Animals and People*, West Lafayette, IN: Purdue University Press.

Boitani, L., Francisci, F., Ciucci, P., and Andreoli, G. (1995) 'Population Biology and Ecology of Feral Dogs in Central Italy', in J. Serpell (ed.), *The Domestic Dog: Its Evolution, Behaviour and Interactions with People*, Cambridge: Cambridge University Press, pp. 217–44.

Budiansky, S. (1992) *The Covenant of the Wild: Why Animals Chose Domestication*, London: Weidenfeld and Nicolson.

Candea, M. (2010) ' "I Fell in Love with Carlos the Meerkat": Engagement and Detachment in Human–Animal Relations', *American Ethnologist* 37: 241–58.

Clutton-Brock, J. (1987) *A Natural History of Domesticated Mammals*, Cambridge: Cambridge University Press.

Coetzee, J.M. (1999) *The Lives of Animals*, A. Gutmann (ed.), Princeton, NJ: Princeton University Press.

Coppinger, R. and Coppinger, L. (2001) *Dogs: A Startling New Understanding of Canine Origins, Behavior, and Evolution*, New York: Scribner.

Crist, E. (1999) *Images of Animals: Anthropomorphism and Animal Mind*, Philadelphia: Temple University Press.

Crofoot, M.C., Lambert, T.D., Kays, R., and Wikelsk, M.C. (2010) 'Does Watching a Monkey Change Its Behaviour? Quantifying Observer Effects in Habituated Wild Primates Using Automated Radiotelemetry', *Animal Behaviour* 80(3): 475–80.

Darwin, C. ([1872]1965) *The Expression of the Emotions in Man and Animals*, Chicago: University of Chicago Press.

Davis, H. and Balfour, D. (eds), *The Inevitable Bond: Examining Scientist-Animal Interactions*, Cambridge: Cambridge University Press.

De Queiroz, K. (2005) 'Ernst Mayr and the Modern Concept of Species', *Proceedings of the National Academy of Science* 102.(1) (3 May): 6600–7.

Dewsbury, D.A. (1995) 'Studies of Rodent-Human Interactions in Animal Psychology', in H. Davis and D. Balfour (eds), *The Inevitable Bond: Examining Scientist-Animal Interactions*, Cambridge: Cambridge University Press, pp. 27–43.

—— (1997) 'Rhetorical Strategies in the Presentation of Ethology and Comparative Psychology in Magazines after World War II', *Science in Context* 10: 367–86.

Dubos, R. (1968) *So Human an Animal*, New York: Charles Scriber's Sons.

Estep, E.Q. and Hett, S. (1995) 'Interactions, Relationships and Bonds: The Conceptual Basis for Scientist-Animal Interactions', in H. Davis and D. Balfour (eds), *The Inevitable Bond: Examining Scientist-Animal Interactions*, Cambridge: Cambridge University Press, pp. 6–26.

Fausto-Sterling, A. (2000) *Sexing the Body: Gender Politics and the Construction of Sexuality*, New York: Basic Books.

Foster, H.L. (1980) 'The History of Commercial Production of Laboratory Rodents', *Laboratory Animal Science* 30: 793–8.

Griffiths, H., Poulter, I., and Sibley, D. (2000) 'Feral Cats in the City', in C. Philo and C. Wilbert (eds), *Animal Spaces, Beastly Places: New Geographies of Human-Animal Relations*, London: Routledge, pp. 56–70.

Haraway, D. (2008) *When Species Meet*, Minneapolis: University of Minnesota Press.

Hare, B., Wobber, V., and Wrangham, R. (2012) 'The Self-domestication Hypothesis: Evolution of Bonobo Psychology is Due to Selection against Aggression', *Animal Behaviour* 83(3): 573–85.

Hediger, H. (1964) *Wild Animals in Captivity*, New York: Dover.

Hemsworth, P.H. (2003) 'Human-Animal Interactions in Livestock Production', *Applied Animal Behaviour Science* 81: 185–98.

Hosey, G.R. (2005) 'How Does the Zoo Environment Affect the Behaviour of Captive Primates?', *Applied Animal Behaviour Science* 90: 107–29.

Iredale, S.K., Nevill, C.H., and Lutz, C.K. (2010) 'The Influence of Observer Presence on Baboon (*Papio spp.*) and Rhesus Macaque (*Macaca mulatta*) Behavior', *Applied Animal Behaviour Science* 122(1): 53–7.

Jack, K.M., Lenz, B.B., Healan, N., Rudman, S., Schoof, V.A.M, and Fedigan, L. (2008) 'The Effects of Observer Presence on the Behavior of *Cebus capucinus* in Costa Rica, *American Journal of Primatology* 70: 490–4.

Kerby, G. and Macdonald, D.W. (1988) 'Cat Society and the Consequences of Colony Size', in D.C. Turner and P. Bateson (eds), *The Domestic Cat: The Biology of Its Behavior*, Cambridge: Cambridge University Press, pp. 67–82.

Knorr-Cetina, K.D. (1983) 'The Ethnographic Study of Scientific Work. Towards a Constructivist Interpretation of Science', in K.D. Knorr-Cetina and M. Mulkay (eds), *Science Observed: Perspectives on the Social Studies of Science*, London: Sage, pp. 115–40.
Latour, B. (1990) 'Drawing Things Together', in M. Lynch and S. Woolgar (eds), *Representations in Scientific Practice*, Cambridge, MA: MIT Press, pp. 19–68.
Leyhausen, P. (1988) 'The Tame and the Wild – Another Just-So Story?', in D.C. Turner and P. Bateson (eds), *The Domestic Cat: The Biology of Its Behavior*, Cambridge: Cambridge University Press, pp. 57–66.
Lynch, M. (1985) *Art and Artifact in Laboratory Science: A Study of Shop Work and Shop Talk in a Research Laboratory*, London: Routledge.
Macdonald, D.W. and Carr, G.M. (1995) 'Variation in Dog Society: Between Resource Dispersion and Social Flux', in J. Serpell (ed.), *The Domestic Dog: Its Evolution, Behaviour and Interactions with People*, Cambridge: Cambridge University Press, pp. 199–216.
Martin, P. and Bateson, P. (1986) *Measuring Behaviour: An Introductory Guide*, Cambridge: Cambridge University Press.
Marvin, G. (2006) 'Wild Killing: Contesting the Animal in Hunting', in The Animal Studies Group (eds), *Killing Animals*, Urbana: University of Illinois Press.
McClintock, M.K. and Adler, N.T. (1978) 'The Role of the Female during Copulation in the Wild and Domestic Norway rat (*Rattus norvegicus*)', *Behaviour* 67: 67–96.
Ord, T.J., Martins, E.P., Thakur, S., Mane, K.K., and Borner, K. (2005) 'Trends in Animal Behaviour Research (1968–2002): Ethoinformatics and the Mining of Library Databases', *Animal Behaviour* 69: 1399–413.
Peniche, G., Vaughan-Higgins, R., Carter, I., Pocknell, A., Simpson D., and Sainsbury, A. (2011) 'Long-term Health Effects of Harness-mounted Radio Transmitters in Red Kites *(Milvus milvus)* in England', *Veterinary Record* 169(12): 311.
Peplow, M. (2004) 'Lab Rats Go Wild in Oxfordshire', *Nature News* (3 February), available online at www.nature.com/news/2004/040203/full/news040202-2.html, accessed 1 February 2012.
Rader, K. (2004) *Making Mice: Standardizing Animals for American Biomedical Research, 1900–1955*, Princeton, NJ: Princeton University Press.
Ramsden, E. and Adams, J. (2009) 'Escaping the Laboratory: The Rodent Experiments of John B. Calhoun and Their Cultural Influence', *Journal of Social History* 42: 761–97.
Rogers, L. (1997) *Minds of Their Own: Thinking and Awareness in Animals*, St Leonards: Allen and Unwin.
Segerdahl, P. (2012) 'Human-enculturated Apes: Towards a New Synthesis of Philosophy and Comparative Psychology', in L. Birke and J. Hockenhull (eds), *Crossing Boundaries: Investigating Human-Animal Relationships*, Leiden: Brill, pp. 139–60.
Sherwin, C.M. (2004) 'The Influence of Standard Laboratory Cages on Rodents and the Validity of Data', *Science in the Service of Welfare. Animal Welfare (Supplement)* 13: 9–16.
Slater, P.J.B. (1985) *An Introduction to Ethology*, Cambridge: Cambridge University Press.
Taylor, N. (2012) 'Animals, Mess, Method: Post-humanism, Sociology and Animal Studies', in L. Birke and J. Hockenhull (eds), *Crossing Boundaries: Investigating Human-Animal Relationships*, Leiden: Brill, pp. 37–50.
Thornton, A. and Clutton-Brock, T. (2011) 'Social Learning and the Development of Individual and Group Behaviour in Mammal Societies', *Philosophical Transactions of the Royal Society B: Biological Sciences* 366: 978–87.
Tiptree J. (1978) 'The Psychologist Who Couldn't Do Awful Things to Rats', *Star Songs of an Old Primate*, New York: Ballantine Books, pp. 227–54.
Whatmore, S. (2002) *Hybrid Geographies: Natures Cultures Spaces*, London: Sage.
Whitman, W. (1900) 'Song of Myself', *Leaves of Grass*, Philadelphia: David McKay, available online at www.bartleby.com/142/, accessed 1 February 2012.
Wieder, D.L. (1980) 'Behavioristic Operationalism and the Life-World: Chimpanzees and the Chimpanzee Researchers in Face-to-Face Interaction', *Sociological Inquiry* 50: 75–103.
Wilson, D.A.H. (2002) 'Experimental Animal Behaviour Studies: The Loss of Initiative in Britain 100 Years Ago', *History of Science* 40: 291–320.
—— (2011) 'British Animal Behaviour Studies in the Twentieth Century: Some Interdisciplinary Perspectives', in C. Blazina, G. Boyra, and D. Shen-Miller (eds), *The Psychology of the Human-Animal Bond: A Resource for Clinicians and Researchers*, New York: Springer, pp. 25–44.
Wolfe, C. (2003) *Animal Rites: American Culture, the Discourse of Species, and Posthumanist Theory*, Chicago: University of Chicago Press.

4

Wherever I lay my cat?

Post-human crowding and the meaning of home

Tora Holmberg

To be one is always to *become with* many.

Donna Haraway, *When Species Meet*

Urban animals are often the subject of complaints; they transgress our sense of cultural order, while roaming uncontrolled in the city. But they are similarly turned into objects of care, conservation practices, and bio-political interventions. Controversies over such animals are thus to be expected, when these two 'frames of meaning' (Goffman 1974) collide. This chapter discusses a specific area of controversy, namely the management of cats in urban settings. On the basis of interviews with animal police, animal welfare inspectors, so-called 'animal hoarders', cat rescuers, and shelter workers, as well as complaint filings and other documents, I investigate a number of debates related to the process of 'becoming with many' cats and the norms of appropriate pet keeping in the city. The phenomenon raises a number of questions: How are 'homeless', 'feral', and other categories produced, and with what consequences? What is the role of 'home' and other places in defining and handling various categories of urban cats?

I use two probing concepts – or figurative tropes (Haraway 1997) – throughout this chapter. The first is that of 'home', in particular the lack of home. Christopher Jencks (1994) states that in order to understand homelessness as a phenomenon, we need to scrutinize the meaning of 'home'. The home is historically the prime place to raise future citizens. It is packed not only with ideas of purity, shelter, exclusivity, and family values, but also belonging. The politics of human homelessness is well covered from a social science perspective (Hutson and Clapman 1999; Jencks 1994; Kennett and Marsh 1999). Essentially, the literature points out that various explanations of homelessness – whether structural (lack of housing, poverty, welfare state cuts, etc.) or individual (drug addiction, unemployment, mental illness) – suggest their own solutions and interventions. It is often emphasized that homelessness is less a state than a process, where a 'cycle of homelessness' can be identified (Wolch *et al.* 1988: 443). People move in and out between shelters, prisons and other institutions, friends, back to the shelter, and so on. Moreover, public shelters constitute 'hybrid organisations' (Hopper 1990), unable to solve conflicting demands – that is, to officially serve as institutions for acute housing needs, while in reality providing more or less permanent housing. From a critical point of view, it has been argued that the shelter system is a node in the apparatus of homelessness, co-constructing and allowing this social problem to continue.

When it comes to feline homelessness, there is a growing volume of literature on the shelter movement, especially in the US. Its history of caring for strays, its connection to the humane societies of the late nineteenth century, and its role in formulating the social problem are all intriguing (Irvine 2003). The problem has been framed as animal surplus – in the US around four million cats are killed annually by shelter workers alone (Patronek *et al.* 1996) – but where to place the blame for this situation has varied over time. Currently, irresponsible owners are on the agenda, but many shelters hesitate to overtly blame owners (Irvine 2003). Thus, animal rescue and adoption are integral parts of a 'negotiated order' (Balcom and Arluke 2001) regarding homeless cats. I borrow the term 'hybrid organisation' to characterize the cat shelter, an organization in which conflicting practices such as advice and courses on behavioural problems, veterinary care, neutering, euthanasia, and adoption all take place under the same roof. The national context of the study is Swedish, where there are countless shelters; every little town has its cat enthusiasts. Some shelters have contracts with the local police, according to which they are housing 'adoptable' homeless cats for a fee. My first question in this chapter is: what is the meaning of home in the context of cat homelessness?

The second concept to probe is that of 'crowding'. The human urban population is increasing worldwide, and in Western countries at least, the number of companion animals is on the rise, too. Thus urbanization and its consequences are not purely human processes (Wolch 1998). Crowding as a phenomenon is well covered in urban studies, and originally it meant the increased density in population due to urbanization (Churchman 1999). However, in the 1970s the concept of crowding also encompassed subjective experiences – originally stress and other negative consequences were mainly highlighted (Stokols 1972; Gove *et al.* 1979). Nowadays, crowding is not often used as an analytical concept, but essentially the same core is addressed when analysing effects of urbanization. Micro-sociology has been deployed in order to understand how people manage to live closely together, and the environmental, psychological, and health-related benefits of crowding have become more and more acknowledged. For example, in urban planning the terms 'smart growth' and 'compact cities' are used to highlight and implement the 'good sides' of higher density. When it comes to official norms of proper housing, 'crowdedness' as a concept has changed historically (Ekstam forthcoming), and the standards for measuring the lower limits of crowdedness are somewhat arbitrary (Ytrehus 2000).

Many actors within the housing sector now choose to limit the number and breed of nonhuman inhabitants allowed. In Sweden, the Animal Welfare Act (1988: §16) stipulates that you are not allowed to house more than nine dogs or cats in a household without a special breeding permit, something which also applies to cat shelters. In the following, I ask how the figure of crowding – or the spatial quality of a collective in a certain social setting – is involved in the understanding and management of urban cats.

In this chapter I approach the concepts of homelessness and crowding as essential to understanding how the 'cat-egorizations' of cats – their 'virtual identity' (Goffman 1971) – play out in relation to home and other places. For these purposes, after going through the theoretical and methodological frames, I will provide two examples of how feral cats are constructed: in relation to practices concerning urban homeless cats and crowded streets, and in the context of hoarded cats and crowded homes. Towards the end, I will return to the questions outlined in the beginning: how are 'crowding' and 'home' produced and made meaningful by various actors, and with what consequences?

Theoretical staging

Sociological interest in human-animal relations has increased since the 1990s (Arluke and Sanders 1996; Franklin 1999; Franklin and White 2001; Peggs 2012; Sanders 1999). I will, however, focus particularly on urban human-animal relations, which pose a specific set of problems. Urban theory is in essence anthropocentric, and urbanization processes have been theorized, as cultural geographer Jennifer Wolch points out, as acting on 'empty spaces', not attending to nonhuman inhabitants (1998: 119). However, cities are occupied by all kinds of animals that contribute to the urban spirit – *anima urbis* (Wolch 2002: 722). Human relations to other animals are part of the construction of urban identities as well as places, and human society rests on the use of other animals (Griffiths *et al.* 2000). When analysing human/animal relations in the city, the nature/culture boundary is a given.

While 'society' consists of humans and their interactions, their institutions and cultures, other animals often fall into the nature category (Sabloff 2001). The nature/culture distinction is also reflected in the binaries wild/domestic (see Birke in this volume). But certain animals do not easily fit into this taxonomy, and the ordering of animals in and through urban space gets further complicated by the notion of the 'feral', typically addressing non-native populations of species (see Franklin in this volume). Since, as pets, cats occupy what can be called a 'liminal space' (Fox 2006), processes of exclusion and inclusion, when they occur, can be expected to be rather complex. But also non-pets, such as pigeons, can be regarded as subverting ordering; they are wild birds that eat from our hands and populate the streets, so no longer regarded as wild, yet they are not thought of as domestic either (Jerolmack 2013: 230). Liminal creatures and other anomalies, as we have learned from Mary Douglas, can be perceived as threatening cultural order and sorting systems and thus triggering closure and other normalizing strategies. Yet through their multiple meanings they also work as potentially subversive actors and contribute to social change (Douglas 1966). One way of dealing with liminality and boundary-crossing in human/animal relations is through the lens of post-humanism, which refers to a temporal space – after humanism – but mainly to a theoretical space (Wolfe 2009). It asks: what happens if we move humans from the centre of inquiry? In short, post-humanism constitutes a challenge to Humanism with a capital 'H' – as anthropocentric and connected to the dominance and exploitation of nature and other animals. Within Humanism, 'human' is a category that you need to qualify for, and thus there are numerous examples of people that historically fell outside the category (slaves, women, Jews, refugees, aboriginal people, etc.).

Giorgio Agamben (2004) writes about the 'anthropological machine', through which the 'human' constantly delineates the 'beast' and mirrors it in the other in order to define itself. For example, people who live in the company of many animals and possessions not valued by the social majority – so-called hoarders and clutterers – seem to challenge certain norms concerning what it is to be human and can become animalized in popular discourse. They are portrayed as 'pack rats' and their behaviour is often explained by their social isolation, although many are not alone at all but have numerous human and nonhuman friends. Moving beyond these apparent contradictions, Donna Haraway's (2003) theory regarding human/animal relations beyond a dichotomous understanding is based on the working relation between human and dog, on cooperation and responsibility, and one in which the relation itself is the most defining element. Humans and other animals are engaged in overlapping 'contact zones' (Haraway 2008) that constitute each other. Even though we are not dealing with equal relations, the embracing of what Haraway calls 'significant otherness' (2003) may lead to new insights and practices.

This significant otherness is created within the ideal typical construction of 'companion species' – related in continual interaction and co-evolution. However, significant otherness also includes its ever present shadow: the insignificants' disconnection. Significant otherness is thus about alliances, conflicts, and distance between species, about being 'messmates' in the mutual but not equal relations of naturecultures (Haraway 2008). Becoming in this co-evolutionary dance is 'always becoming with – in a contact zone where the outcome, where who is in the world, is at stake' (Haraway 2008: 244). But what of those without?

Cats, our most common pets, are characteristically defined in different ways based on both their relationship to humans (homeless, feral, domesticated) and the environment they are moving in (domestic, indoors, farm, city). The cat's place is very much in the home (Grier 2006), but this is not universally the case. The Scandinavian cat was probably domesticated about 1500 years ago (Broberg 2004), yet we do not know much about its background. Gunnar Broberg (2004), who has written the cat's history, argues that their paws step so lightly that they seldom leave any imprint on written history (Broberg 2004: 21). But we know that cats in pre-modern society lived in barns and on ships, and were primarily kept for their outstanding ability to catch mice and rats. Not until the nineteenth and twentieth centuries did the cat, symbolically and materially, enter the home (2004: 339), which also allowed for other, more intimate relationships as pets (Grier 2006: 62). Although many cats today in fact spend most of their lives indoors, many of them live, at least part time, outside of our control and knowledge. Wandering through neighbourhoods, they create their own feline space and relationships with people other than their own (Tipper 2010). Throughout history, the cat is also surrounded by a lush symbolism – unfaithful, deceitful, sensual, frivolous, but also enchanting and associated with the dark forces – and the magic lives on (Broberg 2004: 46). In fact, what contributes to our fascination with cats and their ubiquitous presence may be the cat's seeming ability to embody any character (Rogers 2006). Thus, cats have been as tremendously popular in literature and visual culture as they are in real life.

However, this popularity is not always beneficial for the cats themselves and apart from myths and beliefs about the filthy, treacherous, and magical pussycat, she is also genuinely vulnerable in society. Cats are common – according to a survey, there were 1.3 million in Sweden in 2006 (SCB 2006) – but have low status. The consequence is that many people discard or kill their cat when they find it inconvenient. Each municipality has its cat association and shelter, caring for and relocating homeless cats. The Dog and Cat Supervision Act came into effect in 2008, and along with the Animal Welfare Act (1988) it regulates the care of these animals with minimal standards of what is regarded as good pet keeping. It is difficult to find statistics, but sources refer to 100,000 homeless cats in Sweden. Nevertheless, it is clear that the 'cat issue', especially in urban areas, is a growing problem. At the same time we are witnessing a parallel enhancement of the cat's status; consumerism leaves nothing untouched. Catteries, cat shows, and cat cafés, along with growing cat food and cat insurance industries, tell a clear story: we care more about our feline companions today (Rogers 2006).

Yet is this kind of care always beneficial to us, or to them? And how might a deeper ambivalence relate to the distinctly urban problems of homelessness and crowding? The project this chapter draws from investigates the understanding and management of urban controversies over other animals. I have performed interviews with Swedish police officers working with issues regarding animals, animal welfare inspectors and veterinarians from county boards, shelter workers and activists involved in animal rescue, and people who have been subjected to official complaints (so called 'hoarders'). Twenty interviews have been conducted in four Swedish cities. When quotes appear in the chapter, they have been anonymized and translated into English. To show their connection to broader social trends,

I have also drawn on texts and documents of various kinds, including laws and policy documents, animal welfare protocols, newspaper articles, TV programmes, and social media.

Homelessness and crowded streets

According to my interviews with police and animal welfare inspectors in Sweden, the 'problem of unwanted pets' (Irvine 2003), in particular cats, is one of the top three most common urban sources of animal controversy, along with animal hoarding and dangerous/aggressive/unleashed dogs. It is also something that consumes resources and time and, needless to say, involves much suffering. The problem is in essence that cats, compared to other pets, inhabit a rather strange place in society:

> Police. So the cat problem is rather big and it . . . I don't know, I think it is a matter of attitudes. A cat, you let them out like this.
> Tora. Uh-huh.
> P. People just think that the cat is supposed to run free, and, eh . . . if I had to decide, you'd only be allowed to have cats indoors . . . in the urban environment.
> T. In the city?
> P. Yeah, absolutely. I mean, it's not stranger than if you have a guinea pig or a hamster, you don't let it out in the morning and collect it in the evening. But . . . and the dog, you don't let it out, but you take it out for a walk. And if you think that the cat needs to get out, then you'll just have to walk with it, right. So here we have a big problem, actually, and . . . we haven't come up with a really good way of making everyone understand this.
> T. Uh.
> P. Because it's enough that some people are careless [laugh] then soon there will be many cats, 'course the cat shelters are overcrowded.

The police officer quoted here tries to make sense of the cat overpopulation problem, and his explanation is that people let their cats out to wander around, and, implicitly, they also neglect to neuter their animals. If we would treat cats like dogs, for example, there would simply not be any problems – at least, not with cats crowding the streets of the city. There is however a complicating issue here: the cats that run around do not always belong to an owner. Thus, the police need to distinguish between different kinds of cats which form three different 'cat-egories': the lost, the homeless, and the feral cat. These are interesting since they come with an assemblage of different jurisdictions, formal definitions, and interventions. The first category is the lost cat, salient when discussing a range of urban-related animal controversies handled by the police:

> Police. Yes, and then lost animals, of course. You . . .
> Tora. Right, that people bring in or . . .
> P. Yes, yes, often that people bring in, or that they, they call and say that 'we . . . this cat, it's come and lived here now for three weeks in our carport and we have put up notes at ICA [local grocery store] and everywhere and no one has called, so we . . . this one seems abandoned'.

Cats, dogs, rabbits, and snakes are sometimes lost, or can be assumed to be lost, and they are taken into the police station or straight to a shelter. Pets that are earmarked or equipped with a microchip can easily be identified and then returned to their owners quite quickly. This category is regulated through the Lost Property Act and handled exclusively by the police. Other actors are of course the cat and the (absent) owner. Note how the cat in the quote above is assumed to be lost somewhere close to where it was found, like a wallet or a key. The movements of cats are seldom acknowledged, although we sometimes read stories of cats who get lost in one location and are found years later, somewhere completely out of place. The *Guardian*, for example, reported the following about Willow:

> A cat that went missing in Colorado five years ago was found wandering in Manhattan, and will soon be sent on a plane to reunite it with its former owners, an animal pound spokesman said. Workers at the pet shelter traced Willow the cat back to a family in Colorado, thanks to a microchip embedded in the animal's neck that they checked with a scanner, said Richard Gentles, spokesman for Animal Care and Control of New York City.
>
> *('Missing' 2011)*

The second 'cat-egory' is the homeless cat or, in rare cases, the dog (dogs are seldom homeless in Sweden, in part due to almost a century of dog tax and marking regulations). If the lost cat is not returned to an owner, it is categorized as homeless.

> Police. The homeless cats, on the other hand, they can have run away or they can have been dumped. And they . . . we try to, to, pass them on to a new home.

Homeless cats are said to be adoptable, or it is at least the aim of people's interventions to make them such. Thus, homeless cats are not shot on the spot, but examined by a veterinarian and placed in a shelter for future adoption (for more on adoption practices, see below). This category involves the cat, the police, the shelter, and the future owner, and is regulated by the Dog and Cat Supervision Act (2007: 1150).

The categories 'lost' and 'homeless' are thus produced through their fitting different jurisdictions, and it can be financially advantageous and easier on the workload to place the cat in the 'right' category. I asked how the police officer thinks the system works today:

> Police. Well, I think eh . . . if we talk about cats for a while, it's a curse, it almost knocks you out, right. There are wild cats and abandoned cats and lost property all the time. [. . .] They bring in cats and, eh, they call us and say that they have taken in a cat, and so they book them as lost property . . . and then we are responsible for caring for them, for ten days as it's written.
>
> Tora. Uh.
>
> P. That's 500 bucks [spänn = SEK] a cat. And bills of 20 to 30,000 arrive now and then. [. . .] Regarding the cats, there are three different jurisdictions to turn to, right. The Animal Welfare Act, the Lost Property Act, and then the Supervision Act. We have to toughen up, we can't book everything as lost property. Is it marked, the cat, then you would probably consider it as a lost property then, is there an owner, then it can be the Animal Welfare Act. But if they are not marked and they are a little skinny and a little blowsy and ear mites and the like, then you can like [sigh] suppose that they are abandoned.
>
> T. Uh.
>
> P. Formerly it was written that they, eh, should be, could be suspected to be abandoned *or* feral. Now it says . . . no, *and* it was then, now it's *or*. [. . .] So it makes it a lot easier and can be 'suspected' to be [feral], that's like very low on the scale of suspicion.

Details in the writing of the law give cues about how to define the cat according to the categories available, cues that the police officer in this instance needs to be able to translate and interpret in order to make things work smoothly – in short, to make the procedure accountable. The homeless cat is defined as one born in a home with an owner, but who has been abandoned or run away, thus becoming an object of suspicion and corrective efforts. And, as the quote above shows, the boundary between 'homeless' and 'feral' is a slippery one, yet equally important, since the responsible actors as well as the future fate of the cat hinge on the category:

> Police. We try to find some way of drawing the line . . . between these two concepts. What is a homeless cat and, and what is a feral cat, because *feral* cats you are not allowed to enclose, right. This has been stated by the Ministry of Agriculture . . .
> Tora. Hm.
> P. That it, it's quite demanding, psychologically, to the animals. Ehm . . . so when we trap them, there are these cat populations that *are* feral, right, clearly.
> T. They are like born in the wild?
> P. Yes, they are born outdoors.
> T. Mm.

Note how the interviewer and the interviewee both try to establish whether the cat has been born in the wild, thus producing origin as a key to categorization. However, concerning the feral cat, this origin must be deduced from a number of signs, typically based on a number of visual cues:

> Shelter worker. It's sometimes a little hard to see because you cannot get close to that cat. And it's still the most reliable, or it's not so reliable either, but if you look at a cat that's lived outdoors for a long time, it often looks rather rough. You can see it by the walk, it's often quite sway-back too. Because they, they walk low.
> T. Mm.
> SW. Because they, they are used to . . . they try to make themselves as small as possible. That is a rather reliable sign.
> T. Mm.
> SW. I'd say.

The feral cat looks a bit rough; it walks low and has a sway-back. But there may be other cues as well. Based on a number of markers – whether the cat is chipped and looks healthy, or behaves frightened – it can be understood to be either homeless or feral. Interestingly, the fact that shelters are crowded with cats is used as an argument to kill the feral ones, but there are also other arguments in play. Some interviewees argue that it is not in the best interest of the feral cat to be housed with humans; since they are not tame, they dislike human contact. Whether the feral cats can be tamed is a matter of controversy between the authorities and shelter workers, as well as between different shelters. In short, are they tameable or not?

> Shelter worker. . 'Course some work with saving *all* cats.
> Tora. Uh.
> SW. Or all animals. We don't have, we cannot do that. We take care of these cats that we believe will get a home. That have a chance.
> T. Yeah, right, yeah . . . yeah.

Given that there is a limited amount of space at the shelter, it is better, according to this interviewee, to bet on the ones that do have a chance – that is, the tame ones – and let the others be killed. This view is contested by others, claiming that it is not fair to blame the cat for its bad manners:

> Shelter worker 2. So but, we would like to [sigh] give all cats a chance. Just because you've had a shitty life to begin with, right?
>
> Tora. Yeah.
>
> SW2. It shouldn't mean that you don't get a chance, everyone gets a chance. 'Cause we've had cats that, the first year they have been . . . well, quite feral, you know, they have spat, you, you haven't even been able to look at them. And then suddenly one day, it's turned around and they have become a lap cat and today they are the cuddliest cats. When they finally realize that we are not dangerous, then . . . they can become, those are the finest cats, I think.

In both of the quotes above, the bottom line seems to be that of the cat's 'homeability'; however, the second shelter worker talks about the process of taming as a possible option, eventually resulting in a 'lap cat'. When asked about some good examples, the same shelter worker mentions a particular feral cat who was successfully re-homed.

> Shelter worker 2. And it's so damn nice because we still get e-mails from her and the like, where she kind of tells us how it's going, and the cat, the cat sleeps in her bed, the cat plays with the poodle and the cat is out walking in the garden and so on.
>
> Tora. *laughs.*
>
> SW2. It truly works very well. It's like, even those [feral] cats can actually, in the right environment become . . . quite good cats.

The feral cat is lost in a moral sense, but with a little bit of matchmaking, even the most troublesome character might – according to the interviewee – eventually become a 'good cat'. The successfully transformed feral cat in this example sleeps in a bed, plays with other members of the family, and occasionally strolls in the garden. She is wholeheartedly tamed.

Summing up, the categorization of anomalies, with its consequences in regards to the cats' future fate, points to a construction of the 'normal' case. Normality is here produced as a contrast to the anomalies of the lost, the homeless, and the feral, i.e. the house cat, the cat living in a home with one responsible owner. The handling of these anomalies aims at normalizing the established order: either through returning the cats to their owners, killing them, or sheltering them for future adoption. The above analyses of the classification practices suggest what in ethno-methodological terms is called 'indexicality' (Lynch 1993: 19). This simply means that there is no way of defining once and for all the true or objective meaning of a word; its meaning is always negotiated and defined by its context, or rather by the context that people deploy in order to create meaning in the world (Coulon 1995: 16). However, group members transform, through a number of measurements, the indexical categories into objective ones (Kumlin 2011). In this case, lost, homeless, and feral are defined and made meaningful in the context of the juridical regulations, the actors involved, and, not least, the place of the action. Consequently, 'urban', 'street', 'shelter', as well as 'home', 'lap', and 'garden' play important roles in constructing meaning around the cat. In the next section, I will investigate another apparently liminal case, an additional anomaly: the homeless cat with a home.

Hoarding and crowded homes: homeless cats with a home

A major source of urban controversy and a focal point for official intervention concerns animal hoarding. In Swedish, this phenomenon is called 'animal collecting', still a somewhat contested term, and many officials are reluctant to use it, in part because of its stigmatizing effects. Hoarding is not easily defined and certainly not an issue easily managed (Holmberg forthcoming). It often involves large numbers of animals – most often cats or dogs, but sometimes reptiles, birds, rodents, and rabbits – living in overcrowded homes, under very poor conditions (Arluke *et al.* 2002). The hoarders themselves are demographically often women, over 40, and single, but there is great variation (Patronek 1999). In Sweden, the authority that handles these issues is the County Administrative Board and its animal welfare inspectors, who implement the Animal Welfare Act (1988). The situation is often brought to the attention of the authority by neighbours or others who complain about noise or smell, indicating the mistreatment of animals. The phenomenon concerns norms of living as well as of pet keeping and, as in the case above regarding homelessness, definitions and ideas of 'home' and 'crowdedness' are prevalent.

For the historical and cultural reasons outlined above, cats are the most likely animals involved in such situations in Sweden. If the mistreatment of the cats is not severe, the person under investigation gets to correct the situation him or herself, given a certain time frame. But most often, the animals are immediately taken into custody or euthanized. Quite often, it is not the medical but the social condition of the cat that is decisive for the decision to euthanize – the cat's lack of social skills, when it shows fear, aggression, or even apathy. When it comes to handling cats that are actually taken from hoarders, they are often so antisocial and generally in such a bad situation that they have to be put down:

> Animal welfare inspector. Because then, eh, when we had 170 cats, we put 150 down because they were far too wild [. . .] but, there was a veterinarian, right, who judged who was to be euthanized and who was fit to keep on living.

This quote alludes to a well-known fact, and one that has been documented elsewhere (Arluke and Killeen 2009): animals living in households that consist of too many other animals become antisocial. This is also mentioned in the human crowding literature, which states that stress from high density may lead to all kinds of antisocial behaviour. In the context of hoarding, the cats are said to become impossible to handle; in fact, they become feral:

> Animal welfare inspector. But in other cases then like . . . taking into custody . . . it's about situations where there is such a terrible mess that, that, then we usually don't . . . prescribe, mostly anyway, that people get rid of the animals because we have experienced that they are not capable of it, even if they say that they are going to, they are not capable. So then it is normally about custodies. And depending then on the condition of the animals they either get redeployed or they are put to death. It depends. On what condition they are in.
>
> Tora. Right, because sometimes . . .
>
> A. If we make the judgement that they are fit for re-homing, or if they are completely feral, inbred, sick . . .
>
> T. Uh-huh.
>
> A. Or whatever it can be. [. . .] And right, I am a bit more into the deal that, that eh . . . you should put to death. Not all, but the cats that show such behaviour and are in such

condition. I think that in such cases you should put them to death. For several reasons. Partly because of the cat itself, too, but also because of the difficulty of . . . of what it means to find a new home.

Normally, it is said, the welfare inspectors immediately decide to take the cats into custody, since the people themselves seldom have the capacity to clear up the mess. Apart from often having physical and medical problems, the cats taken from hoarders are considered feral and this is part of the reason they are put to death. The informant adds another reason for the strategy of killing these cats: it is hard to find new homes for these antisocial, possibly inbred, and certainly not very pet-like creatures.

Here we have an interesting anomaly: the feral cat that has been born in a home. Following the previous section in which feral cats were defined by their origin, how is it that cats living in a home can be conceptualized as feral and acted upon as such? One clue can be found in the general disgust for filth that is portrayed both in interviews with authorities and, not least, in the media (Arluke *et al.* 2002; Holmberg forthcoming). Here is an excerpt from an official animal welfare inspection report filed with a County Administrative Board:

> Strong odour of, among other things, ammonia hit the inspectors when they entered the co-op apartment. The house consists of upper and ground floor, and basement, and the cats have access to all three levels. The floors of all levels were untidy and several stains from cat urine can be spotted on the floors. Cat vomit was seen in a living room sofa. In the kitchen, there were four litter trays with cat litter and excrement. The basement was extremely cluttered with furniture, boxes et cetera and heavily polluted by cat excrement. Several cats were located among all the things and they were therefore difficult to control. One cat was repeatedly heard coughing. The upper floor was also stained with cat excrement. Here were 4 trays with litter and cat excrement. The whole house smelled heavily primarily from cat excrement and urine.
>
> *(County 2011)*

Smell, urine, excrement, a sofa soaked with vomit, and cats all over the place in an uncontrolled manner: this is certainly a contrast to the good home, with a single cat lying on the sofa, purring with delight. One of the cat owners that I interviewed, who had had several house calls from the animal welfare authorities, agreed on the inspectors' description of her home as dirty:

> Tora. Well, according to the protocol in the . . . the biggest problem was that they thought it was dirty . . .
>
> Cat owner. Mmm.
>
> T. That it smelled bad. What did you think, kind of, what do you say about this?
>
> CO. I agree.
>
> T. What was the biggest problem, according to you?
>
> CO. I had become so negligent, right, with cleaning, in particular and . . . eh, I had well, come home at three or four in the morning from here [working place], so I was asleep, and so I hadn't had time to clean the litter trays when they arrived. So they were quite full by then, 22 cats have time to . . . but I had like, had 12 trays running, I think, right then. So, there were trays alright. And I had food and water everywhere so that . . . and they noted that, right, that there were. And I always keep fresh litter trays kind of, so that you can switch quickly and the like, so that . . . I know the business.

As stated earlier, numbers are of course important cues for defining hoarders. So are the ways cats are treated in terms of violations of the norms stated in the Animal Welfare Act regarding food, water, housing, health care, etc. However, I would like to argue that in addition, it is the reported lack of homeability – the breaking of norms of normal homes in terms of filth, odour, and lack of control – that allows the authorities to treat these cats as feral and to eventually euthanize them. Of course, a filthy home is in itself not enough to warrant taking the animals into custody.

> Police. Often you'd like to take the animal because of the environment, that they live in a lousy pad and it's dirty and there's glass on the floor and it smells bad and, but there is food and water . . . a cat doesn't care if it's dirty, you know.
>
> T. Uh.
>
> P. That's the way they reason [laugh] at the County Administrative Board, but they don't care, it is us who think it's disgusting.
>
> T. Uh.
>
> P. Custodies like that, they don't get sanctioned, I've learned that. But you have to, eh, right, well the vets, they write that the cats are . . . thin and have, for example, ear mites, that they haven't looked after the animal, then, then sanctions are put in.

The animals need to suffer objectively, that is, it needs to be documented that the individual animal is in a bad state; it is not enough that the animal is at risk due to a hazardous environment. The police and animal welfare authorities, when they find it appropriate, navigate through the various resources available, meaning, for example, taking the cat to a reliable veterinarian who can document such suffering.

One last issue regarding home and crowdedness is that, when it comes to homeless pets, hoarders not only contribute to the surplus and crowding of cats at the shelters, they also take on abandoned pets from others. One police officer told me a story of how he had taken ten dogs into custody, and when one of the largest evening papers wrote about the case, more than 60 people phoned and wanted to adopt the dogs. Even though it is not allowed, the policeman checked the records of the more promising candidates, and it turned out that two were previously convicted of animal abuse. He concludes the story by saying: 'This is typical of animal collectors, you know.' A similar theme appears in a reality show on hoarders, in which it is stated that:

> For some animal lovers, their good intentions of providing a loving home for their pets have gone awry. They may have started out with just a few pets, but then those animals begin to reproduce. Or perhaps they initially set out to rescue abandoned animals and soon find themselves overwhelmed with the responsibility.
>
> *(Animal Planet 2012)*

This phenomenon of hoarders taking on abandoned cats, or even claiming that they run a private shelter, is documented in research from the US (Arluke and Killeen 2009; Frost and Staketee 2010). One could say that the hoarders contribute to the economy of homeless cats in several ways. First, they take on cats that no one else wants. Second, by failing to control the reproduction of cats, they produce large numbers of animals that eventually need to be cared for by shelters. As in the discourse of homeless humans, we could here refer to a 'circle of homelessness' (Wolch *et al.* 1988), in which cats sadly enough can be taken from abusive homes and rescued to a shelter, just to be re-homed with another hoarder.

Summarizing this section, the category of the feral cat is negotiated and made accountable in the context of the crowded homes of hoarders. The meaning of home as well as crowdedness is decisive if we wish to understand how cats that clearly belong to an owner can be produced as feral. The homes of hoarders are not very homelike – in fact, simply not homes at all.

Normalizing living: crowded shelters and sacred homes

It has been recognized that there are at least two ways of defining homeless people: either in terms of lacking a home – as houseless – or by their asocial behaviour (Swärd 1998). Definitions are essential, since they can be linked to explanations, notions of responsibility, and interventions (Sahlin 1992: 51). In the struggle over definitions, different stakeholders and actors introduce and support different versions of reality. A municipality, for example, may be held responsible for caring for homeless individuals – both human and feline ones – and consequently suffer financially. In my data, for example, shelter workers claimed higher numbers of homeless and feral cats than did the authorities. Thus, the existing figures cannot be seen as raw statistics mirroring reality, but as tools in the struggle over definitions (Swärd 1998). My argument here is not to say that cat home issues – including homelessness and shelters – are the same as for humans. There are some defining differences regarding blame and stigma, not to mention the moral discourse on poverty and individual responsibility. Few people would blame the cat for being immoral and living in the street as an individual choice. However, symbolically and institutionally, there are many overlaps.

Home is a place connected to putting down roots; identities are constructed in relation to where we live, where we belong (Duyvendak 2011). In the case of the homeless and feral cat, belongingness is obviously the dimension of home most strongly emphasized. The shelter movement, in turn, aims at bringing the homeless – in terms of being without a home, and sometimes the homeless in terms of antisocial behaviour – to temporary housing at a shelter or in a foster home, while waiting for permanent 'homing'. The animal shelter movement started out early in the late nineteenth century in the US and the UK but much later in Sweden. Its opponents argue that shelters make homelessness permanent through a certain culture. Irresponsible pet owners become excused when cats and dogs may be re-homed (Irvine 2003), and the home as a pure, sacred, and moral place can be saved. Likewise, narratives of hoarders and the official interventions of rescuing, along with media outrage over filth (and less often neglect), restore the ideal home as a place for peace.

Crowding, my second figurative trope, aims at capturing the understanding of the uncontrolled and the collective transformation of individuals in a certain setting (Holmberg forthcoming). As such, it resonates with the notion of vermin. It is well known that the presence of other animals can 'de-civilize' urban places, which in turn calls for such normalizing strategies as sanitization. Both wild and domestic animals can, when transgressing certain boundaries, become vermin and thus objects for extermination. The boundaries can be of geographical or numerical kinds, but it can also be a matter of not fitting into the picture, of being matter out of place or time. I would like to add crowding to this list of 'verminizing' dimensions. One normalizing strategy used for feral cat colonies by a few authorities, but more often by cat advocates, is Trap-Neuter-Release. It is used in order to control the reproduction (so that the colonies do not become overcrowded) while letting the cats maintain their wild existence. An alternative is collective cat 'ownership', as practised in many communities, in which a number of neighbours care for cat colonies with food and sometimes veterinary services. This suggests that the one-cat–one-owner model is not the only possible one. The point here is not to favour any one practice, but rather to indicate the possibility of alternative modes of urban politics.

Returning to the theoretical outline, what can this case study tell us about 'becoming with many' in the urban landscape? Sharing the lives of many refers to a messy terrain where many actors come together and struggle over definitions and outcomes. Moreover, 'the politics of conviviality', in Hinchliffe and Whatmore's (2006) terms, of living together in a more-than-human world, has an epistemological dimension to it. Thus, in order to live well together, we as social scientists need to understand, and make understandable, processes of inclusion and exclusion of humans and other animals.

References

Agamben, G. (2004) *The Open: Man and Animal*, Stanford, CA: Stanford University Press.

Animal Planet (2012) *Confessions: Animal Hoarding*, available at http://animal.discovery.com/tv-shows/confessions-animal-hoarding/about-animal-hoarding/understanding-animal-hoarding.htm, accessed 1 October 2012.

Arluke, A. and Killeen, C. (2009) *Inside Animal Hoarding: The Barbara Erickson Case*, Lafayette: Purdue University Press.

Arluke, A. and Sanders, C.R. (1996) *Regarding Animals*, Philadelphia: Temple University Press.

Arluke, A. *et al.* (2002) 'Press Reports of Animal Hoarding', *Society and Animals* 10: 113–35.

Balcom, S. and Arluke, A. (2001) 'Animal Adoption as Negotiated Order: A Comparison of Open versus Traditional Shelter Approaches', *Anthrozoös* 14: 135–50.

Broberg, G. (2004) *Kattens historia* [The History of the Cat], Stockholm: Atlantis.

Churchman, A. (1999) 'Disentangling the Concept of Density', *Journal of Planning Literature* 13.4: 389–411.

County Administrative Board (2011) *Control Report.*

Coulon, A. (1995) *Ethnomethodology*, London: Sage.

Douglas, M. (1966) *Purity and Danger. An Analysis of Concepts of Pollution and Taboo*, London: Routledge.

Duyvendak, J.W. (2011) *The Politics of Home. Belonging and Nostalgia in Western Europe and the United States*, London: Palgrave McMillan.

Ekstam, H. (forthcoming) *Life in a Small Space – A Feminist Phenomenological Analysis of the Experience of Compact Living*, PhD Thesis, Uppsala: Uppsala University.

Fox, R. (2006) 'Animal Behaviours, Post-human Lives – Everyday Negotiations of the Animal-Human Divide in Pet-keeping', *Social and Cultural Geography* 7.4: 525–37.

Franklin, A. (1999) *Animals and Modern Cultures. A Sociology of Human-Animal Relations in Modernity*, London: Sage.

Franklin, A. and White, R. (2001) 'Animals and Modernity: Changing Human–Animal Relations, 1949–98', *Journal of Sociology* 37.3: 219–38.

Frost, R.O. and Staketee, G. (2010) *Stuff: Compulsive Hoarding and the Meaning of Things*, New York: Houghton Mifflin Harcourt.

Goffman, E. ([1971] 1990) *Stigma. Notes on the Management of a Spoiled Identity*, London: Penguin.

—— (1974) *Frame Analysis: An Essay on the Organization of Experience*, Cambridge, MA: Harvard University Press.

Gove, W.R., Hughes, M., and Galle, O.R. (1979) 'Overcrowding in the Home: An Empirical Investigation of its Possible Pathological Consequences', *American Sociological Review* 44.1: 59–80.

Grier, K.C. (2006) *Pets in America. A History*, Chapel Hill: University of North Carolina Press.

Griffiths, H., Poulter, I., and Sibley, D. (2000) 'Feral Cats in the City', in C. Philo and C. Wilbert (eds), *Animal Spaces, Beastly Places. New Geographies of Human-Animal Relations*, London: Routledge, pp. 56–70.

Haraway, D.J. (1997) *Modest_Witness@Second_Millenium. FemaleMan©_Meets_OncoMouse,*™ London and New York: Routledge.

—— (2003) *Companion Species Manifesto*, Chicago: Prickly Paradigm Press.

—— (2008) *When Species Meet*, Minneapolis: University of Minnesota Press.

Hinchliffe, S. and Whatmore, S. (2006) 'Living Cities: Towards a Politics of Conviviality', *Science as Culture* 15.2: 123–38.

Holmberg, T. (2011) 'Controversial Connections. Urban Studies beyond the Human', in P. Segerdahl (ed.), *Undisciplined Animals. Invitations to Animal Studies*, Cambridge: Cambridge Scholars Publishing.
—— (forthcoming) 'Sensuous Governance: Making Sense of Urban Animal Hoarding'.
Hopper, K. (1990) 'Public Shelter as "a Hybrid Institution": Homeless Men in Historical Perspective', *Journal of Social Issues* 46.4: 13–29.
Hutson, S. and Clapman, D. (1999) *Homelessness: Public Policies and Private Troubles*, London: Cassell.
Irvine, L. (2003) 'The Problem of Unwanted Pets. A Case Study in How Institutions "Think" about Clients' Needs', *Social Problems* 50.4: 550–66.
Jencks, C. (1994) *The Homeless*, Cambridge, MA: Harvard University Press.
Jerolmack, C. (2013) *The Global Pigeon*, Chicago: University of Chicago Press.
Kennett, P. and Marsh, A. (1999) *Homelessness: Exploring the New Terrain*, Bristol: Policy Press.
Kumlin, T. (2011) 'Harold Garfinkel och den omedelbara sociala ordningen', in J. Lindblom and J. Stier (eds), *Det socialpsykologiska perspektivet* (The Social-Psychological Perspective), Lund: Studentlitteratur.
Lynch, M. (1993). *Scientific Practice and Ordinary Action: Ethnomethodology and Social Studies of Science*, Cambridge: Cambridge University Press.
'Missing Colorado Cat Found in New York' (2011) *Guardian*, 15 September, available online at www.guardian.co.uk/world/2011/sep/15/missing-colorado-cat-found-manhattan, accessed 22 June 2013.
Patronek, G. (1999) 'Hoarding of Animals: An Under Recognized Public Health Problem in a Difficult to Study Population', *Public Health Reports* 11.4: 81–7.
Patronek, G.J., Glickman, L.T., Beck, A.M., McCabe, G.P., and Ecker, C. (1996) 'Risk Factors for Relinquishment of Cats to an Animal Shelter', *Journal of the American Veterinary Medical Association* 209: 582–88.
Peggs, K. (2012) *Animals and Sociology*, Basingstoke: Palgrave McMillan.
Rogers, M.K. (2006) *Cat*, London: Reaktion Books.
Sabloff, A. (2001) *Reordering the Natural World. Humans and Animals in the City*, Toronto: University of Toronto Press.
Sahlin, I. (1992) 'Hemlöshet – reflexioner kring ett begrepp', in M. Järvinen and C. Tigerstedt (eds), *Hemlöshet i Norden* (Homelessness in the Nordic Countries), NAD-publikation 22, Helsingfors: Nordiska nämnden för alkohol – och drogforskning, pp. 95–122.
Sanders, C.R. (1999) *Understanding Dogs. Living and Working with Canine Companions*, Philadelphia: Temple University Press.
Statistiska Centralbyrån (SCB) (2006) *Förekomst av sällskapsdjur – främst hund och katt – i svenska hushåll*, Stockholm: SCB Report.
Stokols, D. (1972) 'On the Distinction between Density and Crowding: Some Implications for Future Research', *Psychological Review* 79.3: 275–7.
Swärd, H. (1998) *Hemlöshet. Fattigdomsbevis eller välfärdsdilemma?* (Homelessness. Proof of Poverty or Welfare Dilemma?), Lund: Studentlitteratur.
Tipper, R. (2010) 'Moving Encounters with Pets: The Secret Life of Vince the Cat', paper presented at the conference Animal Movements, Moving Animals, Centre for Gender Research, Uppsala University.
Waddington, D.P. (2011) 'Public Order Policing in South Yorkshire, 1984–2011: The Case for a Permissive Approach to Crowd Control', *Contemporary Social Science* 6.3: 309–24.
Wolch, J. (1998) 'Zoopolis', in J. Wolch and J. Eme (eds), *Animal Geographies: Place, Politics and Identity in the Nature-Culture Borderlands*, London: Verso.
—— (2002) 'Anima urbis', *Progress in Human Geography* 26.6: 721–42.
Wolch, J., Dear, M., and Akita, A. (1988) 'Explaining Homelessness', *Journal of the American Planning Association* 54.4: 443–53.
Wolfe, C. (2009) *What Is Post-humanism?*, Minneapolis: University of Minnesota Press.
Ytrehus, S. (2000) 'Interpretation of Housing Needs? A Critical Discussion', *Housing Theory and Society* 17.4: 166–74.

5

On a wing and a prayer

Butterflies in contemporary art

Giovanni Aloi

> Everyone's frightened of sharks, everyone loves butterflies.
>
> Damien Hirst, 2005

> Butterflies are an easy target, making their captivity and destruction much more criminal.
>
> Mat Collishaw, 2009

On 16 November 2009, Space Shuttle Atlantis took off with a crew of more than astronauts on board. This time, the human ambition for breaching the boundaries of the unknown was shared by four caterpillars of the Painted Lady butterfly, which tagged along in the darkness of the universe (Pathak 2010). For these insects the trip wasn't entirely a matter of leisure, as the eyes of over 180,000 students and teachers on planet Earth kept spying on their metamorphic development in a zero-gravity environment. This educational project, designed by Baylor College of Medicine's Center for Educational Outreach, also encouraged teachers to set up hatching chambers with four Painted Lady caterpillars in classrooms to enable real-time comparison between the insects on Earth and those in space. The University of Colorado beamed images of the insects in space via YouTube, providing students with a direct connection to the mission and prompting reflections on the effect lack of gravity has on bio-rhythms.

Of course, this wasn't the first time animals had left Earth's magnetic field. Experimentation with nonhumans in space began in 1947. Back then it was a small swarm of fruit flies that enjoyed the unique experience in an experiment aimed at gathering information on the effect of radiation on biological functioning, and the fruit flies returned to earth alive (Burgess and Dubbs 2006). However, the same cannot be said for a subsequent experiment that involved a primate. Albert II the monkey was sent into space in 1949 but died when his capsule crashed to Earth due to a parachute failure (NASA 2004). Since then, humans have routinely launched all sorts of animals into space, including mice (1950, 1960, 1966, 1972, 2010), dogs (1951, 1957, 1959, 1960, 1966), monkeys (1959, 1967, 1990), a rabbit (1959), chimps (1961), guinea pigs (1961), frogs (1961, 1990), rats (1961–2, 1966, 1980), a cat (1963), spiders (1972, 2000), and newts (1980), whilst a US biosatellite launched in 1967 contained fruit flies, parasitic wasps, flour beetles, and frog eggs, along with bacteria, amoebae, plants, and fungi (Burgess and Dubbs 2006).

It is rather important to note that unlike previous experiments involving animals in space, the 2009 launch featuring butterflies was expressly designed to invite the participation of wide audiences from around the world. It is perhaps not a coincidence that the butterfly was chosen, as its aesthetic appeal has made it a universal, transhistorical, cross-culturally recognizable symbol of beauty. Following a lengthy absence from the artistic scene, which began with the dwindling of classical still-life painting at the end of the eighteenth century, butterflies have recently re-emerged in the work of a number of contemporary artists but have yet to capture a consistent place in the human-animal studies discourse. Why, after such long absence, are butterflies reappearing in art now and why did they almost entirely disappear from the history of modern painting? Are butterflies of the twenty-first century cultural victims of their own beauty, with nothing left to say, or is there more to their reappearance in art?

It seems at present that mainstream artists are at the forefront of a 'butterfly revolution' that is not informed by the human-animal studies agenda but that nonetheless entails more than a superficial approach to form and colour. Through the contemporary revival of butterflies in art, some artists have begun to define a new *butterfly aesthetic*, which, I will argue, operates through a complex reprising of artistic and scientific representations for the purpose of mobilizing the viewer's perspective on these animals. This chapter will examine the work of prominent and highly controversial artists Damien Hirst and Mat Collishaw in order to illustrate the intricacies involved in their aesthetic choices for the purpose of revealing that much more lies underneath what may simply seem an aesthetics of animal violence. As such, this essay should not be read as condoning animal cruelty but as a genuine attempt to understand highly influential works of art that are very likely to have an impact on the perception of wide audiences.

Butterflies and representation: intermingling discourses

Are there ways in which these artworks can be conceived of that may be productive to a human-animal studies agenda? Hirst and Collishaw are amongst the most internationally popular artists of the contemporary scene and as such their work regularly receives great exposure in the media, outside the strictly artistic circuit. Their practice is not informed by human-animal studies texts, but nonetheless, their work is frequently and consistently associated with the conspicuous emergence of animals in contemporary art. My argument stems from the genuine feeling that Collishaw's and Hirst's treatment of animals may too be considered within the umbrella of the 'postmodern animal' theorized by Steve Baker in 2000, even if the artists' intentionality does not originate from human-animal-studies concerns. The 'botchedness' of these animal bodies suggests that something rather interesting may be at play.

For the purpose of better understanding the complexities presented by the new butterfly aesthetic of Collishaw and Hirst, I will first look at the presence of these insects in the discourses of Western art. A retrieval of these discourses, which have shaped the signification of butterflies in our representational culture, will contribute to a new understanding of certain recurrent aesthetic traits and defining approaches which contradistinguish Collishaw's and Hirst's butterflies.

Very few animals and even fewer insects can command the levels of ecstatic contemplation that butterflies do. As a result, amongst the unfairly discredited world of invertebrates, butterflies occupy a truly unique iconographical niche. In Western culture, early representations of butterflies appeared on sarcophagi of classical antiquity as a symbol of the human soul's migration from the body following death. It is not a coincidence then that the Greek word *psyche* simultaneously means soul and butterfly (Impelluso 2003). Thereafter, the symbolic representation of the human soul imbued in the butterfly was appropriated by Christian imagery to more appropriately embody the idea of resurrection based on the observation that a butterfly's life cycle cannot be fulfilled without the dramatic process of metamorphosis in which symbolically the caterpillar dies and returns to life in a new, spiritually charged body capable of flying (Impelluso 2003). As discussed by David Gallagher,

> Metamorphosis is an inexhaustible malleable device for portraying the fantastic, the oniric, to relate morality [. . .] and to provide religious solutions to readers who face critical moments of transition in life, as where the butterfly is invoked as a symbol of the transmigration of human souls following death to the after-life beyond.
>
> *(Gallagher 2009: 8)*

Interestingly, non-Western cultures, like the Haida of North America as well as Aboriginal Australians, have also extensively and independently developed symbolic imagery relying on butterflies for the representation of human souls, and they are not alone in championing the insect for this purpose. In the tradition of the Navajos, who adopted the butterfly as a totemic animal, the insect symbolizes fertility and life. In the iconography of Japanese art, butterflies symbolize transient joy, womanhood, and female vanity (Werness 2003).

It is again in the Christian-imbued symbolism of still-life painting of the Northern Baroque that we encounter butterflies playing the role of *memento mori*, a signification assigned by the widely spread misconception that butterflies only live for one day. The flickering flight and vivid colours that most Lepidoptera display on their wings have of course largely contributed to the crystallization of this representational association with the ascension into the spiritual, as 'its delicacy expresses the fragility of life and fleet passage of life and death' (Werness 2003: 63).

Although metamorphosis constitutes a pivotal point in our fascination with butterflies, this process is not exclusive to them; many other arthropods undergo radical changes through their life cycles. However, the metamorphic process of butterflies is called 'complete' (in opposition to 'partial') metamorphosis, indicating that the larval and adult stages of the same insects are irreconcilably different in a number of essential biological features. If we take into consideration that other insect orders such as coleoptera, diptera, and hymenoptera also undergo complete metamorphosis, and that even insects experiencing partial metamorphosis like dragonflies present a rather dramatic metamorphic process in which the nymph spends up to three years underwater before emerging in its adult form as one dexterous flying insect, we are then left wondering, why haven't dragonflies caught the human imagination as much? The answer to this question lies in the visibility of the insect itself: it is the colours and intricacy of the patterns on the wings of butterflies (something dragonflies lack) that capture human imagination and subsequently draw attention to the complex life cycle which supports the production of further symbolic levels.

It was indeed the beauty of butterfly wings that prompted Ulisse Aldrovandi to draw them in detail in his *De Animalibus Insectis*, an opera entirely dedicated to insects, a first of its kind, published in 1602. It was the fascination with the symmetry and the familiarity provided by endless repetition of wing patterns that brought Thomas Moffet to compile his *Theatrum Insectorum* in 1634. It was again the beauty of butterfly wings that guided compositional arrangements in the pioneering illustrations of Maria Sibylla Merian in late 1600.

As Janice Neri notices, 'Merian depicted butterflies and moths in either the flat or profile view, a technique that highlighted the patterns and colours of the insects' wings' (2011: 155). And it is because of the colours and patterns of butterfly wings that the insect became almost synonymous with 'collecting of naturalia' during the nineteenth century (Salmon, Marren, and Harley 2000). Perhaps even more than collecting minerals, shells, and ferns, butterfly collecting encapsulated the essence of the Victorian enthusiast lacking professional scientific training but burning with the will to possess the wonderful creations of God (Bernard 1997). It is unclear how much this butterfly-fever contributed to the dwindling numbers that occurred over the last century. Nevertheless, the unregulated killing, preserving, and displaying of butterflies that took place then was an unprecedented phenomenon which strongly contributed to the defining of the modern conception of the butterfly, no longer as a symbol of the human soul, nor a living religious metaphor for Christian dogmas, but as a desirable object of nature to hunt, catch, set, and display.

It is at this stage pertinent to propose that the objectification of butterflies had in some ways already begun through the representational work of naturalists like Aldrovandi. His ambition to catalogue living beings prompted him to draw them in minute detail on the pages of voluminous bestiaries. It is on the surface of these pages that attempted to make sense of nature through classification that butterflies began to lose their spiritual, symbolic dimension. Once fixed as permanent images to be admired, the scientific transfiguration of nature into drawings inevitably oversimplified nature itself, not only from a metaphysical perspective but, in this case, from a metaphorical one too.

The flattening of nature

According to Foucault, it is during the Classical Age (1650–1800) that the writing of history began to be understood as something that required a meticulous examination of things themselves in order to transcribe knowledge in 'smooth, neutralized, and faithful works' (2003: 142). This is where we find the construction of new epistemological sites such as herbaria and collections of plates in which plants and animals were flattened into a system of representational values that aimed at simplifying and clarifying the newly visible world. The shift that stripped butterflies of spiritual, symbolic attributes is therefore to be found in the newly acquired optic of natural history, which emerged as a new discursive formation at the beginning of the Classical Age. In its nomination of the visible, natural history proposed a vision of nature in which living beings made of flesh, feathers, chitin, scales, and hides were translated into language through a process of limiting and filtering, bridging the distances between object, observing gaze, and language itself (144–7). It is through this process of translation that we find the emergence of sight, further assessed towards the end of the seventeenth century by John Locke as the paramount epistemological tool of new scientific disciplines for its peculiar ability to provide proof acceptable to a majority.

Most importantly for this discussion of butterflies and representation, it is relevant to take into account that the episteme of the Classical Age predominantly enabled an understanding of the world as made of orderable surfaces:

> The plant and the animal are seen not so much in their organic unity as by the visible patterning of their organs. They are paws and hoofs, flowers and fruits, before being respiratory systems or internal liquids. Natural history traverses an area of visible, simultaneous, concomitant variables, without any internal relation of subordination or organization.
>
> *(Foucault 1966, 2003: 149)*

This emphasis on the observation of surfaces became pivotal to the rise of interest in butterfly wings, and due to the specific morphology which contradistinguishes these insects, butterflies lent themselves more than other animals to a seemingly 'faithful translation' fitting the materiality of the natural history page. The flatness of butterfly wings as the insect rests or feeds on flowers suggested a readiness for objectification through representation on paper that other animal bodies did not provide; in other words, the morphology of the animal happened to match the peculiar physicality of the privileged medium through which the epistemological locus of the Classical Age configured and supported itself. Furthermore, the flatness peculiar to butterfly wings would have been also readily associated with the morphology of the botanical world. This association became also somewhat consequential to the fact that botany preceded zoology in the formation of discourses of natural history (149). After bringing botany to the fore of taxonomic research, the prominence of vision as the epistemological tool of the Classical Age also, by association with specific media, propelled the artistic as well as scientific interest in butterflies, for, like plants, butterflies appeared to be all exterior. The body of a butterfly was relatively irrelevant to the systems of classification of this episteme: the wings were essential.

Through this process, nature became therefore silent and still at the hands of botanists, zoologists, and artists who engaged in the arduous task of translating the living into the newly founded language of natural history. It is also through this process of translation that we find the scientific and artistic discourses intermingling in the production of new visual imagery, and it is at this stage worth noting that, since the Renaissance, scientists and artists began to work together in order to catalogue nature in the most reliable and objective way possible. As reported by Neri, botanist Carolus Clusius took great care that the craftsman would not 'indulge their whims as to cause the drawing not to correspond accurately to the truth' (2011: xviii). The antagonism between the scientific ambition for establishing a kind of objectivity and the tendency classical art nurtured for idealizing nature began to be at odds with each other.

The purely scientific approach to illustration of the sixteenth and seventeenth centuries and the still-life genre produced since 1600 diverged in their essentially opposed epistemological aims. The main aim of natural history illustration became accuracy and clarity whilst still-life painting proceeded to entangle the insect in the order of the commodity, placing it amongst other desirable objects. Both representational tropes simultaneously objectified butterflies through different and seemingly opposed strategies: the scientific one removed the natural context in which the insect was found, replacing it with a neutral off-white background, whilst still-life painting constructed a stage onto which the insect became enmeshed in a complex stratification of largely implicit exchange-values and spiritual symbolisms. On the off-white backgrounds of natural history plates, the latter entanglement was entirely omitted as was the relationship between the insect and the rest of the natural world (a condition partly challenged by the work of Maria Sibylla Merian). Nevertheless, scientific illustration and still-life painting could be read as different methods contributing to a rationalization of nature, one that attempts to preserve the beauty of what fades and that which is by nature elusive.

Despite their divergence, an intersection between the two representational genres can be identified at the point in which the appearance of butterflies in still-life painting was clearly informed by a number of visual innovations introduced by the emerging scientific discourse and its need of translating nature into a manageable visual language through artists who drew directly from natural history illustrations in order to paint their insects on canvas (Neri 2011). As the craze for butterfly collecting reached its apex during the nineteenth century and the practice became part of the discursive formation of colonialism, representational strategies in the scientific realm shifted. The invention of photography in 1826 began to further bridge the distance between the insect and its illustrational referent.

The optics of scientific enquiry shifted again as the medium provided new visions of nature. The desire to possess nature was fulfilled by the newly acquired image of objectivity that photography proposed as 'truth', and, as a result, illustration lost its privileged place in the scientific discourse. 'A photograph was deemed scientifically objective because it countered a specific kind of subjectivity: intervention to aestheticize or theorize the seen' (Daston and Galison 2007: 135). Of course what photography could not yet provide during the nineteenth century was an adequate reproduction of the coloration of butterfly wings, reducing the insects to monochromatic shadows of their living or preserved selves. Photographic exposure-times were also not fast enough to capture flying insects like butterflies, and black and white images would more often be taken from prepared specimens. These shortcomings became more frustrating with the increased availability on the markets of extremely colourful tropical specimens, justifying in a sense the collecting of these insects, for no illustrative substitute could have captured iridescent colours on wings. Amongst Victorian insect collectors, the 'Aurelians' (as those pursuing butterflies and moths called themselves) outnumbered collectors of all other orders. The hobby apparently become so popular that it 'rank[ed] as a field sport from the 1870's until World War I' (Hancock 2007: 214).

It is impossible to establish if the amount of butterfly collecting that took place in the latter part of the nineteenth century had a major impact on the population of butterflies in Europe. The authors of *The Aurelian Legacy: British Butterflies and Their Collectors* believe that British populations of butterflies were dramatically downsized instead by substantial changes in the countryside and pollution levels rather than collecting (Salmon *et al.* 2000). However, regardless of its predominant cause, a study compiled by the Butterfly Conservation and the Centre for Ecology and Hydrology has recently reported that today 72 per cent of UK butterfly species have declined in abundance and that the distribution of 54 per cent has been reduced, implicating consistent loss of suitable habitat as the root cause (Boettcher 2011).

It is therefore through the intertwining discourses of the mythological, the religious, the scientific, and the artistic that we find butterflies acquiring and losing specific signifying charges. The mythological discourse attributed to the butterfly the symbolic value of the human soul, which was in turn appropriated by discourses of Christianity as a spiritual symbol of resurrection. The intermingling of the religious discourse with the artistic one provided an exuberance of spiritually charged butterfly representations in the still-life paintings of the Classical Age, a phenomenon which owed to the scientific discourses that were introduced, amongst others, by Aldrovandi. Having advanced claims that the representation of butterflies in art was greatly informed by the shifting optics of science first and subsequently by the substitution of the representational for the real during the Victorian period and early 1900s, the

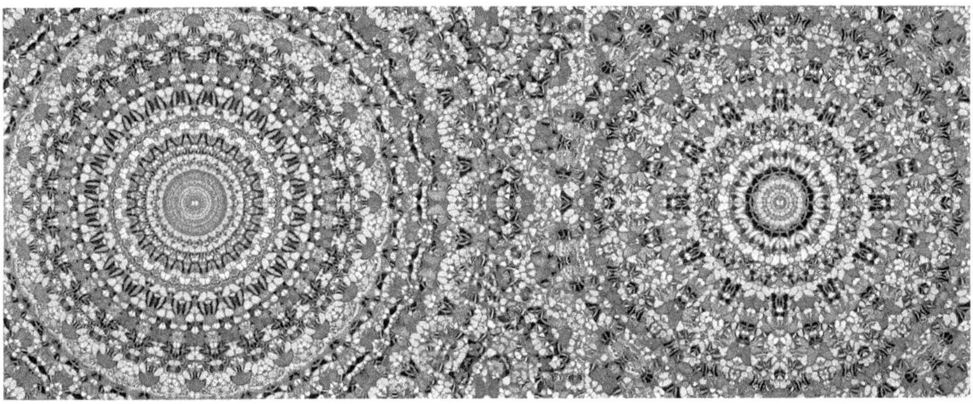

Figure 5.1 Damien Hirst, *I Am Become Death, Shatterer of Worlds*, 2006. Butterflies and household gloss on canvas, 2134 × 5334 mm | 84 × 210 in. Photographed by Prudence Cuming Associates. © Damien Hirst and Science Ltd. All rights reserved, DACS 2014

Figure 5.2 Mat Collishaw, *Insecticide II*, C-type photo on Dibond, 130.5 × 150.8 cm. © Mat Collishaw. Courtesy of the artist

second part of this chapter will explore the representational tropes adopted by contemporary artists Hirst and Collishaw for the purpose of suggesting that the presence of butterflies in contemporary art may be a symptom of a broader – and not simply aesthetic – state of affairs.

Symmetry and the accident

Considering how, over the past hundred years, butterflies have become more readily available to our gaze via photographic reproduction than in nature, the question arises as to how this has altered our relationship with this insect and to the visual more generally. Scientific illustration now capitalizes on the ability that digital photography has to capture crisp detail; and, as the optic of contemporary science has once again shifted, the interest for comparative morphology in Lepidoptera has become mainly the domain of amateurish field-naturalist guides. Nonprofessional illustrated books on butterflies have abandoned the flattened iconography of the Classical Age in favour of a more 'naturalistically constructed' language of wildlife photography. Even when staged in a photographic studio or taken in a butterfly house (which is indeed the case for most butterfly images taken since the 1990s), audiences have come to privilege imageries which artificially 'return' insects to a simulacrum of their ecosystems in opposition to seeing them flattened as specimens against the off-white background of early scientific illustrations. Likewise, Natural History museums around the world have become attuned to this change in visual sensibility, largely removing their butterfly collections of prepared specimens from display and replacing them with seasonal butterfly houses in which visitors can instead observe live insects (Aloi 2007).

As previously mentioned, since the 1990s butterflies have also been the subject of two popular artists whose work emerged through the now legendary phenomenon known as the YBA (Young British Artists). The movement saw a number of young and irreverent artists pushing the boundaries of contemporary art to unprecedented extremes for shock value. Two of the YBA, Damien Hirst and Mat Collishaw, have turned their attention to butterflies at different stages respectively in their careers, most notably Hirst in 1991 with the exhibition titled *In and Out of Love*; in 1997 with a series of circular paintings featuring butterflies; and in the early 2000s with spectacular large-scale butterfly-wing collages. More recently, in 2009, Collishaw unveiled a controversial series of photographic images titled *Insecticide*.

Both artists have in the past been accused of sensationalizing their work through the use of violent imagery, something that might be true, but such accusation somewhat also distracts and confines discussion of the works to the limitations imposed by the ethical and moral. Informed by human-animal studies, I will instead argue that much of contemporary art requires viewers to overcome surface-level shock value in order to reveal the complexity of animal-human interfaces, which helps to more productively understand the presence of butterflies in their art. This does not necessarily equate to condoning the killing of the animals involved, but it seems that, in order to stop audiences in their tracks and induce thinking about the value of the lives of butterflies, this may be the only strategy that presently works. Overcoming the initial shock factor, which usually leads to unconditionally negative responses, may enable viewers to pose questions for the purpose of better understanding our positioning in relation to a number of human-animal issues.

The Nietzschean vaporization of reality, leading to the impossibility of distinction between the 'real' and the 'apparent' world articulated in *The Birth of Tragedy* and *Twilight of the Idols*, reconfigured Western metaphysics around the social construction of the 'I' as moral illusion which projected onto the world itself. Assuming that 'truth' is nothing more than illusion and only individual perceptions exist, then, postmodernist discourse proposed a limitation of the mediation provided by the artist for the purpose of providing a more direct engagement with objects and concepts. Albeit not a compulsory prescription, those artists engaging with the topic of the animal have generally more readily embraced the opportunity to stage unmediated encounters, if not with the live animal, then with preserved animal bodies and surfaces. It is therefore not a coincidence that it is within this paradigmatic set that we witness the rise of photography as privileged medium in offering the opportunity for the abrasiveness of certain encounters to be more than just allusions.

In the first instance it is worth noting that both Hirst and Collishaw present in their works butterfly bodies that have been extensively tampered with. In opposition to the still-life paintings of the Classical Age, the artist's preoccupation does not lie in achieving perfection in the faithful rendition of the animal. This role is now fulfilled by digital photography in the documentary genre. Neither artist's employment of butterflies shares anything with the value associated with the pristine embodied by the scientific specimen. Both artists have in the past engaged with professional scientific imagery in their work; therefore the decision to opt for a different representational route in this instance is deliberately intentional. What is retained of the still-life painting genre in both artists' work is the sense of stillness and suspension.

In the paintings that formed the *In and Out of Love* set and those that subsequently became known as 'butterfly paintings', which surfaced in the late 1990s (and that are still made today as a line of production in Hirst's studio), butterflies are stuck on the surface of monochrome canvases. Their wings, spread flat in the fashion of natural history illustration, are however splattered with paint at the edges, denying a careful scientific killing and summoning, instead presenting the aftermath of the insects' flight onto a canvas primed with fresh, sticky paint. Collishaw's *Insecticide* series proposes a similar 'aesthetic of the accident', albeit presented through the photographic medium, in which butterflies and moths appear horribly crushed and disfigured whilst suspended in midair against a black background.

In the work of both artists, the presence of the insects imbedded in stark backgrounds outlines the staging of peculiar representational planes, which in their neutral timelessness simultaneously deny the taking place of the proposed event. Within these flattened, minimalist, atemporal realms butterflies are caught in suspension, suggesting the possibility of coexistence of life and death in one single moment. The oxymoronic coexisting of life and death that characterized the still-life painting genre of the Classical Age is re-proposed therefore in these works through an allusion to a destructive event that has taken place in the immediate past, one that is so immediate that its spilling into the present becomes an unsettling factor in the fruition of the works: we are not perpetrators of the violence but are implicit witnesses in its fresh aftermath.

This tension is further reinforced by the ambiguity that insects in general present when observed whilst perfectly still. The insects' chitinous bodies don't discolour and they harden soon after death, acquiring a rigour that most often can be misunderstood as livingness. As Hirst noted, 'they don't rot like humans' (2006: 83). An encounter with a dead insect may require therefore a little prodding to be able to conclude that death has really occurred, whilst an encounter with a dead bird or mammal will immediately reveal itself for what it is. Most importantly, in the specific case of butterflies, it is the wings that remain pristine after death (their eyes cloud over rather quickly). As Hirst puts it:

> Then you get the beauty of the butterfly, but it is actually something horrible. It is like a butterfly has flown around and died horribly in the paint. The death of an insect that still has this really optimistic beauty of a wonderful thing.
>
> *(83)*

Through the 'accidents' which foreshadowed the instants framed by Hirst and Collishaw the elusive event which preceded the image has transfigured the butterflies into aesthetic objects in turn produced by the extreme tension which compresses together life, death, beauty, and horror. These butterfly bodies are caught in a moment of passage into intellectual as well as formal transcendence. What Hirst and Collishaw propose is 'a still life on the edge': a condensed image dangerously suspended between life and death, which relies on the notion that 'life does not endure, and the artistic image, in its formal perfection, is the painful agony of the final shock' (Cicelyn 2006: 16).

The friction between the undisputed beauty of the insect and the horror of the suspended, unjustified, and gratuitous death staged by both artists directly challenges past practices of catching and preserving. The moment of the insect's death is erased through the transition granting eternal life in the form of a scientific illustration (butterflies had to be killed to 'pose' for artists and scientists), still-life painting, or cabinet specimen. This is the main defining trait of the flattening of nature outlined in the previous section of this chapter. The concealment of death, achieved through the setting of the symmetrically spread open wings as if in flight, is revealed here in the deliberate derailing of such aesthetic devices. These works bring our attention to the irrefutable fact that the killing of the object of desire has often been a key stage of the epistemological processes of science and art alike; the elegant concealing of such practices has more markedly been art's prerogative. As Maria Sibylla Merian reveals in a letter, the killing of butterflies for the purpose of collecting has never been a pleasant practice:

> If one wishes to [. . .] kill butterflies quickly, then one must hold the point of a darning needle in a flame, thus making it hot or glowing red, and stick it into the butterfly. They die immediately with no damage to their wings.
>
> *(quoted in Neri 2011: 161)*

The value of flawlessness

In the collecting of butterfly specimens the intact preservation of the edges and patterning of wings was, and still is today, of paramount importance. The grading system for charting specimens' conditions is one of the most important reference points for traders and collectors, A1 [ex-pupae] being the highest possible level of perfection a specimen could display, and B being the lowest, where the specimen presents visible damage to wings and body. It is interesting to note that the actual killing of butterflies is verbally omitted even in the technical jargon used to outline the grading system, as we read that the A1 [ex-pupae] butterfly is 'considered "the ultimate" pristine condition. Such a specimen is "mint/proof" with untouched virtual flawlessness. It was most likely dispatched very soon after emergence and is perfect in every way' (Garthe 2007). In this sentence the word 'dispatched' has clearly replaced 'killed' and after that the stretching and drying of the wings.

As seen, the idea of 'perfection' which gave rise to the specimen logic enabled the production of myriad illustrations and paintings in which butterflies are set in the classical pose with spread flattened wings, and it also played a key role in the aesthetics of the entomology cabinet. This iconographical choice acquired popularity during the Classical Age as its clarity enabled the curious gaze to inspect the object in the most favourable condition. Whilst Hirst directly references this formal standardization for *In and Out of Love* and subsequent 'butterfly paintings', Collishaw violently disrupts this normative canon by obliterating the symmetry, which plays a key role in the 'preserved butterfly aesthetic'. As I will argue, this iconoclastic gesture is pivotal to the production of a new and subversive *butterfly aesthetic* in contemporary art. In a sense, Hirst's and Collishaw's images can be understood as functioning in analogous ways to the 'botched taxidermy' concept developed in *The Postmodern Animal* by Steve Baker in which classical aesthetics are subverted through the display of an abrasive aesthetic wrongness aiming at engaging the viewer on a critical level (2000).

It may be surprising to some, but the importance of symmetry is paramount to the phenomenology related to these works of art. Architect György Doczi identified the Golden Section in nature, art, and architecture, revealing that 'the proportioning of butterfly wings varies from species to species, yet there is a unity in these diversities, a unit created by [. . .] proportional limitations, with the 1:2 = 0.5 proportion added corresponding to the octave-diapason' (1981: 89). The encoding of the Golden Section in the wings of butterflies further explains the reasons behind the longstanding fascination humans have with this insect. The proportional aesthetic value embodied by the Golden Section was at the root of the harmony of Egyptian and Greek architecture and sculpture, and it became the essential structural factor in classical art since the Renaissance and until the end of the nineteenth century (Griffing 2007). The Golden Section exerts a peculiar aesthetic power on Western cultures through the production of endless repetition of imbedded classical harmony and balance, which we have become able to instinctively perceive without the need for calculations or further empirical enquiry. The predictability of response which the Golden Section triggers through the repetitive, reflective system of symmetry enhances a sense of stability in the aesthetic paradigm, enabling spontaneous responses of relevance in human and nonhuman animals alike. As Danto notes:

> This is a growing field of experimentation, but it suggests, there being nothing more 'practical' than sex, that dear old natural selection accounts for aesthetic preferences which the clever Greeks introduced into their art, at which, even when the will is out of play because we know them to be statuary, we enjoy looking with the same prurient eyes we cast on one another.
>
> *(1997: 97)*

We can therefore more clearly understand how the obliteration of symmetry in the images created by Collishaw operates as a violent destabilizer of desire. His butterflies and moths have been crushed and their body parts rearranged in such way that the harmonious, proportional balance of the Golden Section is irreparably compromised whilst through this process a different kind of beauty emerges, one that requires dramatic unbalance in order to manifest itself: the sublime.[1]

In *The World as Will and Idea*, Arthur Schopenhauer discusses the idea of symmetry in relation to the metaphysics of music, arguing:

> [w]hen music, as it were in a fit of desire for independence, seizes the opportunity of a pause to free itself from the control of rhythm, to launch out into the free imagination of an ornate *cadenza*, such a piece of music divested of all rhythm is analogous to the ruin which is divested of symmetry and which accordingly may be called, in the bold language of witticism, a frozen *cadenza*.
>
> *(1819: 240–1)*

It is important to note here that Schopenhauer, writing in the early nineteenth century, does not attribute negative value to the obliteration of symmetry, whilst his summoning of the ruin as a comparative element of the visual realm implicitly suggests the emerging sensitivity of the Romantic period and its penchant for tangible manifestation of the destructive effects of decay. In this light, Collishaw's butterflies and moths, more than Hirst's, can be understood as transcendental aesthetic objects charged with the agency of *sublime ruins*. It is at this stage worth noting that one of the most important functions played by the sublime during the Romantic period was that of acting as a vehicle for the spiritual. The sublime extrapolated the spiritual from the exclusive dominion of the Christian church, redistributing it into nature.

As noted by Robert Rosenblum, Romantic poet Thomas Moor expressed the relocation of divinity in nature that, 'far from the traditional rituals of Christianity, could be inspired in many Romantics not only by torrential cataracts and vertiginous abysses, but also by the opposite extreme of an uncommon stillness and silence' (1975: 20). It is therefore in the compositional choices of Collishaw that we find the adoption of the Romantic magnification of scale which allows for the microcosms of nature to become imbued with spirituality as much as the waterfalls, cliffs, and stormy sea-scenes of the 1800s.

In order to advance his argument for the legacy of Romantic art in modern painting, Rosenblum discusses the instance of an early work by Mondrian, the painting of a chrysanthemum. In light of the analogy previously exposed between botany and the illustration of butterflies in scientific plates, Mondrian's flower could be here used to further understand the sublime charge of Collishaw's work in the attempt of understanding how it operates in imbuing the natural with the spiritual. Rosenblum argues that the isolation of the flower, as it stands against a plain neutral dark background, its centrality in the composition, and its magnification confer 'a symbolic aura that far transcends only aesthetic contemplation' (1975: 178). These are all aspects we can easily reconcile with the treatment of Collishaw's butterflies. Placed against a dark, atemporal background, each of the butterflies from the *Insecticide* series is printed on a 182.9 × 182.9 cm photographic sheet, generating a sensational, hyperrealist, overwhelming effect. The process of magnification which allows one to peer into the microcosms of nature was a pivotal Romantic strategy aiming at distorting the world in search of spirituality through the immensely vast and infinitely small. Like Collishaw, Mondrian too focused on decay and the passage from life to death in the portrayal of his flowers, for he would frequently create two painted versions of the same subject seen from the same angle: one showing the flower in its fullest and freshest stage and the other arrested in its decaying process.

The sublime prophecy

In 2003, Hirst came under fire for another body of work involving butterflies. Generally known as 'butterfly wing paintings', the series counts a number of differently sized and shaped canvases on which butterfly wings from a multitude of tropical species have been detached from the insects' bodies and arranged to create the overall effect of a stained glass window. In this case too, as with Collishaw, all works in the series are considerably large pieces, measuring 244 × 153 cm in *Devotion* (2003), 213 × 213 cm in *Amazing Revelation* (2003), or 214 cm in diameter for the round canvases of *Rapture* and *It's Great to Be Alive* (2003). Size, as will be later discussed, is instrumental to the enactment of the sublime in these pieces. All of Hirst's compositional arrangements present levels of entropic harmony structured around symmetrically repeated geometrical patterning. What at first appears to be chaotic and random quickly reveals itself as perfectly ordered, modularly and speculatively referenced within itself. These are images in which the symbolic tensions between life and death explored in the still-life genre have been exasperated.

The work of symmetry previously discussed with Collishaw is at play here again through the Golden Section. The overall visual impact provided by the paintings automatically solicits beauty of a classical kind where what one effectively gazes at, having overcome the hypnotic charge of kaleidoscopic patterns, are dismembered butterfly wings. In other words, the removal of the wings from the butterfly bodies displaces the natural encoding of the Golden Section in the butterfly morphology which is simultaneously recovered through the arrangements of the wings on the surface of the canvas. The geometrized arrangement is here elusive for, in a game of kaleidoscopic, optical illusions, harmonic compositions continuously emerge and dissolve, repositioning themselves on the canvas, outlining endless symmetrical patterning. Beauty and horror coexist again in these ambiguous pictorial dimensions, proposing a tension difficult to emotionally manage: how can so much beauty be produced by so much death? Hirst's works, like Collishaw's, pose questions about the effective value of beauty in our culture and more specifically in relation to nature. Simultaneously, they question beauty's assumed inherent historical positivity in opposition to the absolute negative 'ugly' counterpart that classical art has for centuries attempted to completely remove from sight. Both artists in their specific ways attempt to destroy significant form as tendencies experienced in Gothic and Baroque art.

The dichotomous relationship between fear and attraction which is discussed by Burke in *A Philosophical Enquiry into the Origin of Our Ideas of the Sublime and Beautiful* is here further problematized by a reversal of the dynamics at play, since the painting is captivating beyond expectation due to the differential work of the Golden Section. Paradoxically, we feel compelled to look at what terrorizes us. However, the horror we experience may not so much originate in the nature of the subject we look at – torn butterfly wings – but by the inescapable questioning of how morally appropriate it may be to find pleasure in dismembered butterfly wings. How many butterflies had to die for Hirst's vast canvases to come to life, and how justified is the result in consideration of the required death?[2] The sublime is here directly invoked by quantity as well as by the sensation of greatness. As noted by Jean-François Lyotard, the mathematical sublime relates to magnitude and formlessness of what is perceived, conveying the infinitude of an unlimited extent and absence of boundaries (1994).

The number of units contained in the magnitude escapes us, preventing any mode of rationalization: these images constitute the antithesis of early scientific illustration. The specimen logic is completely sabotaged in these representational planes: the longstanding tradition in which the scientific gaze informed the work of artists has been interrupted in this contemporary manifestation of butterflies. Through the working of what is defined as the mathematical sublime, the image dwarfs us, in this specific case not simply in a sensorial capacity but simultaneously in a cultural one. The clash between the poetics of collecting and the systematic pragmatism required by the curiosity of the gaze for nature's work is here painfully exposed through the reminder that any act of epistemological value of nature involves (and has always involved) an inescapable violence on the animal. What terrorizes us is not therefore just what we see in front of us, but it is what we see in the reflections of ourselves: the dramatic contradictions involved in what it means to be a human animal amongst other animals. The 'negative pain' Burke called delight provides a fissure for the diversion of rational, continuous thought that aims at concealing the functioning of our metaphysical relationship with the insect.

The resurgence of the sublime in contemporary art may constitute a more complex and urgent phenomenon than initially auspicated, one that may be linked to the emergence of capitalist phases of crisis or thresholds proposing new directions for the future. Historically, the Romantic understanding of the sublime in the arts emerged during the height of the Industrial Revolution, which constituted the birth of modern capitalism. If we give credit to the theory that sees the sublime discourse forming as a counterweight to the violent reconfiguration of cultural values imposed by capitalist society, we could therefore read the work of Collishaw and Hirst as symptomatic of the capitalist crisis which has shaped cultural life during the first decade of the new millennium, like the events leading up to and of 11 September 2001 and the subsequent wars in Iraq and Afghanistan. As argued by Luke White:

> If we find ourselves tangling with the sublime again today, the reason for this might be our embrace within a capitalist modernity whose form of capital has come once more to bear uncanny resemblances to the imperial, hyper-liquid and perplexingly spectral capital of the eighteenth century.
>
> *(2010)*

From this perspective, it may be assumed that it is not a coincidence that both Hirst's 'butterfly paintings' and Collishaw's *Insecticide* have trauma at their originating root. The rearranging of torn butterfly wings for the purpose of outlining decorative patterns alludes to torture and castration, whilst Collishaw's crushed Lepidoptera summons the overbearing swiftness of an unexpected, fatal blow. From this perspective, the sublime exerted by these works of art attempts to return us to a mystic world of spiritual dimensions in which the natural body is the repository of a connection to a nature from which we feel more than ever alienated. This too is a paradoxical proposition, which dangerously invites us to find salvation in what we have already perhaps irremediably destroyed: a prayer for a future that may be already doomed.

Through the subversion of classical convention and the obliteration of the Golden Section from nature's form, Hirst and Collishaw propose a new representational paradigm in which the death of butterflies is at play just as much as it was during the Classical Age and thereafter in Victorian times. The substantial difference is that here the death is not concealed. Whilst scientific and painterly representation of butterflies produced between the sixteenth and the nineteenth centuries aimed at concealing the death of the insect – something inevitably necessary for the observation required by the painterly and illustrative act – these images make us painfully aware of death's work as required by the production of beauty and knowledge. The obliteration and reconfiguration of symmetry and balance produced in these works proposes a departure from the figurative realm, pushing representation as closely as possible to abstraction but without ever fully dissolving.

In so doing, the artists partially reconcile butterflies with the range of spiritual associations explored in the first section of this chapter. They attempt to bypass the Victorian obsession for possessing the perfect specimen in order to return a spiritual aura to an insect that has been voided of all auras by the objectification of the scientific gaze. Paradoxically, therefore the butterflies are re-spiritualized through an act of destruction.

In the attempt to reconfigure the spiritual aura of butterflies, Hirst and Collishaw have fragmented the butterfly body, each providing an anti-scientific imagery in which the animal is completely extrapolated from wildlife as well as any natural-history image repertoire. The staging of the passage from life to form becomes therefore much more than that from life to death. It summons the precinct on which art in general functions, exploring the threshold between life and art.

From the purely artistic perspective of abstract art, the shift from naturalistic representation of butterflies to the obliteration of natural symmetry in search of alternative quasi-abstract propositions echoes the history of transcendental experiences with which the abstract art of Mondrian, Kandinsky, and Rothko was imbued. The obliteration of the animal body and its newly acquired ability to function as a vehicle for a kind of spirituality relying on the sublime's dynamics poses therefore questions about the current cultural role played by butterflies and nature at large. Although I am not here arguing for the reconfiguration of a spiritual approach to nature, one is left to wonder what is at stake for a culture that is unable to reconnect with nature in ways other than the scientific.

More alarmingly, the latest incarnation of the sublime we are experiencing in contemporary art may cast a dark of doom. As Luke White argues,

> A key aspect of the recent resurgence of the sublime as a subject for study has undoubtedly been its relevance as an aesthetic of terrible nature, at a moment when, with growing fears about environmental catastrophe, nature has reappeared as a limit to human power, progress and wealth, something which even threatens to destroy us.
>
> *(2010)*

It is the imposition of these limits by nature itself and the seemingly impossible restorability of a balance between nature and humanity that grant these works the extremely unsettling quality that they exude. The images highlight a need for seeking the answer to ultimate questions outside the canons of religion and scientific enquiry. They are a symptom of the facts that science and religion may have both somewhat failed to answer some pivotal questions and that, through the entering of a new capitalist crisis, some questions that have become more urgent than ever demand the implementation of radically new perspectives in order to be appropriately answered.

Notes

1 Collishaw's butterflies are ethically sourced and the butterflies are already preserved when his intervention takes place.
2 Hirst's butterflies are also ethically sourced through fair trade butterfly farming.

References

Aloi, G. (2007) 'The Contemporary Uneasiness with Entomology Displays', *Antennae: The Journal of Nature in Visual Culture* 2: 26–31.
Always, Forever, Now (2005), reduced edition, London.
Baker, S. (2000) *The Postmodern Animal*, London: Reaktion.
Bernard, V.L. (1997) *Victorian Science in Context*, Chicago: University of Chicago Press.
Boettcher, D. (2011) 'Butterflies 72 per cent of UK Species in Decline', *BBC News*, 7 December, available online at www.bbc.co.uk/news/science-environment-16054895, accessed 10 January 2012.
Burgess, C. and Dubbs, C. (2006) *Animals in Space*, London: Springer.
Burke, E. (2008) *A Philosophical Enquiry into the Origin of Our Ideas of the Sublime and Beautiful*, Oxford: Oxford University Press.
Cicelyn, E. (2006) 'The Agony and the Ecstasy', in D. Hirst (ed.), *Damien Hirst, The Agony and the Ecstasy: Selected Works from 1989 to 2004*, English edition, Napoli: Electa.
Danto, A.C. (1997) *After the End of Art*, Princeton, NJ: Princeton University Press.
Daston, L. and Galison, P. (2007) *Objectivity*, New York: Zone Books.
Doczi, G. (1981) *The Power of Limits: Proportional Harmonies in Nature, Art, and Architecture*, London: Taylor and Francis.
Foucault, M. (1966/2003) *The Order of Things*, London: Routledge.
Gallagher, D. (2009) *Metamorphosis: Transformations of the Body and the Influence of Ovid's Metamorphoses on Germanic Literature of the Nineteenth and Twentieth Centuries*, Amsterdam: Rodopi.
Garthe, B. (2007) *Garthe's Insect Gradebook*, available online at www.garthesinsectgradebook.20megs free.com/index.html, accessed 23 March 2012.
Griffing, L.S. (2007) *The Golden Section: An Ancient Egyptian and Grecian Proportion*, Bloomington, IN: Xlibris.
Hancock, M.W. (2007) *Boffin's Books and Darwin's Finches: Victorian Cultures of Collecting*, Lawrence: University Press of Kansas.
Hirst, D. (2006) *I Want to Spend the Rest of My Life Everywhere, with Everyone, One to One*, D. Hirst (ed.), *Damien Hirst, The Agony and the Ecstasy: Selected Works from 1989 to 2004*, English edition, Naples: Electa.
Impelluso, L. (2003) *Nature and Its Symbols*, Los Angeles: Getty Museum.
Kant, E. (2008) *Critique of Judgment*, Oxford: Oxford University Press.

Locke, J. (1999) *An Essay Concerning Human Understanding*, State College: Pennsylvania State University Press.

Lyotard, J.-F. (1994) *Lessons on the Analytic of the Sublime*, E. Rottenberg (trans.), Palo Alto, CA: Stanford University Press.

Miall, N. (2009) 'The Butterfly Hunter', in M. Collishaw, *Insecticides*, London: Haunch of Venison, pp. 2–9.

NASA (2004) *History of Research in Space Biology and Biodynamics at the Air Force Missile Development Center, Holloman Airforce Base, New Mexico, 1946–1952*, available online at http://history.nasa.gov/afspbio/top.htm, accessed 31 January 2009.

Neri, J. (2011) *The Insect and the Image*, Minneapolis: University of Minnesota Press.

Newman, B. (1948) 'The Sublime Is Now', in *The Tiger's Eye* 1.6 (December), New Haven, CT: Yale University Press, pp. 51–3.

Nietzsche, F. (1998) *Twilight of the Idols*, D. Large (trans.), Oxford: Oxford University Press.

—— (1967) *The Birth of Tragedy and The Case of Wagner*, W. Kaufmann (trans.), New York: Random House.

Pathak, D. (2010) 'Space Station Butterfly Experiment Grabs International Interest', *RedOrbit*, 2 February 2010, available online at www.redorbit.com/news/space/1817390/space_station_butterfly_experiment_grabs_international_interest, accessed 23 January 2012.

Rosenblum, R. (1975) *Modern Painting and the Northern Romantic Tradition: Friedrich to Rothko*, London: Thames and Hudson.

Salmon, A., Marren, M., and Harley, P. (2000) *The Aurelian Legacy: British Butterflies and Their Collectors*, Berkeley: University of California Press.

Schopenhauer, A. (1819) (1948) *The World as Will and Idea*, London: Routledge and Keegan Paul.

Truman, J.W. and Riddiford, L.M. (1999) 'The Origins of Insect Metamorphosis', *Nature* 401.6752: 447–52.

Werness, H.B. (2003) *Animal Symbolism in Art*, London: Continuum.

White, L. (2010) 'Damien Hirst's Shark: Nature, Capitalism and the Sublime', *Tate's Papers* 14, available online at www.tate.org.uk/research/tateresearch/tatepapers/10autumn/white02.shtm, accessed 1 August 2011.

6
'This ain't agriculture'

Bernard E. Rollin

Consider the story told to me by one of my colleagues in Animal Science at Colorado State University. This man told of his son-in-law, who had grown up on a ranch, but could not return to it after college because it could not support him and all of his siblings. (Notably, the average net annual income of a Front Range – i.e. eastern slope of the Rocky Mountains – rancher in Colorado, Wyoming, or Montana is about $35,000!) He reluctantly took a job managing a feeder-pig barn at a large swine-factory farm:

> One day he reported a disease that had struck his piglets to his boss. 'I have bad news and good news,' he reported. 'The bad news is that the piglets are sick. The good news is that they can be treated economically.' 'No,' said the boss. 'We don't treat! We euthanize!' He proceeded to demonstrate by dashing the baby pigs' heads on the side of the concrete pen, and then throwing the still-twitching piglet into a garbage heap. The young man could not accept this. He bought the medicine with his own money, clocked in on his day off, and treated the animals. They recovered, and he told the boss. The boss's response was 'You're fired!' The young man pointed out that he had treated them with his own time and money, and was thus not subject to firing. He did, however, receive a reprimand in his file. Six months later he quit and became an electrician. He wrote to his father-in-law: 'I know you are disappointed that I left agriculture, Dad. *But this ain't agriculture!*'

I

It is arguable that the development of human civilization was directly dependent on the creation of a secure and predictable food supply. Such a food supply freed peoples from the uncertainties and vagaries of depending on hunting and gathering, and enabled the establishment of communities. Predictability regarding food was assured by the development of both plant and animal agriculture, which operated synergistically. Cultivation of crops and plants secured human ability to depend on (barring catastrophes of weather) foods of plant origin and a steady and local source of animal feed. Animal agriculture in turn provided a source of labor for crop production, as well as a predictable reservoir of animal protein for human consumption. The secure food supply ramified in the ability to develop manufacturing, trade, commerce, and, in Hobbes's felicitous phrase, the 'leisure that is the mother of philosophy', construed in the broadest sense as speculative thought, science, technological innovation, art, and culture.

A presupposition to the development of both plant and animal agricultures was the concept of sustainability; that is, assurance that the conditions and resources necessary to them were indefinitely renewable. As children, many of us learned about balanced aquaria. If we wished to keep a fish tank where the fish lived and we didn't want to keep tinkering with it, we needed to assure that the system in question was as close to a 'perpetual motion' machine as possible, a system that required little maintenance because all parts worked together. That meant including plants that produced oxygen and consumed carbon dioxide, enough light to nourish the plants (or rather plants that thrived in the available light source), water that was properly constituted chemically, scavengers to remove wastes, and so on. When such a system worked, it required minimal maintenance. If something were out of balance, plants and animals would die, and require constant replacement. The fish tank aims at being a balanced ecosystem and thus represents a model of traditional approaches to cultivation of land, wherein one sought to grow plants that could be grown indefinitely with available resources, could conserve and maximize these resources, and the plants would not die out or require constant enrichment. Hence the beauty of pastoral agriculture, where pasture nourished herbivores, and herbivores provided us with milk, meat, and leather, and their manure enriched the pasture land in a renewable cycle.

Cultivation of land evolved locally with humans. If one did not attend to the constraints imposed by nature on what and how much could be grown in a given region, the region would soon cease to yield its bounty, by virtue of salinization, or depletion of nutrients or overgrazing, or insect infestation. Thus, over time, humans evolved to 'farm with nature' in a way that involved animal husbandry both as a rational necessity and an ethical imperative. Local knowledge, accumulated over a long period of trial and error, told us how much irrigation was too much; what would not grow in given soils; what weeds left standing protected against insects; where shade and windbreaks were needed, and so on. Thus accumulated wisdom was passed on – and augmented – from generation to generation, and was sustainable, meaning that it required minimal tweaking or addition of resources. The genius of agriculture was to utilize what was there in a way that would endure. If the land did not thrive, you did not thrive. Traditional agriculture, then, was inherently sustainable; by trial and error over a long period of time it evolved into as close to a 'balanced aquarium' as possible.

Not surprisingly, precisely isomorphic logic applied to sustainability in animal production. The maxim underlying continued success in rearing animals was *good husbandry*, which represented a unified synthesis of prudence and ethics. What I term 'husbandry' here meant, first of all, placing the animals into the optimal environment for which they had been bred, and where they could maximally fulfill their physical and psychological needs and natures. Having done so, the herder and later farmers then augmented animals' ability to survive and thrive by watching over them – by providing protection from predators, food during periods of famine, water during times of drought, shelter during extremes of climate, assistance in birthing, medical attention, and generally ministering to whatever needs the animals had.

So powerfully ingrained was this imperative in the human psyche that when the Psalmist searches for a metaphor for God's ideal relationship to human beings, he can do no better than seize upon the conceit of the Good Shepherd. The Shepherd serves as far more than merely a herdsman, but more as a guardian and protector of the sheep under his aegis:

> The Lord is my shepherd; I shall not want.
> He maketh me to lie down in green pastures.
> He leadeth me beside the still waters.
> He restoreth my soul.
>
> (Psalm 23, Old Testament)

We want no more from God than what the Good Shepherd provides for his flock. As we know from other passages in the Old Testament, a lamb on its own would live a miserable, nasty, and short life by virtue of the proliferation of predators – hyenas, raptors, wolves, bears, lions, foxes, jackals, and numerous others. With the care and ministrations of the Shepherd, the animal lives well until such time as humans take its life, in the meantime supplying us with milk, wool, and, in the case of some domestic animals, the labor that became indispensable to the working of land for crops.

The power of this symbiotic image cannot be overestimated in the history of Western civilization. In Christian iconography, for thousands of years, Jesus is depicted both as Shepherd and as lamb, a duality built into the very foundations of human culture. The pastor, a word harking back to *pastoral*, tends to his flock; the members of his congregation are his sheep. And when Plato discusses the ideal political ruler in the *Republic*, he deploys the shepherd-sheep metaphor: the ruler is to his people as the shepherd is to his flock. *Qua* shepherd, the herder exists to protect, preserve, and improve the sheep; any payment tendered to him is in his capacity as wage-earner. So too the ruler, again illustrating the power of the concept of husbandry on our psyches.

Animal agriculture was indispensable to the subsequent development of society. Husbandry is the *ancient contract* that was a precondition for the entire cultural-evolutionary process. In one of the most momentous ironies in the history of civilization, this ancient contract with the animals, as well as with the Earth, in terms of sustainability, contained within it the seeds of its own undoing. It was by virtue of a secure and predictable food supply that humans could proceed with trade, manufacturing, invention, and the general flourishing of cultures.

By the late nineteenth century, industrial proliferation and innovation had reached a point where sustainability and good husbandry seemed to be no longer essential presuppositions of civilization. The ancient contract, which we may characterize as husbandry with regard to animals and stewardship with regard to the land, was the presuppositional bedrock upon which economics, art, and culture rested. Yet with the profound hubris akin to that of an Icarus who challenged inherent human limitations, with blind and abiding faith in the humanly crafted tools that repeatedly show themselves as impotent in the face of natural disaster, we thumbed our noses at both morality and prudence. As the ancients crafted the tower of Babel, so we began to overreach the constraints imposed on us by the natural world. In both crop and animal agriculture, the ancient values of sustainability, stewardship, and husbandry inexorably gave way to modernist values of industrialization, productivity, and efficiency. The symbiotic partnership between humans and the Earth, and between humans and animals, was rapidly transmuted into patent exploitation with no respect for or attention to what priceless elements were lost.

Although my major focus in this chapter is animal agriculture, I would be remiss if I did not at least sketch the developments in crop agriculture that paralleled the loss of animal husbandry. Cultivation of land evolved locally with humans. If one did not attend to the constraints imposed by nature on what and how much could be grown in a given region, the region would soon cease to yield its bounty, by virtue of salinization, depletion of nutrients, overgrazing, and insect infestation. The genius of agriculture was to utilize what was there in a way that would endure. As in animal husbandry, if the land did not thrive, you did not thrive. Traditional agriculture, then, was inherently sustainable; by trial and error over a long period of time it too evolved into as close to a 'balanced aquarium' as possible.

Also, as in the case of animal husbandry, humans broke their contract of stewardship over the earth as soon as they were able, given the tools of technology. No longer was agriculture directed at sustainability. Instead it was driven by a mantra of productivity: suck as much yield from the earth as possible with the help of technology, sometimes in the name of profit and sometimes in the name of 'feeding the world', using huge tractors for tillage, irrigation systems, fertilizers, pesticides, and herbicides. What had historically been gentle intercourse with the earth became rape and plunder. And yield grew.

But at what cost? Science and technology enabled extraction of greater crop yields than hitherto imagined, but, as it is said, there is no free lunch. Instead of farming depending on idiopathic wisdom, local knowledge, know-how – passed from generation to generation – it was now conceived of as technology, as applied science, as nomothetic or law-like, and in principle applicable to any locale. As in animal agriculture, capital and machinery supplanted knowledgeable labor, farms got bigger and bigger, and 'get big or get out' became their mantra. Food was plentiful, and thus cheap.

With these gains came major costs, albeit costs that were not immediately obvious, attended by long-term liabilities. But costs they were. If you forget about having a balanced aquarium, then you must pump resources in regularly to compensate for the loss of balance. The new agriculture required a great deal of fuel to run the machinery, make the chemicals. Massive amounts of water were also required. Only 75 percent of modern agriculture's consumption of water is replenished. And these chemicals, fertilizers, pesticides, herbicides, and fuels left residues polluting air and water, leading to disturbance and death of fragile ecosystemic health. Growing crops required land, so forests and rain forests were cleared, uprooting and impoverishing endogenous ecosystems, annihilating species, losing wild plants whose significance to the balance of nature was unappreciated. Repetitive tillage led to soil erosion and depletion, degradation, and pollution. Toxic residues abounded. Excess irrigation depleted soil nutrients, requiring yet more fertilizer. Monoculture – cultivation of one cultivar, that which is most profitable, and most productive, but also most susceptible to devastation (by putting all of one's eggs in one bucket) – replaced traditional crop diversity.

Powerful and plentiful agricultural chemicals had negative health effects on workers and consumers. Indiscriminate use of pesticides predictably and inexorably led to inadvertent selective breeding of super-pests highly resistant to these chemicals, even as the massive use of antibiotics in confinement animal agriculture both to promote growth and to mask the effects of bad husbandry led to the evolution of antibiotic-resistant pathogens. In addition, pesticides non-selectively killed off both target pests and their more widely beneficial natural enemies.

As farmers' debt-load increased in the industrialized world, increasing numbers of small farmers lost their farms, unable to afford the infrastructure required. And as they went out of business, so died the little communities they inhabited, and their cultures, and their way of life. And, with unsurpassed and bitter irony, in the United States, the land-grant universities chartered in the nineteenth century to help small farmers and the rural communities they constituted came to contribute to their demise, by developing the very technologies that caused the problems we have described, and they were funded to do so by the US Department of Agriculture (USDA). Farming as a way of life became agribusiness, the grand monoliths dominating the food supply that Thomas Jefferson feared as inimical to democracy. And agribusiness funded the science that perpetuated agribusiness, leaving no niche for the small farmer. So bad has this become that when I did contract research for the USDA on farm animal welfare and cited western ranchers' views of their animals, and the sort of concern for their well-being evidenced earlier, I was told by a high official that 'western ranchers are not real agriculturalists – they are a bunch of damn Romantics.'

The sentence just quoted is tragically emblematic of the degree to which animal husbandry has been ridiculed as antiquated and supplanted by an industrial model. For it is undeniably the case that ranchers in the western United States are the last remaining vestige of the husbandry approach to animal agriculture that was historically dominant. What did the officials cited above mean by calling them 'Romantics'?

In the first place, income is not the primary rationale for ranching. As mentioned above, the average net income of a Front Range rancher is $35,000 per annum. Having lectured to well over 20,000 ranchers in my career on their home grounds, I am well aware of the fact that their 'way of life' is far more important to them than income. These people, while geographically isolated (the nearest neighbors may be 100 miles or more away), willingly trade sociality for proximity to nature. Rancher families are extraordinarily close-knit, and even small children work closely with their parents from a very young age. The overarching value is a fierce commitment to their animals and to allowing the animals to live under the pastoral conditions dictated by their natures. I cannot begin to count the number of times I have heard ranchers declare that 'if I had to raise animals like the chicken and pig people do, I would get the hell out of the business.' In my experience, a higher percentage of ranchers, over 90 percent, would affirm that animals have the right to live their lives in accordance with their natures than would members of the urban public!

And, most eloquently attesting to their commitment to husbandry, the vast majority of ranchers have spent more on saving the life of a marginal calf, be it in money or time, than the calf is worth. When I point out the economic irrationality of spending $20 to produce a 'widget' that one sells for $10, the angry response I invariably elicit is: 'but these are not widgets; they are living things for which we are responsible!' It is no wonder that the USDA, an agency that is greatly complicit in the industrialization of agriculture, mocks and derides these people as 'Romantics'. The bottom line is that, short of such minor changes as riding fence with ATVs or motorcycles rather than horses, Western cow-calf ranching is unchanged from what it was in the nineteenth century and so has grown out of step with the (animal) exploitative and (farmer) self-defeating notions of progress touted by the USDA.

While intensification of agriculture clearly eroded and indeed virtually eliminated husbandry-based animal agriculture, it also dealt serious and indeed mortal blows to its sustainability. Consider, for example, pastoral rearing of cattle or pigs. Under these conditions, overgrazing or exceeding the land's carrying capacity soon made manifest one's exceeding of sustainable numbers, so that self-interest, indeed, survival of the farmer, corrected such mistakes. Under technological, intensive, agricultural conditions, technological innovations or 'fixes' concealed going beyond sustainability until a great deal of damage was done. Whereas pastoral agriculture represented a 'balanced aquarium' in the case of grazing and waste disposal – animal waste served as fertilizer and helped nourish forage – confinement agriculture does not utilize animal waste, and disposal of such waste becomes a major problem, rather than an asset, as is evident in the huge and formidable problems of waste disposal encountered by the swine industry in the southeastern United States.

Critics of my work have argued that I present a far too rosy picture of preindustrial, husbandry agriculture, implying that everyone in agriculture treated their animals well. That is certainly not my intention. In fact, the logo of the ASPCA (American Society for the Prevention of Cruelty to Animals), the first animal welfare organization in the United States, shows the founder of the organization, Henry Bergh, staying the hand of a carter who has whipped his horse to the ground. Obviously, there was animal abuse well before the advent of industrial agriculture due to short tempers, cruel individuals, and ignorance. One would be both stupid and utopian to deny that. My point, however, was that success in preindustrial animal agriculture was *conceptually* (if not always *de facto)* based ideally on maximizing animal interests in pursuit of producer interest. No one operating rationally in a husbandry-based system would intentionally bring harm to the animals if only for reasons of self-interest.

II

If everything we have stated is correct, how did industrial agriculture come to supplant husbandry agriculture? In the mid-twentieth century, husbandry agriculture was replaced by industrial agriculture in many countries. Ruth Harrison (1964), in her famous book *Animal Machines*, described the horrid conditions that she observed on intensive farms in England. Chickens crammed into tiny cages, tethered veal calves, and sows that were not able to turn around. Her book was translated into seven languages, and it served as an impetus for the formation of the Bramble Committee, which wrote the five freedoms, the pioneering compact of rights for animals under human control. Her book caused the public to become concerned about the intensification and lack of husbandry that was occurring on English poultry, veal, and pig farms. And it helped to contextualize how industrialization of animal agriculture in the US occurred for a variety of understandable and even laudable reasons that are worth recounting. When industrial agriculture began, roughly in the 1940s, the US was confronted with a variety of new challenges related to food. In the first place, the great economic Depression and Dust Bowl (the dust storms that swept the US and Canadian prairies in the 1930s, caused by severe drought and topsoil erosion from plowing) had soured many people on farming and, even more dramatically, had raised the specter of starvation for the American public for the first time in US history. Vivid images of bread-lines and soup kitchens drove the desire to assure a plenitude of cheap food. By the late 1960s and 1970s, the US had large-scale industrialized animal agriculture with much bigger units than those in Europe.

Because better jobs were to be found in cities, displaced rural people flocked to them in hopes of a better life, creating a potential shortage in agricultural labor. Correlative with the growth of cities and suburbs came encroachment on agricultural land for various forms of development, raising land prices and moving acreage once available for agriculture out of that pool. Many people who would otherwise have been happy with a slow, rural way of life also were exposed to greater sophistication by virtue of military service in World War I and World War II, and thus were dissatisfied with an agrarian existence. Recall the song popular after World War I, 'How ya gonna keep 'em down on the farm (after they've seen Paree)?'

Ironically, by the late 1960s many urban people were yearning for the 'simple life' represented by the small farm. Today, in developed countries, there is a huge interest in buying animal products from small, local family farms. Urban people are yearning for a return to animal husbandry. Along the way, demographers predicted a precipitous and dramatic increase in population, which turned out to be accurate. With the success of industrialization in new areas, notably Henry Ford's application of the concept of the assembly line to the production of the automobile, it was probably inevitable that the concepts of industrialization would be applied to agriculture. (After witnessing the Chicago stockyard operations, Ford himself had already characterized slaughterhouses as 'disassembly lines'.) Thus was born an industrial approach to agriculture, with machines taking the place of labor. The traditional Departments of Animal Husbandry in agricultural schools symbolically marked this transition by changing their names to Departments of Animal Sciences, a field defined in textbooks as 'application of industrial methods to the production of animals'.

In this transition, the traditional bedrock values of agriculture – husbandry, sustainability, agriculture as a way of life, not only a way of making a living – were transmuted into values of efficiency and productivity. With human labor replaced by machinery, in turn requiring large amounts of capital, farm units grew larger, eventuating in the mantra of the 1970s: 'get big or get out'. Agricultural research stressed producing cheap and plentiful food, and moved in unprecedented directions. With animals confined for efficiency and away from forage, much research was directed towards finding cheap sources of nutrition, in turn leading to feeding such deviant items to animals as poultry and cattle manure, cement dust, newspaper, and, most egregiously, bone-meal to herbivores, the last of which is believed to have created Bovine Spongiform Encephalopathy (BSE) or 'Mad Cow Disease'. Animals were kept under conditions alien to their natural needs for the sake of productivity. Whereas small-scale, sustainable animal agriculture stressed putting square pegs into square holes, round pegs into round holes, while producing as little friction as possible, industrialized animal agriculture forced square pegs into round holes by utilizing what I have called 'technological sanders', such as antibiotics, hormones, extreme genetic selection, air-handling systems, artificial cooling systems, and artificial insemination to force animals into unnatural conditions while they nonetheless remained productive, in the narrow sense of profitable to investors.

Consider, for example, the egg industry, one of the first areas of agriculture to experience industrialization. Traditionally, chickens ran free in barnyards, able to live off the land by foraging and to express their natural behaviors of moving freely, nest-building, dust-bathing, escaping from more aggressive animals, defecating away from their nests, and, in general, fulfilling their natures as chickens. Industrialization of the egg industry, on the other hand, meant placing the chickens in small cages, in some systems with six birds in a tiny wire cage, so that one animal may stand on top of the others and none can perform any of their inherent behaviors, unable even to stretch their wings. In the absence of space to establish a dominance hierarchy or pecking order, they cannibalize each other and must be 'debeaked', producing painful neuromas since the beak is innervated. The animal is now an inexpensive cog in a machine, part of a factory, and the cheapest part at that, thus totally expendable. If a nineteenth-century farmer had attempted such a system, he would have gone broke, with the animals dead of disease in a few weeks. Some genetic lines of pigs and chickens are so highly selected for egg and meat production that they have less disease resistance. Pigs have become so susceptible to disease that some farmers have installed anti-bacterial filters to take germs out of the air that enters the building. This is a technological sander taken to the extreme.

The steady state, or enduring balance of humans, animals, and land, is lost. Putting chickens in cages and cages in an environmentally controlled building requires large amounts of capital and energy and technological 'fixes'; for example, to run the exhaust fans to prevent lethal buildup of ammonia. The value of each chicken is negligible so one needs more chickens; chickens are cheap, cages are expensive, so one crowds as many chickens into cages as is physically possible. The vast concentration of chickens requires huge amounts of antibiotics and other drugs to prevent wildfire-like spread of disease in overcrowded conditions. Breeding of animals is oriented solely towards productivity, and genetic diversity – a safety net allowing response to unforeseen changes – is lost. Bill Muir, a genetics specialist at Purdue University, found that commercial lines of poultry have lost 90 percent of their genetic diversity compared to noncommercial poultry. Dr. Muir is extremely concerned about the lack of genetic diversity (Lundeen 2008). Small poultry producers are lost, unable to afford the capital requirements; agriculture as a way of life as well as a way of making a living is lost;

small farmers, who Jefferson argued are the backbone of democratic society, are superseded by large corporate aggregates. Giant corporate entities, vertically integrated, are favored. Manure becomes a problem for disposal, and a pollutant, instead of fertilizer for pastures. Local wisdom and know-how essential to husbandry is lost; what 'intelligence' there is is hard-wired into 'the system'. Food safety suffers from the proliferation of drugs and chemicals, and widespread use of antimicrobials to control pathogens in effect serves to breed – select for – antibiotic-resistant pathogens as susceptible ones are killed off while resistant ones multiply. Above all, the system is not balanced – constant inputs are needed to keep it running, to manage the wastes it produces, and to create the drugs and chemicals it consumes. And the animals live miserable lives, for productivity has been severed from well-being.

One encounters the same dismal situation for animals in all areas of industrialized animal agriculture. Consider, for example, the dairy industry, once viewed as the paradigm of bucolic, sustainable animal agriculture, with animals grazing on pasture giving milk and fertilizing the soil for continued pasture with their manure. Though the industry wishes consumers to believe that this situation still obtains – the California dairy industry ran advertisements proclaiming that California cheese comes from 'happy cows' and showing the cows on pastures – the truth is radically different. The vast majority of California dairy cattle spend their lives on dirt and concrete, and in fact never see a blade of pasture grass, let alone consume it. So outrageous is this duplicity that the dairy association was sued for false advertising, and a friend of mine, a dairy practitioner for 35 years, was very outspoken against such an 'outrageous lie'.

In actual fact, the life of a dairy cow is not a pleasant one. In a problem ubiquitous across contemporary agriculture, animals have been single-mindedly bred for productivity, in the case of dairy cattle, for milk production. Today's dairy cow produces 3 to 4 times more milk than 60 years ago. In 1957, the average dairy cow produced between 500 and 600 pounds of milk per lactation. Fifty year later, it is close to 20,000 pounds (Colorado Foundation for Dairy Agriculture 2005; NASS 2006). From 1995 to 2004 alone, milk production per cow increased 16 percent. The result is a milkbag on legs, and unstable legs at that. A high percentage of the US dairy herd is chronically lame (some estimates range as high as 30 percent), and these cows suffer serious reproductive problems. Whereas in traditional agriculture, a milk cow could remain productive for 10, 15, or even 20 years, today's cow lasts slightly longer than two lactations, a result of metabolic burnout and the quest for ever-increasingly productive animals, hastened in the US by the use of versions of the hormone bovine somatotropin (BST, also known as bovine growth hormone [BGH]) to further increase production. Such unnaturally productive animals naturally suffer from mastitis, and the industry's response to mastitis in portions of the US has created a new welfare problem by docking of cow tails without anesthesia in a futile effort to minimize teat contamination by manure. Still practiced, this procedure has been definitively demonstrated not to be relevant to mastitis control or lowering somatic cell count (Stull *et al.* 2002). (In my view, the stress and pain of tail amputation coupled with the concomitant inability to chase away flies may well dispose to more mastitis.) Calves are removed from mothers shortly after birth, before receiving colostrum, creating significant distress in both mothers and infants. Bull calves may be shipped to slaughter or a feed lot immediately after birth, generating stress and fear in these baby animals.

The intensive swine industry, which through a handful of companies is responsible for 85 percent of the pork produced in the US, is also responsible for significant suffering that did not affect husbandry-reared swine. Certainly the most egregious practice in confinement swine industry and possibly, given the intelligence of pigs, in all of animal agriculture is the housing of pregnant sows in gestation crates or stalls – essentially small cages. The recommended size for such stalls, in which the sow spends her entire productive life of about four years, with a brief exception we will detail shortly, according to the industry is 0.9 m high by 0.64 m wide by 2.2 m long – this for an animal that may weigh 275 or more kilograms. (In reality many stalls are smaller.) The sow cannot turn around, walk, or even scratch her rump. In the case of large sows, she cannot even lie flat, but must remain lying on her sternum. The exception alluded to is the period of farrowing – approximately three weeks – when she is transferred to a 'farrowing crate' to give birth to and nurse her piglets. The space for her is not greater, but there is a 'creep rail' surrounding her so the piglets can nurse without being crushed by her postural adjustments.

Under extensive conditions, a sow will build a nest on a hillside so excrement runs off; forage an area covering 2 kilometers a day; and take turns with other sows watching piglets and allowing all sows to forage. With the animal's nature thus aborted, she goes mad, exhibits bizarre and deviant behavior such as compulsively chewing the bars of the cage; she also endures foot and leg problems, and lesions from lying on concrete in her own excrement.

III

These examples suffice to illustrate the absence of good welfare in confinement. Rest assured that a long litany of issues could be addressed. In general, all animals in confinement agriculture (with the sometime exception of beef cattle who live most of their lives on pasture and are 'finished' on grain in dirt feed lots, where they can actualize much of their nature) suffer from the same generic set of affronts to their welfare absent in husbandry agriculture:

1. *Production diseases*: by definition, a production disease is one that would not exist or would not be of serious epidemic import were it not for the method of production. Examples are liver and rumenal abscesses resulting from feeding cattle too much grain, rather than roughage. The animals that get sick are considered more than balanced out economically by the remaining animals' weight gain. Other examples are confinement-induced environmental mastitis in dairy cattle, weakness caused by beta-agonists to increase muscle mass in pigs, and 'shipping fever' in beef cattle. There are textbooks of production diseases, and one of my veterinarian colleagues calls such disease 'the shame of veterinary medicine' because they should be working to eliminate such pathogenic conditions, rather than treating the symptoms.
2. *Loss of workers who are 'animal smart'*: in large industrial operations such as swine factories, the workers are minimum wage, sometimes illegal, often migratory workers with little animal knowledge. Confinement agriculturalists will boast that 'the intelligence is in the system' and thus the historically collective wisdom of husbandry is lost, as is the concept of the historical shepherd, now transmuted into rote, cheap, labor.
3. *Lack of individual attention*: under husbandry systems, each animal is valuable. In intensive swine operations, as referenced earlier in the anecdotes illustrating industrial versus traditional husbandry, the individuals are worth little. When this is coupled with the fact that workers are no longer caretakers, the result is obvious.
4. *The lack of attention to animal needs determined by their physiological and psychological natures*: as mentioned earlier, 'technological sanders' allow us to keep animals under conditions violating their natures, thus severing productivity from assured well-being.

It is absolutely essential to stress here that there was no malicious attempt to ride roughshod over animal welfare by those who created and perpetuated these systems. The failure to account for animal welfare in these new systems basically arose from a conceptual error.

Let us recall that traditional agricultural success or good husbandry depended upon proper treatment of animals; generally if the animals suffered the producer lost out. Under such circumstances, it was perfectly reasonable (subject to some refinements) to use animal productivity as a good (rough) criterion for welfare. The people who developed confinement agriculture continued to use this criterion, failing to mark the singular difference between what is good for industry and husbandry. In husbandry conditions, animal suffering would *usually* lead to loss of productivity. Under industry, however, the technological sanders mentioned sever the close connection between productivity and welfare, insofar as the animals may suffer harm in ways that do not impinge on economic productivity. In some cases, such as the modern dairy cow, great productivity has been univocally selected into the cow's genetics, in ways that actually create new welfare problems, such as metabolic 'burnout', foot and leg problems, and reproductive problems. In one study, the most productive cows had a higher percentage of swollen legs (Fulwider *et al.* 2007). Again it is worth noting that, in many intensive dairies, a cow lasts for only two lactations.

Animal welfare is not the only problem inadvertently occasioned by the industrialization of animal agriculture. The PEW Commission on Industrial Farm Animal Production, on which I was privileged to serve, recently issued a report on all of the multifarious practical and ethical problems to which industrial agriculture has given rise.[1] In addition to seriously compromising the welfare of all animals raised in confinement, modern agriculture has been responsible for unprecedented amounts of environmental despoliation, including pollution of waterways and ground water, deterioration of air quality for large areas surrounding confined animal feeding operations, and problems resulting from the inability of such operations to dispose of animal waste in a sustainable manner. When floods hit North Carolina, for example, lagoons containing hog waste overflowed and burst, covering large areas with waste and with dead animals.

Rural communities made up of small farmers were another victim of industrialization of agriculture. It is virtually universal that large, well-capitalized production systems invariably out-compete small operations. Eloquent evidence of this can be found in the fact that, since 1980, the US has lost over 85 percent of the small hog producers who flourished before 1980. Today, five large corporate entities produce 90 percent of the pigs in America. Many formerly flourishing communities, based in production of hogs, are now ghost towns, with small operations bought out for pennies on the dollar, and local people unwilling to work in giant hog factories. Large entities controlling the food supply represent a real and major threat to democratic institutions.

Human health and animal health are also jeopardized by the proliferation of confinement agriculture. A vast number of studies exist documenting this threat. People living within 60 miles of confinement feeding operations, be the systems aimed at swine, poultry, or dairy production, suffer from a wide range of diseases, ranging from respiratory problems to depression. The use of antibiotics, both as a technological sander to mitigate the pernicious effects of close confinement on animal health and as a vehicle for growth promotion in animals, have served as an artificial selection pressure, selecting inevitably for pathogens that carry resistance to antimicrobial drugs to the animal, and/or human pathogens that are impervious to the antimicrobials in question. Researchers at the Johns Hopkins University School of Public Health, conducting investigations of water contamination in the Delaware/Maryland/Virginia area, were amazed and horrified to find the cutting-edge antibiotic vancomycin in the water, a drug that should have been held in reserve to fight highly resistant human pathogens, instead being deployed in and subsequently leaking out of chicken-production systems, in which it was intended to prevent the animals from dying. In addition, huge feeding operations holding vast numbers of poultry or swine serve as incubators for new pathogens such as influenza viruses. Food safety is also adversely affected by animal-product contamination that is disseminated nationally, rather than merely locally.

In addition, the confinement animal agricultural claim that industrial animal production has created a source of cheap food, something allegedly demanded by society, is somewhat misleading, for two reasons. First, while it is true that foods of animal origin are inexpensive 'at the cash register', that is in large part due to the fact that the true costs of industrial agriculture are often 'externalized', or passed on to the public, whether in the form of mitigating pollution costs of industrial agriculture; or road repairs required of long-haul travels to slaughter, processing, and commercial venues;, not to mention significantly increased health care costs disproportionately borne by people living within the immediate environments of confinement feeding operations.

Whereas European societies became aware of the multitude of problems, particularly animal welfare problems, emerging from the major intensification of confinement agriculture again as early as the 1960s, the issue has entered American consciousness only in the past few decades. In Europe, one can track a pretty steady movement in social thought culminating in severe restriction of confinement agriculture by the European Union, particularly in Sweden in 1980, which passed a law abolishing in timed stages most of the confinement systems the United States takes for granted. Nonetheless, increased American awareness of the implications of this agriculture, which has been occasioned by successful Humane Society of the United States-initiated referenda that are abolishing sow stalls, veal crates, and battery cages, as well as by ever-increasing media coverage of agriculture, and by the well-publicized Pew report, have turned animal agriculture into a major societal issue. Indeed, the members of the Pew commission in their final report called for the elimination of high-confinement animal agriculture within a decade. In 2007, this author convinced Smithfield, the largest pork producer in the world, to commit to phasing out sow stalls. By June 2012, 250,000 Smithfield sows will have been moved into open systems out of sow stalls. This, in my view, represents the death-knell for severe gestation crate confinement, as other companies will be forced into eliminating them as well.

The agricultural industry has had little to say in defense of high-confinement systems. After all, these systems were developed as a perfectly legal and putatively laudable response to societal demand for abundant food of animal origin. Therefore, if society demands major animal agricultural systemic changes, some relief, perhaps in the form of tax exemptions, should be accorded to the relevant industries as a matter of fairness. As I have told the animal science industry in numerous talks and papers, they should not see the public as simply rejecting all the work done in the twentieth century to create a cheap, plentiful, and nutritious food supply of animal products. Rather, in devising these systems, neither they nor the society that benefited from these innovations were able to control the unintended consequences of radically turning from husbandry to industry. Now that the problems have clearly emerged, society is asking animal science to 'go back to the drawing board' and devise systems that incorporate solutions to societal concerns for animal welfare, sustainability, environmental health, human and animal health, preservation of rural communities, and all of the other problems that confinement agriculture has been found to carry in its wake.

Correlatively, it may well be the case that resolving the issues arising from intensive agriculture will eventuate in increases in the price of food. The industry certainly thinks this is axiomatic and uses it as an argument against reform. However, there is good reason to believe that increased food prices are not necessarily a consequence of changing to a more humane and sustainable agriculture. For example, Dr. Tim Blackwell, chief swine veterinarian for

Ontario, Canada, and I, in research completed about ten years ago, found that moving from gestation crates to group housing systems actually cuts the cost of capitalizing new sow barns by about 50 percent, since a huge part of the cost of building full confinement barns is in the ironwork entailed in constructing sow stalls.

IV

One significant result of the industrialization of agriculture in the twentieth century has been to turn the choice of production systems in both plant and animal agriculture into moral issues. When all agriculture was played out on a stage of husbandry and sustainability, an ethic was already implicit in the system by virtue of commitments to animals living well and to keeping crop-yield sustainable. Industrial agriculture, however, presents us with a much wider range of moral choices. A paradigmatic example of such choices is given implicitly in the subject of this chapter: what, and to what extent, are we willing to sacrifice for the sake of efficiency and productivity guaranteeing cheap and abundant food?

This is clearly a question for social morality to answer. In discussing ethics, we can distinguish between *personal ethics* and *societal ethics*. In social life, many issues are left to an individual's personal ethics. For example, being a vegetarian or vegan is considered an individual and personal, albeit moral, choice. On the other hand, committing rape, murder, or bank robbery has far-reaching implications for others. Prohibition of the latter type of activities is obviously presuppositional to the possibility of groups of people living together more or less harmoniously. For this reason, every society must prohibit those sorts of activities in a well-publicized *societal consensus ethic*, invariably codified in the legal system. These prohibitions may change with time, as occurred in the past 50 years in the US regarding abortion.

Historically, societal ethics for animal treatment were extremely limited in scope. Although society has always had an articulated ethic regarding animal treatment, that ethic has been very minimalistic, leaving most of the issue of animal treatment to people's personal ethic, rather than to the social ethic. Since at least biblical times, that limited social ethic has forbidden deliberate, willful, sadistic, deviant, purposeless, unnecessary infliction of pain and suffering on animals, or outrageous neglect, such as not feeding or watering. Beginning in the early nineteenth century, this set of prohibitions was articulated in the anti-cruelty statutes of the laws in many societies worldwide. But even in biblical and medieval times, the social ethic inveighed against cruelty. The Old Testament injunctions against yoking an ox and an ass together to a plow, or muzzling the ox when it is being used to mill grain, or seething a calf in its mother's milk, all reflect concern with, and abhorrence for, what the Rabbinical tradition called *tsaar baalei chaiim*: the suffering of living things. In the Middle Ages, St. Thomas Aquinas, while affirming that, lacking a soul, animals enjoyed no moral status, nonetheless strictly forbade cruelty, on the grounds that permitting such behavior towards animals would encourage its spreading to human beings, an insight buttressed by much recent research.

When the primary use of animals in society was husbandry-based agriculture, treating the animals well became a prudential as well as an ethical imperative. Self-interest dictated that agriculturalists treat their animals well to assure maximal productivity. Thus the only social consensus ethic for animals was the ethic of anti-cruelty, designed to capture the behavior of the minority of people unmoved by self-interest and motivated by sadistic and/or psychopathic impulses. The rise of industrialized animal agriculture, however, represented a historically new playing field, where significant animal suffering not motivated by sadism, but rather by efficiency and productivity, aimed at ensuring a plentiful and relatively inexpensive supply of animal products for food. These latter motives were quite distant from cruelty, and thus the societal ethic of anti-cruelty did not apply.

On the other hand, as Western society has become aware of the price paid by animals for new, efficiency-driven production systems, dissatisfaction with resulting animal suffering, even though emphatically not the result of cruelty, gave rise to a demand for a new societal consensus ethic for animal treatment, which has gradually been encoded in the legal system. The society now demands that limits be placed on animal suffering occasioned by industrial agriculture, as described above. This emerging ethic has borrowed from our venerable ethic for human beings, wherein *rights* based in interests fundamental to human nature, such as freedom of speech or religion or protection from torture, serve as checks upon excessive utilitarian use of human beings. Animals too have natures, which, following Aristotle, I have called *telos*: the 'pigness' of the pig; the 'cowness' of the cow. Although, strictly speaking, animals are property in the eyes of the law, and thus are incapable of having rights, something approximating rights can be achieved by limiting property use. Hence, the proliferation in recent years of new laws designed to protect animals – over 2100 across US legislatures in 2004.

V

In the end, society is demanding that our use of animals be constrained by the same notions of minimal decency that followed of necessity from husbandry. While society clearly wishes to continue using animals for food and other purposes, it also clearly wants assurance that animals live decent lives in the context of that use. The sort of situation encountered by the young man from a ranch background recounted at the beginning of this chapter is unacceptable to society in general. Gratifyingly, society is rejecting the great cost to animals attendant upon industrial animal production and is correlatively demanding a significant retreat from agriculture that 'ain't agriculture'.

Note

1 The report *Putting Meat on the Table: Industrial Farm Animal Production in America* may be found online: http://www.ncifap.org/_images/PCIFAPFin.pdf.

References

Colorado Foundation for Dairy Agriculture (2005) *Colorado Dairy Facts.*

Fulwider, W.K., Grandin, T., Garrick, D.J., Engle, T.E., Lamm, W.D., Dalsted, N.L., and Rollin, B.E. (2007) 'Influence of Free Stall Base on Tarsal Point Lesions and Hygiene in Dairy Cows', *Journal of Dairy Science* 90: 3559–66.

Harrison, R. (1964) *Animal Machines: The New Factory Farming Industry*, London: V. Stuart.

Lundeen, T. (2008) 'Poultry Missing Genetic Diversity', *Feedstuffs*, December 1: 11.

NASS (2006) *Milk Production and Dairy Cows.*

Stull, C.L., Payne, M.A., Berry, S.L., and Hullinger, P.J. (2002) 'Evaluation of the Scientific Justification of Tail Docking in Dairy Cattle', *Journal of the American Veterinary Medical Association* 220: 1298–303.

7

Beyond the wild, the feral, and the domestic

Lessons from prehistoric Crete

Kerry Harris and Yannis Hamilakis

Type 'Minoan' Crete or 'Minoan civilization' (the societies of Crete in the third and second millennium BC) in any Google search, and some of the first images you will get depict animals, either on their own, or interacting with humans. It is through animal imagery to a large extent that this cultural formation has become known worldwide, ever since its archaeological discovery or rather archaeological remake and constitution at the start of the twentieth century. The bull-leaping scenes from Knossos, for example, or the bull's head *rhyta* (vessels thought to have been used for libations) often embody Minoan-ness as a whole. To a large extent, it is animals that have constructed and popularized the image of 'Minoan' Crete in modernity. We will argue in this chapter that in prehistory too it was the intricate entanglement of humans with other animals (along with things and environments) that co-shaped the 'Minoan phenomenon'. Moreover, such interaction cannot be easily confined within the 'wild', 'domestic', and 'feral' categories, and it offers in fact some compelling evidence for the problematic effects of these categorizations.

Who has domesticated whom? The archaeology of animals

From the start, the human story is interwoven with that of (nonhuman) animals. From the point at which humans became more or less sedentary, archaeologically termed the Neolithic period, a defining element of life was the consistent cohabitation with and husbanding of a range of animals (and plants). Archaeologically, it is often the presence of the remains of particular 'domesticated' animal species (e.g. cattle, pigs, sheep, goats) in conjunction with specific remains of human practices – the material culture – that identifies as such these early periods of settlement.

Understanding domestication was arguably one of the cornerstones of the development of zooarchaeology, the archaeological subdiscipline that studies the interaction between humans and animals on the basis of their material traces. Domestication is also a topic that is continually being redefined and reconceptualized today. The reassessment of animal domestication within a primarily zooarchaeological rather than zoological framework led to a number of key conclusions; perhaps one of the most important of these conclusions was the recognition that domestication was not likely to have been a single dramatic event but rather a gradual and long-term process, encompassing diverse forms of human-animal interaction, beyond the rigid categories of the 'wild' and the 'domestic'. In addition, it was proposed that domestication should be considered in terms of variations in human behaviour rather than animal morphological divergences. As such, although domestication was recognized as 'a matter of reciprocal relationships', it was the 'human aspect of man-animal relationships that was of primary importance' (Jarman 1976: 86).

Indeed, in conventional zooarchaeological discourse, definitions of domestication are strongly anthropocentric: animals are described as captured, removed from their natural environment, and bred under human control, entirely at human direction. Other, less popular views, however, have emphasized mutuality, at least in the biological sense, in the process of domestication or have even attributed to nonhuman animals the initiative of having 'manipulated unwitting humans into relationships that gave the domesticate great evolutionary advantage at the expense of human fitness' (Zeder 2012: 228). Social approaches have also recognized the active agency of animals and have even attributed to animals 'direct moral status' in their own right (Armstrong-Oma 2007: 18). The recovery and study of animals buried together with humans or receiving formal burial on their own, going as far back as 14,000 years ago (e.g. Morey 2006), has also foregrounded the affective bonds between human and nonhuman animals and has raised the issue of animal personhood in the prehistoric past.

In most discussions of the matter, however, the often undefined 'wild' category is, as noted by O'Connor, '*presumably* left to encompass animals which have little or no contact with humans, animals which cohabit with humans but which are not domesticated and animals which may actively predate humans' (1997: 150, our emphasis). More often than not, non-domestic animals are grouped together under the undefined and subsuming term 'wild resources'. The implication is a treatment of nonhuman animals as objectified and exploitable commodities. In such narratives, wild animals are seen as 'risk-buffering' options, for example, as a fall-back resource in the case of failure in the 'harvest' of domestic animals (Boyle 2006: 19). Moreover, the term 'wild' as a blanket description for 'non-domestic' animals homogenizes the widely varying characteristics of different species and prevents the detailed exploration of species-to-species interactions. Distinguishing between 'wild' and 'domestic' species in zooarchaeology thus remains a standard practice. Such distinctions usually rely on zoological methodologies based on bone morphology, particularly changes in animal size, and a certain amount of inference. Many researchers, however, acknowledge that a wild-domestic dichotomy is too simplistic and cannot satisfactorily account for the diversity of human-animal relations. For example, it has been recently recognized that humans' interactions with other animals, prior to what we describe as domestication, involved diverse practices, including some sort of 'management', lasting for thousands of years, and which did not lead to any morphological changes (Zeder 2008).

Furthermore, anthropological accounts of the social constitution, ontological status, and perceptions of wild and domestic in non-Western contexts have demonstrated the ethnocentric traditions upon which Western zooarchaeological wild/domestic definitions are often based. Perhaps due partly to these dichotomous understandings of domestic and wild, and partly to the methodological difficulties in their identification, feral animals are rarely discussed in detail in zooarchaeological discourse. The predominant descriptions see feral animals as those who 'fall between' the domestic and the wild, animals that have 'escaped the domestication process' and returned to the wild (Russell 2012).

In this chapter, we will problematize the terms 'wild', 'domestic', and 'feral', and we will emphasize the diverse social relationships amongst humans and other animals that these terms encompass. Having noted at the outset the emphasis in zooarchaeology on domestication and domestic animals, here we will focus on the significant diversity that exists in human-animal relationships outside the conventionally understood 'domestic' context and the associated husbandry practices. Rather than assume 'wild', 'domestic', and 'feral' categories to be stable and immutable, we interpret them as potentially permeable and contextually specific. These categories thus can be seen as the embodiment of a particular set of relationships rather than static concepts or structural opposites. Nevertheless, such labels carry with them the inherent limitations of anthropocentrism, obscuring the on-going 'mutual becomings' (Birke *et al.* 2004: 174) that characterize all human-animal interactions. Our qualified use of them here rests on the conviction that to explore their contextually specific constitution can be instructive on its own, as well as an appropriate way to decentre and undermine them.

Humans and other animals in prehistoric Crete: the case of cattle, agrimia, and fallow deer

The Holocene introductions of animals onto Crete is a fascinating subject. During the Late Pleistocene a number of species such as the dwarf hippopotamus, elephant, deer (*Candiacervus*, not *Cervus* or *Dama*), and various species of microfauna were endemic to Crete (see Masseti 2003; and papers in Reese 1996). It seems that these animals, however, did not survive into the Holocene and the cause of their extinction is not known, although a number of natural and anthropogenic factors have been postulated (Broodbank and Strasser 1991; Lax and Strasser 1992). There are, as yet, no Pleistocene faunal remains with evidence for human involvement and no cultural objects in pre-Neolithic levels containing Pleistocene animals, leaving an apparent hiatus in the sequence of human-animal interactions from the Pleistocene to the Holocene (Broodbank and Strasser 1991; Hamilakis 1996a; Mavridis 2003). This break is further illustrated by the lack of species such as sheep, goats, cattle, and pigs prior to the first evidence for human occupation on the island. Given the distances, it is unlikely that these animals would have swum to the island. The most plausible explanation thus is that these animals came along with humans, using the same navigational means. At the time of first human permanent settlement on Crete during the Neolithic period (from *c.* 7000 BC), there were no indigenous large mammals left. Interestingly, and as noted on Mediterranean islands elsewhere, the people who arrived in Crete in the Neolithic were accompanied not only by 'domesticated' animals but 'wild' animals too, such as deer. They travelled together, most likely from Anatolia, on the same boats. The early 'colonization' of the island thus was a multi-species affair. Given this shared voyaging experience, and the close proximity it entailed, can we still call these animals 'wild'?

Whilst we have clear evidence for the presence of sheep, goats, cattle, pigs, and dogs in the earliest human settlement contexts at the site of Knossos (a period termed the Aceramic Neolithic),[1] the trajectory involving various 'wild' species is less clear. Fallow deer and red deer were also introduced at this time, but the very low quantities of the remains suggest perhaps an unsuccessful establishment of breeding populations in these early stages (Isaakidou 2004). Reintroduction of deer species during the subsequent Bronze Age is a more likely scenario, judging by the more frequent and widespread occurrence of deer remains in archaeological deposits dating to this period (see Yannouli and Trantalidou 1999). The Cretan 'wild goat' – the *agrimi* – is first attested in contexts dated to the Late Neolithic period (Wilkens 1996), and from this time onwards it occurs throughout prehistory and history; it still survives in limited numbers in the mountains of West Crete today. In the following discussion, we will focus on the diverse engagements and interactions of humans with three species in particular: cattle (*Bos taurus*), wild goat/agrimi (*Capra aegagrus cretica*), and fallow deer (*Dama dama*), generally described as domestic, feral, and wild, respectively.

Across Europe, cattle appear to be an extremely important animal for Neolithic communities. Indeed, on Crete cattle are present from the time of the earliest permanent settlement (Aceramic Neolithic at Knossos), and it is during the Neolithic period that the greatest quantities of cattle (as attested by their remains in archaeological settlement sites) are known. At Knossos, an intensification in the consumption of cattle occurs, both commensally as 'prestige food for conspicuous on-site butchery, consumption and discard' and 'symbolically' through an increase in bovid figurines (Broodbank 1992: 62); this coincides with a period of rapid settlement growth, population expansion, and 'social reordering', described as the Early Neolithic to Middle Neolithic transition (Tomkins 2008). Zooarchaeological analysis of the Knossos material suggests that cattle would have been kept for transport and ploughing, and a high percentage seemingly were raised for meat (Isaakidou 2004).

During the Bronze Age, there is an apparent increase in the proportion of adult male cattle, and a slight increase in body size, which potentially reflects the breeding of large males for prestige events (Isaakidou 2004: 248). The importance of cattle beyond a 'subsistence' role is evident in the representation of the bull as one of the most important pictorial themes and in a variety of media, especially at Late Bronze Age Knossos. A particularly important theme within this corpus is the depiction of bull-leaping, in which human figures are seen 'somersaulting' over the back of a bull, a theme we tackle in more detail in the next section. It is of interest that at Knossos, at times of intensification in socio-political relationships (rapid population expansion and social reordering in the Middle Neolithic, and increased factional competition in the Neopalatial Bronze Age period), an increase in human-cattle interaction occurs.

In the Knossos assemblage, the size of the cattle is generally smaller than wild cattle (*Bos primigenius*) and comparable to domestic cattle from contemporary sites on the Greek mainland (Isaakidou 2004). Occasionally, however, larger skeletal fragments occur that fall within the size range for wild cattle (Isaakidou 2004; Jarman 1996). Whilst it is highly unlikely that wild cattle would have been an indigenous species to Crete, aurochs (wild cattle) remains have been reported by Nobis (1990, 1996) and Persson (1993); the possibility that importation of some 'wild' individuals from mainland Greece, where they did occur, could be considered. Isaakidou notes that the large bone specimens in the Knossos assemblage do not exhibit the pronounced muscle attachments common on skeletal elements of wild individuals and concludes that there is 'no indication that large specimens enjoyed a radically different lifestyle from smaller counterparts' (2004: 238). This observation, however, refers to 'radically different lifestyles' that lead to changes in bone morphology; in other words, these animals could not have been involved in activities that were arduous and repetitive enough to have left traces on the bones. The assumption thus is that at Knossos a single population of cattle existed and its size variation was due to sexual dimorphism, rather than a separate feral population consisting of animals of larger size (Jarman 1996; Isaakidou 2004). Nevertheless, not only in the Neolithic but also through the Bronze Age, the presence of some large cattle individuals is evident, whether of traditionally defined 'domestic', 'feral', or 'wild' status.

At the Late Neolithic site levels at Phaistos, however, a different picture emerges. Despite a period of contemporaneity between the Knossos and Phaistos settlements, there seems to be an interesting difference in some aspects of human-animal interaction between the two sites. Whilst high quantities of cattle also occur at Phaistos in this period, these animals are described as small to medium sized and used for farm work (Wilkens 1996). The presence of large cattle individuals here, unlike Knossos, is not apparent. Not only is there a seeming lack of large individual cattle at Phaistos, there is also evidence for a human consumption of agrimi at Phaistos that is not seen at Knossos (Wilkens 1996). This evidence for human-agrimi interaction here is significant; it is the first evidence for agrimi in archaeological deposits, and, despite their having been identified in archaeological contexts in subsequent periods, this Phaistos group still constitutes the highest quantity of archaeological agrimi remains recovered to date.

In appearance, the agrimi bears a strong physical resemblance to its Near Eastern progenitor, the Asiatic bezoar goat (*Capra aegagrus*): a distinct black line along the spine and base of the neck, brown flanks, white underbelly, and characteristically large, backward-curving 'scimitar-shaped' horns (Horwitz and Bar-Gal 2006: 126). Because of the physical similarity, it was originally considered to be a subspecies of the bezoar goat; more recent discussions, however, assumed that agrimi was a feral species – escapees from the early domestic goat herds that were introduced onto Crete during the early Neolithic period. Whilst the anthropogenic introduction of agrimi onto Crete is certain, its status as wild, feral, or domestic at the time of its introduction is by no means clear. A recent DNA analysis proposes that, in fact, agrimia were introduced onto the island as wild animals, and it was subsequent interbreeding with domestic goats that affected the genetic profile (Horwitz and Bar-Gal 2006). In such a scenario, 'free-living' goat populations may have been established by the releasing of wild animals onto the island, and agrimia thus represent not feral animals but relics of wild taxa (Horwitz and Bar-Gal 2006).

Although Wilkens (1996) identified the Phaistos agrimi on the basis of its large size and increased robusticity of the bones, the animals are difficult to distinguish from domestic goats on the basis of their skeletal remains, and there is as yet no established methodology to do so. There is, therefore, the possibility of inter-observer variation in the identification of agrimi remains in zooarchaeological material. An important exception, however, is the very distinctive and characteristic agrimi horns; indeed, in the subsequent Bronze Age period the symbolic and perhaps synecdochic significance of agrimi horns is denoted by their frequent placement in 'special deposits'. Such a practice is encountered at the Late Bronze Age site of Nopigeia (West Crete), for example, where, in the base of a ditch containing feasting remains, a partial agrimi skull with its horns intact was carefully deposited on a layer of sand, perhaps as an initiatory deposit with meaningful connotations (Hamilakis and Harris 2011).

Not only does the occurrence of agrimia in archaeological deposits increase during the Bronze Age, but so too does the frequency of deer remains, predominantly fallow deer but also red deer. The prehistoric introduction of deer onto Mediterranean islands is generally attributed to the practice of stocking these islands with 'game', especially where large native species are absent. In the case of Crete, it is likely that fallow deer would have come from Anatolia, whereas red deer could have been brought from the Greek mainland. As noted above, the status of these animals too is somewhat ambiguous. They are usually assumed to be 'wild', but it has been noted that the effort expended to import them suggests they were at least 'semi-domestic rather than fully wild' (Jarman 1996).

Masseti notes that the introduction of cervids onto islands in sufficient numbers to act as a breeding stock argues for 'a high degree of control over the animals and for their sophisticated management' (1996: 17). He further suggests, however, that it is unlikely that deer on the Mediterranean islands were domesticated, but were possibly released and hunted. It is clear that archaeologists face serious difficulties in attempting to explain these human-animal interactions based on dichotomous categorizations and on the implicit logic which sees animals primarily as economic resources, managed by humans on the basis of output maximization and formalist economics in general.

It is generally accepted that during the Bronze Age on Crete an increase in human interaction with 'wild' animals occurs, usually interpreted as a rise in hunting. That said, the role of hunting in prehistoric Crete is rarely discussed in detail in zooarchaeological discourse. This, in part, is perhaps due to the development of conventional zooarchaeology within a theoretical framework in which animals are primarily discussed as commodities and subsistence resources, although a new, social, and non-anthropocentric zooarchaeology is currently emerging (Overton and Hamilakis 2013). Within the conventional zooarchaeological framework, it is assumed that the higher the quantity of remains of a particular species, the greater its significance to a society in economic terms, as a provider of meat, wool, dairy products, and so on. Interestingly for this discussion, the perceived economic importance of an animal is often seen as an indicator of its domestic status, especially for deer. Jarman, for example, notes that

> [o]n Crete [. . .] the deer seem to have been introduced primarily for non-economic purposes, possibly as park animals or for hunting [. . .] [T]hey never occur in numbers sufficient to imply their use as an important economic resource. It is therefore less likely that the Cretan deer were behaviourally domestic than the economically important Cypriot deer.
>
> *(1976: 92)*

Recent analysis of the Cypriot fallow deer, however, suggests that, although it was imported, it was not domesticated but released and hunted (Vigne *et al.* 2011). Conflicting interpretations such as this perhaps highlight the inadequacies of seeing domestication, and human-animal interaction more generally, in purely economic terms and indicate that investigation of the social relationships between human and nonhuman animals might be a more fruitful line of enquiry (e.g. Armstrong-Oma 2010; Orton 2010).

Thus, on Crete, although deer remains become more widespread during the Bronze Age, they still rarely comprise more than 2 per cent of faunal assemblages. These generally low quantities of deer (and wild animals overall) have resulted in their rendering as 'insignificant' to archaeology. Occasionally, as noted above, a symbolic significance is sometimes proposed, for example as 'prestige objects' for the purposes of hunting and consumption by an elite group, in a manner akin to medieval deer parks in North-West Europe (Isaakidou 2004; Jarman 1996). At Bronze Age Knossos, it has been proposed that fallow deer may have been kept in parks and closely monitored by the 'palace' (Isaakidou 2004; Jarman 1996). A denigration of the role of some species such as deer on the basis of the quantitative occurrence of their remains obscures the role of live animals as sentient entities, which could have been valued as living beings and as partners in various social interactions, rather than as 'walking larders', as providers of calories and protein, as potential 'deadstock'.

Furthermore, it is becoming increasingly apparent that in the Late Bronze Age at a number of sites in west and central Crete high quantities of deer do occur, allowing us to explore deer-human interaction in more detail. At Chania, Chamalevri, and Thronos/Kephala, for example, deer appeared to be a significant element in feasting events, the remains of which were subsequently deposited in a seemingly structured or ritualistic manner in 'ceremonial pits' (D'Agata 1997–2000; Hallager 2001; Mylona 1999). The high percentage of deer in these contexts, the apparent evidence for their having been consumed (based on butchery marks present on the bones), and the potentially conspicuous manner in which the remains were deposited (large quantities of feasting debris in pits) suggest that the presence and consumption of deer constituted these commensal events as distinctive, as occasions that punctuated both the temporality of the everyday and the periodic calendar of feasting (Hamilakis 2008; Isaakidou 2007).

In this region, humans and deer were incorporated within a set of practices, intensive and structured enough to leave a significantly durable trace in the archaeological context. Yet what was the nature of the interaction with the living deer? The higher quantities of deer remains in these regions could point to the purposeful development and nurturing of deer herds; alternatively, it may signify a more intensive approach to hunting deer in 'the wild'. The data from Late Bronze Age Chania indicate that the majority of deer in these encounters (which culminated in the death of the animal) were adult animals in the prime of life, with possible evidence for a preference for adult males. In these deposits, bones from all parts of the deer body are present. Either the deer was hunted further afield and the whole carcass was brought back to the site or the living animal was brought into the site and killed, constituting a significant and memorable part of the drama of the event. Either way, the performance of a direct physical interaction with these large, impressive animals was crucial. Through the fragmentation of the animal body and its subsequent consumption in public, feasting, and ceremonial events, a wider range of people beyond the hunters would have taken part in these encounters with deer. The evidence for these practices bears some similarities to a wider 'Mycenaean' (Late Bronze Age southern mainland Greece) ideology of hunting (see Hamilakis 1996b, 2003) and is potentially related to the strong links between West Crete – Chania in particular – and the 'Mycenaean' mainland at this time.

Bronze Age iconographic depictions

That deer and agrimia were certainly imagined and visualized in hunting contexts in the Bronze Age on Crete is attested by their frequent depiction in hunt scenes on carved sealstones and later on painted burial coffins (*larnakes*). In this imagery, agrimia and deer are depicted as wounded by spears and/or attacked by dogs (Bloedow 2003; Younger 1988). However, hunting is not the only iconographic *topos* in which deer and agrimia are present. Other images are interpreted as 'ritual' or sacrificial scenes; for example, a three-sided seal described as showing a 'priest's head between a bow and arrow', a 'running goat' (agrimi), and a 'stylised bucranium' (the sacrifice indicator) is interpreted by Marinatos (1986) as depicting the priest as sacrificer but also as hunter, pursuing the running goat. On a stone rhyton from Zakros, east Crete, sitting agrimia (along with two birds) crown the entrance of a building which has been interpreted as a mountain-top sanctuary (Figure 7.1).

That deer, too, were possibly considered sacrificial animals is indicated by a sealstone described as depicting a deer above a table or altar attacked by a griffin (Marinatos 1986: 12), and a fresco from Ayia Triada in south central Crete (Figure 7.2), which depicts human figures, seemingly women, leading several fallow deer to an altar (Militello 1998) – 'a rare example of this animal in a cultic context' (Morgan 1988: 55). One of the fallow deer wears something that looks like a collar, denoting thus some sort of 'tamed' status. Marinatos proposes that in Late Bronze Age Cretan 'religion' there is an equivalence between hunting and sacrifice, with many 'priests' portrayed as hunters and ritual hunting almost certainly practised (1986: 42).[2] But were these animals led to the altar for sacrifice or were they processing along with humans, a well-known theme in Cretan frescoes? And might the colour difference in the hides of the two fallow deer and the presence of the collar aim to accentuate the individuality of each animal?

Scholars acknowledge that two types of cattle seem to be represented in Bronze Age Cretan iconography: seemingly 'domestic' animals, being smaller and depicted in a variety of poses; and cattle portrayed as larger, horned, and part of bull-leaping scenes (Marinatos 1993; Younger 1995). Others even suggest that there are hardly any scenes involving cattle

Figure 7.1 The Zakros Sanctuary, Rhyton, depicting agrimia, 1450 BC. © Archaeological Museum of Haraklion. Reproduced with permission.

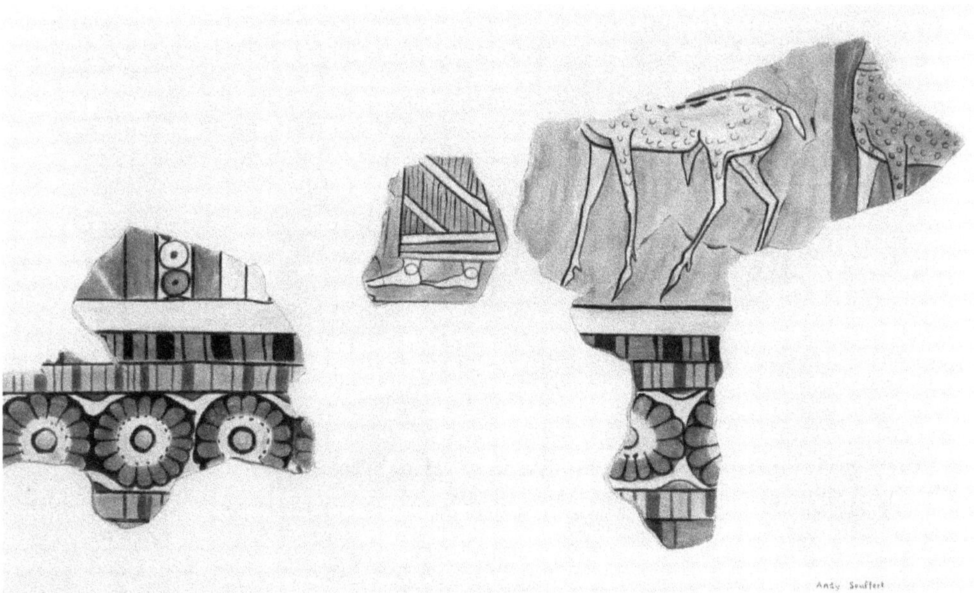

Figure 7.2 Ayia Triada fresco with women and fallow deer, 1430–1300 BC.

in domestic and husbandry routines (Shapland 2009). Perhaps some of the most striking bull images are the depictions of bull-leaping, portrayed in a variety of media, but with most visual impact in the life-size bull-leaping fresco scenes at the Late Bronze Age 'palace' at Knossos, where they are situated 'at some of the most spectacular and important places in the palace' (Hallager and Hallager 1995: 547). Interestingly, bull iconography is concentrated at Knossos and is virtually absent from other sites on Crete.

Whilst bull-leaping iconography has been widely examined, for the purposes of this discussion the link between bull-leaping and bull-hunting is most interesting. Younger (1995: 512), for example, notes that 'many bull-hunting and bull-capture scenes also show informal bull-leaping and bull-wrestling' and proposes that all these practices were linked. There are examples of depictions of bulls with nets across their backs and of bulls being speared, including the depiction, on sealstones, of cattle hunt scenes with human figures, spears, and/or dogs, in the same manner as deer and agrimia described above. A particularly interesting depiction on a small ivory vessel found in a tomb at Katsamba near Knossos (Figure 7.3) depicts a charging bull with a human figure holding onto the horns and somersaulting over its head; two other men run looking back towards the bull, one of which brandishes a spear at the bull's head, the other possibly holding a folded net (Younger 1995: 524). The additional details of flanking palm trees, a flying bird, and rocky background terrain suggest, in this example, an outdoor environment. Most interpretations of bull-leaping scenes generally assume that they depict a sport (Younger 1995), a symbol of power (Hallager and Hallager 1995), a ritual performance (Soar 2009), or a prestige activity of sorts (Shapland 2009, 2013) likely associated with the elite at Knossos and probably taking place within the palace (Soar 2009). These categorizations are not always informative and at times betray modern assumptions. Whilst the iconographic data may represent stylistic conventions relating to fictional and epic narratives, and do not necessarily portray realistic breeds or species, it is clear that

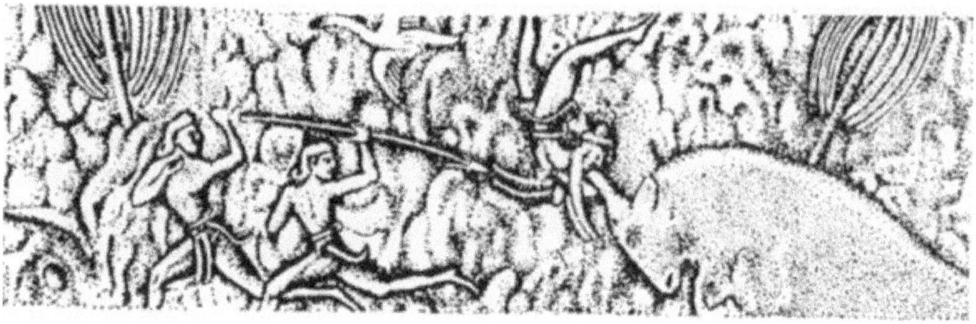

Figure 7.3 Katsamba ivory-vessel depiction with bull and men, 1600–1450 BC.

depictions of specific animal types can occur in a variety of scenes, including contexts perhaps not traditionally associated with particular animals, such as cattle in hunt contexts and 'docile' deer being led to an altar.

Domestic-feral-wild

What can we learn from this evidence? When people first settled permanently on Crete, they brought animals with them. Or, to phrase it differently and in less anthropocentric terms, at the beginning of the Neolithic period a range of animals settled on Crete: cattle, sheep and goats, pigs, dogs, deer, and humans. Archaeologists to date have spoken of the human colonization of Crete in the Neolithic, with little consideration of the other sentient beings that actively shaped this process. When they settled in Crete, humans and 'domestic' animals lived together within the rhythms of daily habitus. Alongside these inter-species interactions within the familiar cycles of daily routines, there was also a need to engage with other animals in less familiar contexts. Interaction with 'wild' animals that made the trip from either Anatolia or mainland Greece, along with humans, satisfied this need, and so did the animals that chose to or were allowed to become 'feral' after they arrived in Crete. One key arena for such interactions and performances in unfamiliar domains was hunting. Hunting was not simply a sport or just a prestige activity for the elites, although it would have certainly acquired these connotations too. It was rather a performative, embodied, and multi-sensorial venture into unfamiliar realms, into other spaces. It involved journeys and travelling in both space and time, as well as the transgression of geographical and symbolic boundaries. Furthermore, it appears that, through time, this engagement with 'the wild' seemingly became a more structured and performative one.

It is no coincidence that particular attention was paid to the careful and deliberate deposition of agrimi horns: the physical feature that perhaps most distinguishes agrimia from domestic goats. These are also the body parts that are crucial in confrontations with other animals of the same species (during the mating season, for example) and with other species, including humans. Living high in the mountains, these 'wild' goat groups would have occupied some of the most inaccessible areas of the Cretan landscape. An encounter with these animals would have necessitated an actual physical journey beyond the familiar agricultural landscapes to the remotest realms. Conspicuous display and deposition of the agrimi horns, as practised in the Bronze Age, would have been a means to express and demonstrate to the wider community this specific human-animal encounter. They would have been material mnemonics of these embodied journeys into the mountains (cf. Jones 1998) as well as hunting trophies, reminders and proofs at the same time of a clash and confrontation between agrimia and humans. In Bronze Age iconography the frequent depiction of agrimia with elements characteristic of this mountain landscape suggests that these animals evoke a certain distance from the immediate, familiar, agricultural locales. In discussing the depiction of an agrimi with female human figures in Cretan Bronze Age Art, Shapland proposes that 'the animal body [agrimi] acts as a marker of difference from the familiar domestic world: in association with a human body it creates a context in which one can interpret the females as deities' (2009: 122).

Even within the familiar domains of domestic and urban spaces, however, humans and other animals were engaging in a series of performative acts, whether in reality or in representational media. Depictions of cattle in the Bronze Age iconography show cattle differentiated by size and also by activity – with large cattle portrayed in bull-leaping iconography. Although Younger (1995) interprets these large cattle as *Bos primigenius* (wild cattle), biogeographically it is unlikely that these animals were indeed wild, unless certain individuals had travelled from mainland Greece. It is more likely, however, that some animals were selectively bred from within the domestic cattle herd for size and perhaps aggressivity, much like the breeding of bulls for bullfighting in contemporary Spain (Marvin 1988). This may explain the fact that in the depictions of bull-leaping, as far as colour and hide patterns are concerned, bulls resemble domestic cattle, not wild ones. If that is the case, then some of these larger (taxonomically domestic) cattle were amongst the ones identified by zooarchaeologists. Isaakidou (2004) notes that there are no zooarchaeological indications that the lifestyle of these larger cattle individuals was different enough from the smaller cattle to lead to a change in bone morphology – a change that could be caused by occupying a more rugged terrain, for instance. The implication is that the larger cattle were not living separately from the smaller cattle in terms of contrasting environmental habitats, as suggested for the goats. Thus perhaps we might conclude that both the larger and the smaller cattle were taxonomically the same, occupied the same place, and perhaps formed part of the same herds. What differentiated them was their different mode of interaction with humans, the development of certain morphological and ethological features such as size and aggressivity, and their participation in different performative contexts, such as bull-leaping.

Shapland sees the depictions of bull-leaping as a 'performance placing the bull outside everyday activities'. He, however, interprets this as a means of changing the human-cattle relation into a prestige activity rather than a mundane one (2009: 122). We too see this activity as one in which the 'everyday' aspect of cattle is transcended, but to call it merely a 'prestige' activity does not exhaust its meaningful connotations. Moreover, the talk of prestige is inscribed into the anthropocentric framework that portrays animals as material or symbolic resources to be used by humans, denying thus the agency of animals themselves, both as actual embodied entities and as depictions.

While Shapland draws a parallel between hunting and bull-leaping, we propose here a direct connection between the two. We suggest that it is perhaps possible to conceive of bull-leaping as a ritual or choreographic performance of bull-hunting. Of particular relevance in this respect is Marvin's suggestion that, through the heightened embodied and sensory experience of hunting, the ordinary is transformed and transcended (2000). Assuming thus that the bull-leaping scenes correspond to actual bull-leaping performances involving large and fierce bulls, we may suggest that these cattle were not separated from the rest on the basis of taxonomy or geographical or genetic distance, but that the status of some of them was transformed, albeit perhaps only temporarily, through their incorporation into a particular type of encounter. Even if these images did not correspond to actual performances, they would have provided the mnemonic props and the background for such performances to be narrated or staged in other ways, making explicit discursive and visual references to past events, replete in mythical, heroic, or celebratory connotations.

These performances created a 'place outside a place'; they produced a heterotopic domain: a different space where the normal conventions of time did not apply (Foucault 1986). Temporality is a key dimension here. Foucault notes that '[t]he heterotopia begins to function at full capacity when men arrive at a sort of absolute break with their traditional time' (1986: 26). The actual or evoked acts of wrestling with supposedly wild bulls in a familiar setting would have transported people to other places and other times. The hunting of wild bulls in their native habitat is not a practice that happened on Crete; it was, however, a practice of the mainland ancestors of the first settlers at Knossos. By referencing hunting thus, these bull-leaping performances would have temporarily produced a heterotopic locus, a space of wilderness structured by ancestral and genealogical time.[3] If bull-leaping is a ritualized performance of bull-hunting, it references and cites earlier, originary times, perhaps the times of the first colonization of Crete. As Knossos is one of the earliest sites occupied by the first, Neolithic settlers on Crete, this may go some way to explaining the apparent long-standing significance of cattle for this particular site. Indeed, the emergence of the monumental 'palaces' of Crete, such as the one at Knossos, celebrated and materialized history (Hamilakis 2013). They were established at these specific locales because of their long occupation, and thus their ancestral and genealogical weight, and the equally long history of embodied and sensorial performances, including the rituals of commensality. The actual or assumed re-enactments of bull-leaping/bull-hunting in these ancestral contexts would thus have operated as mnemonic citations of an even longer originary narrative, an ancestral charter myth.

It is important not to lose sight of the fact that the main protagonist here is the bull himself. Not only is his size in wall paintings deliberately exaggerated compared to human figures, occupying a central position in the iconographic syntheses, but the motif itself is the most widespread animal depiction, especially at Knossos. If in addition to wall paintings we consider the bull's head rhyta and the imagery in seals and other media, then the animal acquires almost totemic dimensions.

The human-deer relationship in prehistoric Crete is by no means less complex. Archaeologists often assume that these were wild species, and yet their arrival together with humans during the first colonization of the island in the Neolithic, in other words their intentional importation and the establishment of breeding populations of deer, speak of a rather tangled and multi-faceted interaction. It is not clear whether these animals existed in a free-roaming state or were managed herds akin to the deer parks of more recent periods. It was not until the Bronze Age that the consumption of deer became more widespread across Crete, at least in a manner in which their remains became incorporated into settlement sites.

Although deer are present in sealstone imagery, compared to depictions of cattle and agrimia these are relatively rare. Only one example, from Ayia Triada, is known of deer in a Cretan fresco medium, described above (Figure 7.2), in contrast to the southern Greek mainland where deer, and hunting more generally, are frequently present in the fresco programmes of the 'Mycenaean' Bronze Age 'palace' sites. Hunting imagery in Mycenaean palatial iconography is often positioned in visually prominent locations and high-status areas, and studies have often highlighted its role as a symbolic device associated with Mycenaean power dynamics. Palatial transactions recorded in tablets from the mainland site of Pylos with Linear B script mention live deer apparently destined for a feast, strengthening further the idea of game parks, possibly involving fallow deer (Palmer 2012). Some of these animals may have been involved in performative, ritualistic hunting events.

Interestingly, the majority of depictions of deer in both the sealstone and the fresco imagery, both on Crete and on the mainland, seem to indicate fallow deer, judging by the spotted coats and palmated antlers. An emphasis on fallow deer over red deer is also evident in the rich zooarchaeological assemblages from the Late Bronze Age West Cretan Chania sites, where fallow deer comprise 70 to 80 per cent of the deer remains. Why this focus on fallow deer? On the Bronze Age southern Greek mainland, despite their relative frequency in the iconography, their physical remains are scarce (Yannouli and Trantalidou 1999). In the context of the Mycenaean 'palaces', 'what counted more than the actual activity was the deployment of the idea of hunting and its representation in the media' (Hamilakis 2003: 244). Thus in the case of fallow deer it could be suggested that the aim was the demonstration of human knowledge of these animals despite, or perhaps because of, their relative scarcity in the local landscape.

At the end of the last Ice Age, the native range of fallow deer was restricted to the Near East, but since the Neolithic period subsequent human transportation resulted in their reintroduction across Europe (Sykes *et al.* 2013), first occurring more intensively on Crete and the southern Greek mainland during the Bronze Age, although, as noted above, their archaeological remains are scarce (with a few exceptions). Helms (1988) has previously noted that practices such as hunting, long-distance travel, acquisition of exotic goods, and skilled crafting are indicators of participation in external geographic, symbolic, and temporal realms, and as such are used as mechanisms for the generation of power in many societies. As the Bronze Age in Greece was a period partly characterized by extensive links across the wider Mediterranean, encounters with fallow deer, perhaps an exotic animal at the time and rarely experienced in the flesh, would have embodied participation in, and knowledge of, such distant geographical realms as the Near East.

But it would be a mistake to see fallow deer operating as static symbolic devices, as proxies for the demonstration of prestige and authority on the part of humans. While the materialization of the link with the East would have been important for humans, it would have been perhaps more important to engage with the fallow deer itself (and its depictions) in a direct, sensorial manner. The ethology of the animal, its assumed docility compared to other deer species, and its preference for large herds facilitates these performative encounters and the staging of various ritualized events. The Ayia Triada fresco portraying fallow deer (Figure 7.2) offers hints that individual animals may have even acquired a personalized status and may have taken part in processions on a par with and alongside humans.

Conclusion

The 'animal turn' in a number of disciplines has certainly been a most encouraging move in our attempts to decentre and undermine the philosophical-cum-political constitution of the individuated, anthropocentric self within capitalist modernity. Unlike most other disciplines in the humanities and social sciences, however, archaeology has engaged with animals from early on and even facilitated the development of a thriving subdiscipline, zooarchaeology. Moreover, despite its anthropocentric heritage, archaeology, having access to the material traces of the interaction between humans and other animals over the last 2.5 million years, harbours the potential to demonstrate the diverse and multi-faceted nature of these interactions and to historicize and thus relativize or even subvert the dominant, contemporary classifications and divisions. Or, to recall Giorgio Agamben, to 'render inoperative' the 'anthropological machine', which would entail nothing less than showing 'the central emptiness, the hiatus that – within man – separates man and animal, and to risk ourselves in this emptiness' (2004: 92).

Nonhuman animals in prehistoric Crete were not simply the passive pieces in chess games played out by humans; in possession of their own agency, their animal-agency, they were rather co-participants in 'colonizing' the island, and some of them may have consented to the various performances and re-enactments, and others not. As in human-human relationships, however, attempts at control and subjugation – in all possible directions – would have certainly been part of the equation. The encounters discussed above represent only a fraction of the diversity of human-cattle, human-agrimi, human-fallow deer, and in general human-other animal relationships. Indeed, the amazing and wonder-full material culture of Neolithic and Bronze Age or 'Minoan' Crete was produced by multiple species, not just humans: cattle that worked as traction animals to generate the subsistence surpluses which allowed many humans to engage in elaborate and time-consuming projects in architecture and artefact manufacture; the sheep, goats, cattle, and pigs (as well as the deer and agrimia) that were sacrificed and consumed in large numbers to provide for the feasting events, so central in social reproduction and in the negotiations of power amongst diverse groups; the thousands of sheep that provided the wool for the elaborate textile industry, for which Late Bronze Age Knossos was so renowned. These inter-species collaborative projects no doubt contain their own complex and fascinating micro-histories, which remain to be written.

Notes

1 The Neolithic period for Crete is divided into chronological phases: Aceramic Neolithic (7000–6500/6400 BC), Early Neolithic (6500/6400–4500/4400 BC), Middle Neolithic (4500/4400–3900 BC), Late Neolithic (3900–3600 BC), and Final Neolithic (3600–3000 BC), and is followed by the Bronze Age (3000–1100 BC). All dates are approximations.

2 In other themes there are depictions of agrimia pulling chariots and agrimia as supposed 'attendants of divine figures' (Hiller 2001); Bloedow (2003) argues on the basis of certain agrimi depictions (e.g. the Zakros rhyton) that in some cases agrimia may have been kept in captivity (1990). Both deer and agrimia are also depicted in 'natural' poses (Younger 1988).

3 On the links between bullfighting and hunting in contemporary Spain, see Marvin 1988.

References

Agamben, G. (2004) *The Open: Man and Animal*, Palo Alto, CA: Stanford University Press.

Armstrong-Oma, K. (2007) *Human-Animal Relationships: Mutual Becomings in Scandinavian and Sicilian Households 900–500* BC, Oslo: Unipub/Oslo Academic Press.

—— (2010) 'Between Trust and Domination: Social Contracts between Humans and Animals', *World Archaeology* 42.2: 175–87.

Birke, L., Bryld, M., and Lykke, N. (2004) 'Animal Performances. An Exploration of Intersections between Feminist Science Studies and Studies of Human/Animal Relationships', *Feminist Theory* 5.2: 167–83.

Bloedow, E. (2003) 'The Significance of the Goat in Minoan Culture', *Praehistorische Zeitschrift* 78: 1–59.

Boyle, K.V. (2006) 'Neolithic Wild Game Animals in Western Europe: The Question of Hunting', in D. Serjeantson and D. Field (eds), *Animals in the Neolithic of Britain and Europe*, Oxford: Oxbow, pp. 10–23.

Broodbank, C. (1992) 'The Neolithic Labyrinth: Social Change at Knossos before the Bronze Age', *Journal of Mediterranean Archaeology* 5.1: 39–75.

Broodbank, C. and Strasser, T.F. (1991) 'Migrant Farmers and the Neolithic Colonization of Crete', *Antiquity* 65: 233–45.

D'Agata, A.-L. (1997–2000) 'Ritual and Rubbish in Dark Age Crete: The Settlement of Thronos/ Kephala (Ancient Sybrita) and the Pre-Classical Roots of a Greek City', *Aegean Archaeology* 4: 45–59.
Foucault, M. (1986) 'Of Other Spaces', *Diacritics* 16.1: 22–7.
Hallager, E. (2001) 'A Waste Deposit from LBA Shrine in Khania (?)', in R. Laffineur and R. Hägg (eds), *Potnia: Deities and Religion in the Aegean Bronze Age: Proceedings of the 8th International Aegean Conference*, Liège: Austin Université Histoire de l'Art et Archéologie de la Grèce Antique; Austin: University of Texas, Program in Aegean Scripts and Prehistory, pp. 175–80.
Hallager, E. and Hallager, B.P. (1995) 'The Knossian Bull – Political Propaganda in Neo-Palatial Crete?', in R. Laffineur and W.-D. Niemeier (eds), *Politeia. Society and State in the Aegean Bronze Age: Proceedings of the 5th International Aegean Conference/5e Rencontre égéenne internationale, University of Heidelberg, Archäologisches Institut, 10–13 April, 1994*, Liège: Université de Liège, Histoire de l'art et archéologie de la Grèce antique; Austin: University of Texas, Program in Aegean Scripts and Prehistory, pp. 547–55.
Hamilakis, Y. (1996a) 'Cretan Pleistocene Fauna and Archaeological Remains: The Evidence from Sentoni Cave (Zoniana, Rethymnon)', in D. Reese (ed.), *Pleistocene and Holocene Fauna of Crete and Its First Settlers*, Madison, WI: Prehistory Press, pp. 231–40.
—— (1996b) 'A Footnote on the Archaeology of Power: Zooarchaeological Evidence from a Mycenaean Chamber Tomb at Galatas, NE Peloponnese', *Annual of the British School at Athens* 91: 153–66.
—— (2003) 'The Sacred Geography of Hunting: Wild Animals, Social Power and Gender in Early Farming Societies', in E. Kotjabopoulou, Y. Hamilakis, P. Halstead, C. Gamble, and P. Elefanti (eds), *Zooarchaeology in Greece: Recent Advances*, London: British School at Athens, pp. 239–48.
—— (2008) 'Time, Performance and the Construction of a Mnemonic Record: From Feasting to an Archaeology of Eating and Drinking', in L. Hitchcock, R. Laffineur, and J. Crowley (eds), *DAIS: The Aegean Feast*, Liège: Université de Liège, pp. 3–22.
—— (2013) *Archaeology and the Senses: Human Experience, Memory, and Affect*, Cambridge: Cambridge University Press.
Hamilakis, Y. and Harris, K. (2011) 'The Social Zooarchaeology of Feasting: The Evidence from the "Ritual" Deposit at Nopigeia-Drapanias', in E. Kapsomenos, M. Andreadaki-Vlazaki, and M. Andrianakis (eds), *Proceedings of the 10th International Cretological Congress (Khania, 1–8 October 2006)*, Khania: Literary Society of Khania 'Chrysostomos', pp. 225–44.
Helms, M.W. (1988) *Ulysses' Sail: An Ethnographic Odyssey of Power, Knowledge, and Geographical Distance*, Princeton, NJ: Princeton University Press.
Hiller, S. (2001) 'Potnia/ Potnios Aigon. On the Religious Aspect of Goats in the Aegean Late Bronze Age', in R. Laffineur and R. Hägg (eds), *Potnia: Deities and Religion in the Aegean Bronze Age: Proceedings of the 8th International Aegean Conference*, Liège: Austin Université Histoire de l'Art et Archéologie de la Grèce Antique; Austin: University of Texas, Program in Aegean Scripts and Prehistory, pp. 293–304.
Horwitz, L.K. and Bar-Gal, G.K. (2006) 'The Origin and Genetic Status of Insular Caprines in the Eastern Mediterranean: A Case Study of Free-Ranging Goats (*Capra aegagrus cretica)* on Crete', *Human Evolution* 21: 123–8.
Isaakidou, V. (2004) 'Bones from the Labyrinth: Faunal Evidence for Management and Consumption of Animals at Late Neolithic and Bronze Age Knossos, Crete', unpublished PhD thesis, University College London.
—— (2007) 'Cooking in the Labyrinth: Exploring "Cuisine" at Bronze Age Knossos', in C. Mee and J. Renard (eds), *Cooking Up the Past: Food and Culinary Practices in the Neolithic and Bronze Age Aegean*, Oxford: Oxbow, pp. 5–24.
Jarman, M.R. (1976) 'Early Animal Husbandry', *Philosophical Transactions of the Royal Society of London. Series B, Biological Sciences. The Early History of Agriculture* 275.936: 85–97.
—— (1996) 'Human Influence in the Development of the Cretan Mammalian Fauna', in D. Reese (ed.), *Pleistocene and Holocene Fauna of Crete and Its First Settlers*, Madison, WI: Prehistory Press, pp. 211–29.
Jones, A. (1998) 'Where Eagles Dare: Landscape, Animals and the Neolithic of Orkney', *Journal of Material Culture* 3.3: 301–24.
Lax, E. and Strasser, T.F. (1992) 'Early Holocene Extinctions on Crete: The Search for the Cause', *Journal of Mediterranean Archaeology* 5.2: 203–24.
Marinatos, N. (1986) *Minoan Sacrificial Ritual*, Stockholm: Paul Åström Förlag.

—— (1993) *Minoan Religion: Ritual, Image and Symbol*, Columbia: University of South Carolina Press.
Marvin, G. (1988) *Bullfight*, Oxford: Blackwell.
—— (2000) 'Natural Instincts and Cultural Passions: Transformations and Performances in Foxhunting', *Performance Research* 5.2: 108–15.
Masseti, M. (1996) 'The Postglacial Diffusion of the Genus *Dama* Frisch, 1775, in the Mediterranean Region', *Supplemento alle Richerche di Biologia della Selvaggina* 25.1: 7–29.
—— (2003) 'Holocene Endemic and Non-endemic Mammals of the Aegean Islands', in E. Kotjabopoulou, Y. Hamilakis, P. Halstead, C. Gamble, and P. Elefanti (eds), *Zooarchaeology in Greece: Recent Advances*, London: British School of Athens, pp. 53–63.
Mavridis, F. (2003) 'Early Island Archaeology and the Extinction of Endemic Fauna in the Eastern Mediterranean: Problems of Interpretation and Methodology', in E. Kotjabopoulou, Y. Hamilakis, P. Halstead, C. Gamble, and P. Elefanti (eds), *Zooarchaeology in Greece: Recent Advances*, London: British School of Athens, pp. 65–74.
Militello, P. (1998) *Haghia Triada I. Gli Affreschi*, Scuola Archeologica Italiana di Atene, Padova: Bottega d'Erasmo.
Morey, D.F. (2006) 'Burying Key Evidence: The Social Bond between Dogs and People', *Journal of Archaeological Science* 33: 158–75.
Morgan, L. (1988) *The Miniature Wall Paintings of Thera: A Study in Aegean Culture and Iconography*, Cambridge: Cambridge University Press.
Mylona, D. (1999) 'Chamalevri Zooarchaeological Material', in Y. Tzedakis and H. Martlew (eds), *Minoans and Mycenaeans: Flavours of Their Time*, Athens: National Archaeological Museum, p. 62.
Nobis, G. (1990) '"Der Minotaurus" von Knossos auf Kreta – im Lichte moderner archäeozoologischer Forschung', *Tier und Museum* 2.1: 15–19.
—— (1996) 'Der Aeroschse oder Ur (*Bos primigenius* Bojanus, 1827) auf Kreta', in D. Reese (ed.), *Pleistocene and Holocene Fauna of Crete and Its First Settlers*, Madison, WI: Prehistory Press, pp. 263–72.
O'Connor, T.P. (1997) 'Working at Relationships: Another Look at Animal Domestication', *Antiquity* 71: 149–56.
Orton, D. (2010) 'Both Subject and Object: Herding, Inalienability and Sentient Property in Prehistory', *World Archaeology* 42.2: 188–200.
Overton, N. and Hamilakis, Y. (2013) 'A Manifesto for a Social Zooarchaeology: Swans and Other Beings in the Mesolithic', *Archaeological Dialogues* 20.2: 111–136.
Palmer, R. (2012) 'Deer in the Pylos Tablets', in *Études mycéniennes 2010. Actes du XIIIe colloque international sur les textes égéens, Sèvres, Paris, Nanterre, 20–23 septembre 2010. Biblioteca di Pasiphae 10*, Pisa and Roma: Fabrizio Serra editore, pp. 357–82.
Persson, O. (1993) 'Ure in Chania auf Kreta', *Tier und Museum* 3.4: 121–3.
Reese, D. (ed.) (1996) *Pleistocene and Holocene Fauna of Crete and Its First Settlers*, Madison, WI: Prehistory Press.
Rehak, P. (1995) 'The Use and Destruction of Minoan Bull's Head Rhyta', in R. Laffineur and W.-D. Niemeier (eds), *Politeia: Society and State in the Aegean Bronze Age: Proceedings of the 5th International Aegean Conference/5e Rencontre égéenne internationale, University of Heidelberg, Archäologisches Institut, 10–13 April 1994*, Liège: Université de Liège, Histoire de l'art et archéologie de la Grèce antique; Austin: University of Texas, Program in Aegean Scripts and Prehistory, pp. 435–60.
Russell, N. (2012) *Social Zooarchaeology: Humans and Animals in Prehistory*, Cambridge: Cambridge University Press.
Shapland, A. (2009) 'Wild Nature? Human-Animal Relations on Neopalatial Crete', *Cambridge Archaeological Journal* 19.3: 109–27.
—— (2013) 'Jumping to Conclusions: Bull-leaping in Minoan Crete', *Society and Animals* 21: 194–207.
Soar, K. (2009) 'Old Bulls, New Tricks: The Reinvention of a Minoan Tradition', in M. Georgiadis and C. Gallou (eds), *The Past in the Past: The Significance of Memory and Tradition in the Transmission of Culture*, Oxford: Archaeopress, pp. 16–27.
Sykes, N.J., Carden, R.F., and Harris, K. (2013) 'Changes in the Size and Shape of Fallow Deer – Evidence for the Movement and Management of a Species', *International Journal of Osteoarchaeology* of the United States of America. 23.1: 55–68.
Tomkins, P. (2008) 'Time, Space and the Reinvention of the Cretan Neolithic', in V. Isaakidou and P. Tomkins (eds), *Escaping the Labyrinth. The Cretan Neolithic in Context*, Oxford: Oxbow, pp. 21–48.

Vigne, J.-D. (1999) 'The Large "True" Mediterranean Islands as a Model for the Holocene Impact on the European Vertebrate Fauna? Recent Data and New Reflections', in N. Benecke (ed.), *The Holocene History of the European Vertebrate Fauna: Modern Aspects of Research*, Rahden/Westf: Verlag Marie Leidorf GmbH, pp. 295–322.

Vigne, J.-D., Carrère, I., Briois, F., and Guilaine, J. (2011) 'The Early Process of Mammal Domestication in the Near East: New Evidence from the Pre-Neolithic and Pre-Pottery Neolithic in Cyprus', *Current Anthropology* 52.S4: S255–S71.

Wilkens, B. (1996) 'Faunal Remains from Italian Excavations on Crete', in D. Reese (ed.), *Pleistocene and Holocene Fauna of Crete and Its First Settlers*, Madison, WI: Prehistory Press, pp. 241–61.

Yannouli, E. and Trantalidou, K. (1999) 'The Fallow Deer (*Dama dama* Linnaeus, 1758): Archaeological Presence and Representation in Greece', in N. Benecke (ed.), *The Holocene History of the European Vertebrate Fauna: Modern Aspects of Research*, Rahden: Verlag Marie Leidorf, pp. 247–81.

Younger, J.G. (1988) *The Iconography of Late Minoan and Mycenaean Sealstones and Finger Rings*, Bristol: Bristol Classical Press.

—— (1995) 'Bronze Age Representations of Aegean Bull Games III', in R. Laffineur and W.-D. Niemeier (eds), *Politeia: Society and State in the Aegean Bronze Age: Proceedings of the 5th International Aegean Conference/5e Rencontre égéenne internationale, University of Heidelberg, Archäologisches Institut, 10–13 April 1994*, Liège: Université de Liège, Histoire de l'art et archéologie de la Grèce antique; Austin: University of Texas, Program in Aegean Scripts and Prehistory, pp. 507–45.

Zeder, M. (2008) 'Domestication and Early Agriculture in the Mediterranean Basin: Origins, Diffusion, and Impact', *Proceedings of the National Academy of Sciences, USA* 105.3: 1597–604.

—— (2012) 'Pathways to Animal Domestication', in P. Gepts, T. R. Famula, and R. L. Bettinger (eds), *Biodiversity in Agriculture: Domestication, Evolution, and Sustainability*, Cambridge: Cambridge University Press, pp. 227–59.

8

Netherworld envoy or man's best friend?

Attitudes toward dogs in the ancient world

Sophia Menache

Human attitudes toward the fauna convey a faithful manifestation of time and space, combining religious and philosophical tenets with ecological and environmental demands. The prevailing attitudes toward animals reflect in this regard just a fragment of the constant process of reflection and contemplation and the resulting attempts to cope with a challenging environment, the rules of which remain unknown or incomprehensible for most human beings. Selected from the rich spectrum of human attitudes toward the nonhuman world, this chapter focuses on dogs, the first domesticated animal and, as such, a continuous associate, if not a real component, of human society. From a chronological perspective, this study devotes much attention to the ancient world, when the basic principles of religion and philosophy were discussed and eventually established.[1] These thematic and chronological demarcations allow some responses to the intriguing attitudes of human beings toward their most faithful friends among the nonhuman habitants on earth.

The perception of the cosmos as a harmonious wholeness encouraged positive attitudes toward animals in Chinese traditions (Seiwert 1984). Additionally, the central concept of non-violence (*ahimsạ̄*) toward all beings left its mark on the integralist relationship with animals that characterizes Hinduism, Buddhism, and Jainism, as well (Figl 1998: 3–4). The Bhagavad Gitã – a 700-verse Hindu scripture attributed to Lord Krishna, the manifestation of God – states that the true spiritual Self, notwithstanding its different forms, is qualitatively identical in all beings. Upon reaching their ultimate beatitude, all beings are then submitted to a process of transmigration and rebirth that includes plants and animals. Equality of spirit and reincarnation thus justify the universal principle of non-injury to all living creatures (Nelson 2006: 181–6).

Such pantheistic principles did not a priori ensure a positive attitude toward all animals, first and foremost to dogs. The commentator Vyãsa, a central and revered figure in most Hindu traditions, asserted that sinful human beings would be reborn as pigs or dogs, since both of these species were regarded as the outcasts of the animal world. He further claimed, 'One who reverts to what has once been renounced is like a dog licking up its own vomit' (quoted in Nelson 2006: 183), a characterization to be found in the Abrahamic traditions as well. Further emphasis was laid on the dogs' main vices, such as promiscuous sexual behaviour and gluttony, which included eating carrion. No wonder, therefore, that by their simple contact dogs were said to pollute the food of Brahmins and, no less threatening, the sacred offerings to the gods. Dogs also served as a kind of punishment in both this world and the world to come: if their lovers were low-caste men, adulterous women would be devoured by dogs (Doniger 2009: 41–3).

On the other hand, some stories recognized the loyalty of dogs and their meritorious behaviour toward human beings, acts that awarded them celestial rewards. In the epic *Mahabharata*, Lord Yama accompanies the Pandavas all the way to paradise in the guise of a dog in order to test the wisdom of Yudhisthira, the eldest brother. Yudhisthira eventually refuses to enter heaven without his dog, which has faithfully followed and protected him. This behaviour ensured his celestial reward. The ambiguity that characterizes the image of dogs in Hindu traditions is not fortuitous and may reflect the historical process from pre-Vedic and Vedic times (1700–1100 BCE to 500–150 BCE) to the *Mahabharata*: though having a utilitarian role as watchdogs and beasts of burden, the traditional Vedic gods – who were associated with dogs – went into decline, bringing about some deterioration in the status and image of canines (Debroy 2008).

For completely different reasons, the attitudes toward dogs in ancient Greece and Rome were characterized by similar ambivalence. Since nature made nothing in vain, Aristotle's *Politica* maintained that animals were created for the sake of human beings, the 'noblest of all living beings'; domestic animals were therefore expected to labour, while wild animals were to be hunted. Cicero, in his *De nature deorum,* went one stage further and claimed that nature as a whole belonged to man, thus excluding any duty or mercy toward animals. This philosophical framework encouraged hostile attitudes toward dogs. Thus, Phocylides saw in the unmanageable, female termagants the offspring of bitches. In his *De natura animalium*, Aelian further identified a basic deficiency in the most essential virtue ascribed to canines, their loyalty, since they had betrayed their owners' confidence by falling asleep during the Celts' attack on Rome in 390 BCE.

The suspicious attitude toward dogs facilitated their use as a means of punishment against the peripheral elements of contemporary society: as illustrated often in literary texts, including the *Iliad, Antigone*, and *Hecuba*, it was common practice in 'Homeric' society (1400–1100 BCE) to throw people to them as a form of execution or to leave them the corpses of persons unworthy of funerary honours – which implies that dogs were not, or were not thought to be, entirely domesticated.[2] The very word for dog, *kuwv* in ancient Greek, became an insulting designation covering human vices such as cowardice, immodesty, and arrogance – all presumed blots on the canine species. Thus, the most infamous creatures in Spartan society, the helots, were compared to dogs (David 1993: 403–8).

However, symbolic associations were not unequivocally negative. Even the appellation of Cynics *(kuvikoi)* given to the followers of Diogenes of Sinope essentially suggests their distinctive traits, such as their rejection of all conventions, their attempt to live on nothing, and their support and exercise of shamelessness (Bonilla 1967: 60). The Stoics, as well, were often represented as dogs – as in the *Satyricon* by Petronius – probably because one of their cardinal virtues was to be truth-tellers. Similarly, *kyniskos* and *kyniska*, the male and female words for puppy in Ancient Greek, were used by the Spartans as persons' names and nicknames.

Beyond the symbolic level, there was also some awareness of the injuries or impairment that dogs could cause. In the case of personal harm, Solon, Plato, and the Twelve Tables (the foundation of Roman law) all recognize the obligation of the dogs' owners to compensate anyone injured by a dog and to transfer the guilty animal to that person for proper punishment. Awareness of the potential danger of dogs brought about the obligatory use of clear warnings of their presence. '*Cave canem*' (beware of the dog) was a common exhortation, perpetuated in archaeological and literary documentation. Still, it seems that the manifestations and ways of transmission of the most dangerous affliction that dogs could cause – rabies – remained unknown. In Homeric society, the word 'rabies' was used in its metaphorical form, to designate fury or uncontrollable wrath. Although Aristotle in his *Historia animalium* was the first author to depict the symptoms and effects of rabies, he still denied its transmission to human beings and limited its effects to animals alone.

The knowledge that dogs may cause harm – archaic as it was – fostered a clear tendency to keep canines away from the world of the living and consign them to the realm of Hades. This netherworld abode for dogs – where they accompanied their owners – became a common belief among Indo-European civilizations. Brewer *et al.* (2001) detail a notable similarity among the Greek dog Cerberus (which welcomes dead souls at the entrance of Hades), Garmr (the Germanic hell-hound), the Vedic sons of Sarama (the dog-messengers of death), and Odin's two wolves (which later became dogs). Dogs were also sacrificed in acts of purification of the family and the house, and they constituted one of the most prominent sacrifices in association with funerary rites.

Aelian reports the habit of the inhabitants of Argos to kill dogs that approached the marketplace during some festive days. In the Spartan *agoge*, the ferocious fight between two bands of youths was preceded by the sacrifice of two puppies to Enyalius, presumably the Laconian name of Ares, the god of war. Dogs howled at the approach of Hecate, a nightmarish lunar goddess of the infernal regions and of witchcraft, their howling also being conceived as an augur of death. Hounds also became companions of Hecate and were sacrificed to her. Like their mistress, they became the overseers of cyclical time, guardians of life, and very influential in the awakening of slumbering vegetation and plant life (Gimbutas 1989). Their link with the world of death turned dogs into unclean animals, whose entry into the temples in Delos or on the Athenian Acropolis was forbidden. The Romans, too, believed that neither dogs nor flies should enter the Temple of Hercules in the *forum Boarium*.

In clear contrast to the negative nature of most of the testimonies mentioned above, there is pictorial evidence that both Greeks and Romans developed definitive breeds of dogs, a clear reflection of the growing attachment to these animals. In sepulchral monuments from the late Archaic to the Hellenistic periods, dogs are depicted as protecting their owners against the living, as well as against the forces of hell. Homeric narrative further presents a more commendable image of dogs, which has permeated Western culture up to this day. Dogs appear as the most helpful auxiliaries of men for hunting and safeguarding purposes in the *Odyssey* and the *Iliad* as well. These tasks confronted them with the most dangerous of the wild animals, such as lions and bears, thus creating a suitable arena in which canines could prove their courage and tenacity. No wonder, therefore, that in the framework of the *oikos* (household) – and the obligation of its head to avenge any harm to the household, slaves and animals included – the killing of a shepherd dog was punished like the murder of the shepherd himself.

Dogs also emerge as the loyal companions of gods and men, and not just in literary and artistic representation. They were employed as guards in the temple of Jupiter on the Capitoline, of Hephaestus on Etna, Athena in Daunia, and Adranus in Sicily. Dogs escorted their owners to social and political gatherings, especially in aristocratic circles. An impressive number of vases show dogs behind a table waiting to be fed (a scene persisting in an immense number of medieval paintings); the *Odyssey* explicitly refers to delicacies generously provided for them by their masters. Hounds were considered private property in Spartan society, and their temporary use by some other citizen was conditioned on the prior consent of the owner, usually a wealthy man. Accordingly, Xenophon approaches hunting-dogs as one of the most important possessions worthy of adorning an estate. Plato considers dogs to be the trustful protectors of sheep and a model for the guardians of the ideal State because they are both gentle and spirited.

Of the many qualities associated with dogs, ancient authors emphasize loyalty, perception (being the sole animals to recognize their names), memory, and devotion to their offspring. Some writers further enlist historical testimony to prove canine loyalty. When the Persian invasion caused the Athenians to evacuate their city in 480 BCE, Plutarch reports the story of Pericles's father, Xanthippus, whose dog swam across the strait by the side of his master's trireme and staggered out on Salamis to die straight away. Pliny the Elder, as well, sees in the dog – together with the horse – the animal that is most faithful to man, a premise he demonstrates through an impressive documentary corpus.

Dogs gradually lost their animal essence and were credited with magical skills that went beyond those of the ordinary human being. Thus, Aelian attributes to dogs the good management of household affairs, especially among the poor. Though acknowledging that they are devoid of reason, he further credits dogs with a presentiment of imminent famine, pestilence, or earthquake. They can also foretell fair weather and the fertility of crops. Canine excellence embraced bitches as well, as faithfully demonstrated by a pregnant bitch that gave birth to nine puppies only after completing her hunting duties. The virtues assigned to canines, furthermore, embraced devotion not only to their masters but also to conjugal fidelity. By his barking, a lap-dog of Sicily caused a husband to capture his adulterous wife in concert with her paramour. Following this substantial attestation of canine virtues, Aelian concludes: 'It fills me with pain that a dog should be shown to have more loyalty, more kindly feeling than man' (1958: 10).

Dogs also proved their merit in times of war. The *canis bellator* or *canis pugnator* was assigned crucial military roles, a practice that probably originated in the Orient. Aelian reports the case of a dog that fought bravely at his master's side in the Battle of Marathon, its meritorious behaviour perpetuated in pictorial representations. In 352 BCE, Philip the Macedonian used dogs in the Thracian war to force the Arbelians to leave the forest where they had found shelter. The same tactic was followed by the Roman consul Pomponious Matho against the Sardes in 230 BCE.

Contrary to digestive taboos that applied to monotheistic societies, dogs were part of the ancient Greeks' menu, and curative effects were attributed to their meat. Puppies were used as scapegoats against intestinal and stomach disorders, the diseases supposedly transmitted to them through physical contact (Burriss 1935: 32–3). From the long list of therapeutic qualities credited to dogs (Gourevitch 1968: 247–58), Pliny the Elder mentions the effectiveness of their blood both as an antidote for poison and as a remedy for itching. It was also believed that canine blood, if placed under the threshold, kept away evil spirits, especially the spirits of the dead that haunted the location and might harm the living members of the family. The flesh of a suckling puppy taken with wine and myrrh was considered an expedient against epilepsy; the ashes of a dog's head, when mixed with wine and honey, against jaundice; and the ashes alone, against burns and chilblains. Bitches enjoyed some therapeutic merits of their own: their proestrus blood was said to cure the bite of a mad dog, and their milk to prevent the growth of hair. In this rather tortuous way, canines thus accompanied ancient Greeks and Romans through every part and parcel of life, in this world as well as in the world to come.

Similar to the utilitarian attitude found in the classical heritage, the acknowledgement of the human race as the work of art of an omnipotent God led all Abrahamic traditions to develop a functional approach to nonhuman beings; like all animals in the Genesis account, they were created for the sake of humans and, as such, were left under the arbitrary rule of men. Such a patronizing attitude, however, did not dictate a clearly defined attitude toward different animals. The extreme antagonism toward dogs in theological writings, moreover, presents a challenge that has not yet encountered satisfactory answers in modern research.

The Old Testament justifies a negative attitude toward dogs because of their connection with carrion and carcasses, a behavioural pattern that ancient societies often perceived as threatening and a source of danger (Douglas 1966: 11). To their essential impurity – which, according to Jeremiah, will burden dogs with carrying the dead on Doomsday – the Book of Proverbs attaches their stupidity. Biblical narrative also uses dogs as a common metaphor to hint at the unfortunate status of men and/or the negative elements of society, whether the enemy, deserters in general, false prophets, or despicable persons. Goliath's sarcastic diatribe to David, 'Am I a dog that thou comest to me with staves?' (I Sam. 17: 43), clearly reflects the low status of dogs in contemporary society. When criticizing the many sins of his contemporaries, the prophet Isaiah compares them to dumb dogs that cannot bark and evince a selfish character. Deuteronomy, hinting at divine repugnance, proscribes any acceptance by the Holy Temple of profits deriving either from male or female prostitution, an intriguing association that appears in the New Testament, as well.

> Thou shalt not bring the hire of a whore, or the price of a dog, into the house of the Lord thy God for any vow: for even both these are an abomination unto the Lord thy God.
>
> *(Deut. 23: 18).*

The practical conclusion is therefore clear, and it justifies the need to keep at a safe distance from the canine race.

Such antagonism to dogs – bizarre as it may appear to modern eyes – was hardly exceptional in the ancient Levant. Besides the examples quoted above from the Graeco-Roman world, dogs were regarded as unclean animals in the Hittite Empire, whose temple personnel were specifically ordered to keep them away – together with pigs – from the offerings to the gods. Yet, though not exceptional in its geopolitical space, the negative approach of biblical narrative toward canines seems rather problematic in light of the prevailing attachment to dogs in agrarian and nomadic societies, for which canines fulfilled guard functions and were highly esteemed, at least from an instrumental angle. One possible explanation for this dichotomy may lie in the Old Testament's essential opposition to any remnant of polytheism, in which dogs were commonly bestowed a prominent role. In the Egyptian pantheon, for example, Anubis, the god of death, had a dog's head; and Isis, the goddess of the netherworld, was represented riding on a dog. Dogs were also involved in healing cults in many ancient cultural centres in Mesopotamia and around the Mediterranean, the Land of Israel included. Large cemeteries of sacred dogs were found in the archaeological excavations at the eastern and southern areas of modern Israel, hinting at the ritual function of dogs in the geographical space of the Chosen People, as well. There is also evidence from the Natufian period (from 10500 to 8300 BCE) of a human skeleton buried with a dog, the man's arm stretched toward his pet, thus reflecting their mutual affection in life (Patton 2006: 27).

The Book of Psalms corroborates the ubiquity of dogs in urban centres and the resulting noise they created there, while the Book of Job refers to shepherd dogs in agrarian communities. Rabbinical literature, as well, indicates that dog-owning had been common practice among Jews in agricultural villages, especially between the third and sixth centuries CE. Daily practices of this kind probably encouraged the peremptory tone of the Babylonian Talmud,[3] which differentiated at least in principle between wild and good dogs. It applied a cautious attitude toward good dogs, which also had to be securely chained and could be freed only at night when only suspicious people walked the streets. On the other hand, the ownership of an evil dog – that is, one that bites and barks – was completely forbidden, since it could turn into a public danger and, as such, cause its owner to violate the biblical prohibition, 'Do not place blood in your home' (Deut. 22: 8).

Daily practices and dictates, as well as pagan remnants, may therefore have justified to some degree the negative approach to dogs in Jewish literature. Still, other domestic animals, cats and sheep among them, were part and parcel of daily life while receiving a prominent role in pagan cults without generating the extreme dislike that faced dogs. The ecological and theological perspectives may introduce additional possibilities of analysis in this regard: the domestication process was far from being completed in the ancient world, and rapidly increasing packs of pariah dogs, with the resulting problem of rabies, overran cities and villages. They created serious security and sanitary problems, thus justifying and even dictating general taboos against such pernicious animals (Serpell 1995: 112–16).

On the other hand, the attachment of human beings to the canine species in actual practice does not suit, if not openly contradicts, divine plans. As peremptorily established in Scripture, God submitted the whole of fauna to the absolute rulership of the human race, while the tolerance of animals was conditioned on their fulfilling human needs. One may therefore assume that pet-keeping was not envisaged by God in creation. On the contrary, God clearly established human rule – people being created in His Own image – over all other creatures upon the face of earth (Patton 2000: 405–6). A close partnership/companionship between humans and dogs thus seemed to contradict divine commands and to some degree challenged the most desired submission of the human race to His designs. Yet the Babylonian Talmud refers to an old tradition according to which God, after exiling Cain from paradise, furnished the first criminal in the history of humankind with a dog, to defend him from attack by savage animals (*Bereshit Rabbah* 22: 12). Domestication in this version appears as an act not only envisaged by God but also initiated by Him to defend the otherwise vulnerable humans against the forces of nature. In a similar vein, the apocryphal *Book of Tobit* refers to Tobit's dog, which faithfully accompanied its owner on his journey to Media, a motif very similar to that found in Hindu traditions with regard to the Pandavas.

The same ecological conditions and similar, if not identical theological traditions encouraged an analogous ambivalent approach to canines in the New Testament and early Christian theology. Following the Old Testament's tenets, such leading Christian theologians as Tertullian, Origen, and St Augustine maintained the total mastery of human beings over animals, thus excluding any familiarity or companionship with them. In his eschatological vision, John perpetuated the connection of dogs to whores. Together with the most despicable strata of human society and as the only representative of the animal kingdom, dogs were excluded from heavenly Jerusalem. Their dishonourable status was corroborated by Jesus, who decreed: 'Give not that which is holy unto the dogs, neither cast ye your pearls before swine, lest they trample them under their feet, and turn again and rend you' (Matthew 7: 6).

As time went by, Christian homilies often presented dogs as tools of divine punishment or as messengers from the afterlife. The motif of the devil in dog form survived into the modern era, and the same code served against witchcraft, the devil, and mad dogs in sixteenth-century prayers (Woods 1959: 14). One typical formula advised holy bread or water to be taken for nine days, together with the recitation of three Paternosters and three Aves in honour of the Trinity and St Hubert, the patron of huntsmen, whose assistance against hydrophobia was to be begged.

In contrast to the negative approach found in the apostolic age, there are some indications of pet-keeping in medieval Christendom, perhaps as a result of the changing environment.

Commenting on the parable of Dives and Lazarus, Ambrose of Milan, one of the four recognized doctors of the Latin Church, blessed dogs 'that represent those who guard the flock and protect it against the wolves [. . .] keep guard for their masters' safety [. . .] and display noteworthy smelling skills' (Springer 1931: 29–32). He reports in detail the genealogies of several dogs – carefully established in patterns similar to those that served fourth-century aristocracy – and birthday parties celebrated in their honour. Acknowledgement of the canine's instincts led Basil the Great to deduce that dogs had achieved the power of reason, while their loyalty toward their owners should embarrass those sinners who are ungrateful to their benefactors. John Scotus Erigena admits that dogs have some advantages over people, such as memory and faithfulness, both virtues exemplified by Argos, Odysseus's dog, recognizing his master after twenty years of absence. Hildegarde of Bingen claims that dogs are able to detect human lechery, as well as to augur events, unfortunate or auspicious, that are about to befall people. As a most indispensable component of hunting expeditions, dogs also formed part of the burial gifts to wealthy men (Ohman 1983). Medieval representations of dogs further reflect their many varieties, both in cities and in rural settings.

The evolution from the evangelical antagonism to dogs to the gradual recognition of their virtues in medieval Christendom corroborates, again, the importance of the historical perspective in analysing the changing attitudes toward canines. The more friendly attitudes, which are clearly detectable in the Central Middle Ages, are an integral part of the secularization process that Western society underwent from the twelfth century onwards. The Pauline ideal of contempt for this world of sin was then questioned by a less radical vision of this world and a greater desire to enjoy its delights. The growth of cities was part of this process, but it did not change the essentially agricultural nature of medieval society or its allegiance to feudal norms, among which the underling's expected fidelity toward his lord played a most critical role. The 'discovery' of canine loyalty thus conjoined with the maturation of chivalry and the lordly code of behaviour, while the unconditional fidelity of dogs toward their masters offered a most suitable pattern for imitation (Menache 2000: 42–60). The intensification of hunting among aristocratic circles simultaneously elevated the socioeconomic status of canines.

In contrast to developments in medieval Europe, the different environmental background – very similar to that of the ancient Hebrews and early Christianity – fostered hostile attitudes toward dogs in Muslim society (Foltz 2006: 157). The widespread belief that dogs, especially black dogs, were in fact a demonic emanation of evil spirits created the ground for a biased approach. Faced with the problem of a plague of stray dogs, Mohammed at first took an uncompromising decision to exterminate 'all dogs'. Afterwards, he mitigated his decree by reasoning that the canine genus was created by Allah and people needed certain species of dogs. The Prophet thus decided to exterminate only black-coated strays, particularly those with light patches, the indisputable mark of the devil. According to Ḥamdullāh al-Mustaufī al-Qazwīnī, 'any place in which the eye of a black dog is buried will fall into ruins' (1928: ix, 34). Freed from divine condemnation were useful dogs that obeyed their master; that is, trained hunting dogs and watchdogs. Muslim doctors of law allowed the possession of such dogs as long as their use could be justified; their killing, furthermore, was punished with heavy fines according to the species and the functions these dogs fulfilled.

Though useful dogs were socially tolerated, they remained unclean with respect to religious practice, since the Prophet once said: 'Prayer is interrupted by a woman, a donkey [. . .] and a black dog' (quoted in al-Damīrī 1857: 430). Everything a dog touched or licked was rendered impure, and the place where it had lain had to be purified with water. Moreover, if a dog drank from a human's utensil, it was imperative to wash it seven times. A dog prowling close to a believer in prayer invalidated his prayer, and its presence prevented angels from visiting a house. Ultimately, any believer who kept a useless and vicious dog lessened his final reward. No wonder, therefore, that the word 'dog' in the Arabic language, *kalb*, became a biting insult and appears pejoratively in numerous proverbial sayings.

The negative, even diabolic image of dogs among Muslims was undoubtedly connected with the danger of rabies, which they were suspected of carrying and which became an endemic phenomenon in the Levant. Islamic scholars developed a remarkable awareness of the disease, its symptoms, and its dangerous consequences. Still, for a long time, a person infected with rabies was considered possessed by madness and treated by exorcism rather than therapy. Ibn Qutayba claims that 'if a rabid dog bites a human being, it happens that it changes him into a barker like itself, renders him pregnant, and impregnates him with little whelps that you see as coagulated blood in the shape of dogs' (1949: 55).

Other early Islamic scholars like Al-Djahiz – whose *Kitab al-hayawan* furnishes the richest documentation on dogs – reflect a more moderate view. He affirms that, in principle, no dog will attack a person who has not previously harmed it. Perhaps to weaken prevailing fears, he reports the story of a boy who, one month after having been bitten by a dog, did not bark and was completely healthy, an assertion particularly important against widespread fears. Al-Qazwīnī, on the other hand, further asserts that a 'mad dog' is one of the 'five scoundrels' and ought to be killed (1928: 34). Everyone it bit was in danger of death for forty days, after which the person was deemed safe. He also claims that becoming afraid of water was an unquestionable sign of rabies and of approaching death. Drinking the blood of a king was held to be the supreme remedy against rabies, as it was against insanity and possession, a linkage that indicates, again, the devilish nature of dogs and the disease they transmitted. More commonly available means of combating rabies were also identified, such as Spanish flies, lentils, and a beverage called 'old drink', which was artificially prepared – all treatments that were kept secret and passed from father to son in some families.

Notwithstanding the widespread awareness of the danger of rabies, there are also some indications of dog-keeping in medieval Muslim society. Thus, Sultan Mohammed al-Nasir (1310–41), who was well known for his love of hunting, maintained eighty different groups of hunting-dogs, which were carefully raised by special officials nominated for this purpose. The Prophet himself was said to have promised a divine reward to an old woman for her act of charity to a thirsty dog. According to Muslim tradition, moreover, a dog by the name of Kitmir was allowed to enter paradise, since he became the symbol of fidelity and was given the gift of speech by Allah. Al-Qazwīnī acknowledges that a dog

> is a faithful beast, patient in undergoing hardships and in enduring hunger, in rendering service and in keeping off enemies. It follows up game with quick intelligence, and although kept hungry it is faithful and will not leave its master, acknowledging the duty of obedience.
>
> *(1928: 34)*

Al-Djahiz further praises dogs' memory and loyalty (their skill in recognizing their owners after long periods of absence), their intelligence, and the services they provide as sentinels. Particularly touching are his references to spoiling dogs, which will never sleep on the ground if they can enjoy the texture of a delicate carpet and will always prefer a pillow to both of these beds.

This short review of the prevailing attitudes toward the canine species in ancient and medieval societies shows a clear gap between two conflicting tendencies: the mythical-theological approach labelled dogs as 'outcasts', possessing the most infamous human vices and patterns of behaviour and thus delegating canines to the netherworld. Their comparison to prostitutes and their characterization as unclean, their very proximity harming the dialogue between the believer and his God, the many problems they may cause, first and foremost rabies – all these prejudices were partially the product of the historic and ecological conditions of Hindus, Greeks, Romans, Jews, Christians, and Muslims. However, such an explanation – though dictated by clear, empirical factors – leaves much room for speculation as to the intensity of religious antagonism to dogs. One may argue in this regard that the proximity between humans and dogs – unique in its intensity – might strengthen human self-esteem, and thus it represented a threat to the ruling theocracy. The priesthood consequently responded with all the catechisms of fear at its disposal, including punishment in both this world and the world-to-come. The representatives of God on earth thus reinforced the well-known dichotomy between purity and pollution, virtues and sins, and used dogs as a projection mechanism that encapsulated all the vices of both humans and nonhumans.

Yet, against the theological-mythical tenets, there is ample evidence of dog-keeping, with all its emotional weight, in all the traditions reviewed in this chapter. Faithfulness, memory, intelligence – all the rich spectrum of human virtues were also attributed to dogs, which in many cases exceeded the skills of the average human being. The rich evidence of dog-keeping in ancient and medieval societies thus clearly challenges the assumption that pet-keeping is a modern phenomenon. The pendulum that characterized the attitudes toward dogs thus justifies the thesis advanced by James Serpell some years ago, according to which dogs are 'an interstitial creature, neither person nor beast, forever oscillating uncomfortably between the roles of high-status animal and low-status person' (1995: 254). One may therefore conclude that a better knowledge of human attitudes toward dogs from a historical perspective will substantially advance a broader understanding of human society itself, with all its dreams and ideals, as well as its weaknesses and fears, concerning our companions throughout the ages.

Notes

1 The imperative to include Muslim attitudes toward dogs dictated research into the medieval period, as well; thus rigid chronological boundaries are not observed in this chapter.

2 Thucydides, however, reports that during times of plague, dogs refrained from touching human corpses (1981: 30).

3 The *Mishnah* (meaning repetition, or to study and review) is the first major written redaction of Jewish oral traditions and a most important work of Rabbinic Judaism. It was redacted *c.* 200 CE by Rabbi Judah the Prince, and its authority is second only to the Bible, which it complemented. Rabbinic commentaries on the *Mishnah* over the next three centuries were redacted as the *Gemara*, which, coupled with the *Mishnah*, constitute the *Talmud* (meaning instruction, learning). In its two versions, one version written in Jerusalem and the other, more influential in Babylon (*BT*), the Talmud covers commentaries from the second to the eighth centuries CE.

References

Aelian (1958) *De natura animalium*, A.F. Schofield (ed.), Cambridge, MA: Harvard University Press.

Bonilla, L. (1967) *Historia y Psicología del Perro*, Madrid: Editorial Tecnos.

Brewer, D.J., Clark, T., and Phillips, A. (2001) *Dogs in Antiquity: Arnubis to Cerberus: The Origins of the Domestic Dog*, Warminster: Aris and Phillips.

Burriss, E.E. (1935) 'The Place of the Dog in Superstition as Revealed in Latin Literature', *Classical Philology* 30.1: 32–42.

Cicero, M.T. (1979) *De nature deorum*, Arthur Stanley Pease (ed.), Cambridge, MA: Harvard University Press.

al-Damīrī, M.M. (1857) *Kitāb ḥayāt al-ḥayawān al-kubrá*, Cairo.
David, E. (1993) 'Hunting in Spartan Society and Consciousness', *Echos du monde classique* 12: 393–413.
Debroy, B. (2008) *Sarama and Her Children: The Dog in Indian Myth*, New Delhi: Penguin Press.
Doniger, W. (2009) *The Hindus: An Alternative History*, New York: Penguin Press.
Douglas, M. (1966) *Purity and Danger: An Analysis of the Concepts of Pollution and Taboo*, New York: Praeger.
Figl, J. (1998) 'The Human Animal Relationship in World Religions', in *The Changing Roles of Animals in Society: Abstract Book of the 8th International Conference on Human-Animal Interactions*, Renton, WA: International Association of Human-Animal Interaction Associations, pp. 2–6.
Foltz, R. (2006) '"This She-Camel of God Is a Sign to You": Dimensions of Animals in Islamic Tradition and Muslim Culture', in P. Waldau and K. Patton (eds), *A Communion of Subjects: Animals in Religion, Science, and Ethics*, New York: Columbia University Press, pp. 149–59.
Gimbutas, M. (1989) *The Language of the Goddess*, London: Harper and Row.
Gourevitch, D. (1968) 'Le chien, de la thérapeutique populaire aux cultes sanitaires', *Mélanges d'archéologie et d'histoire* 80.1: 247–81.
Lewinsohn, R. (1954) *Animals, Men and Myths: An Informative and Entertaining History of Man and the Animals around Him*, New York: Harper and Brothers.
Menache, S. (2000) 'Hunting and Attachment to Dogs in the Pre-Modern Period', in A. Podsberscek, E. Paul, and J. Serpell (eds), *Companion Animals and Us: Exploring the Relationships between People and Pets*, Cambridge: Cambridge University Press, pp. 42–60.
Nelson, L. (2006) 'Cows, Elephants, Dogs, and Other Lesser Embodiments of Atman: Reflections on Hindu Attitudes toward Nonhuman Animals', in P. Waldau and K. Patton (eds), *A Communion of Subjects: Animals in Religion, Science, and Ethics*, New York: Columbia University Press, pp. 179–93.
Ohman, I. (1983) 'The Merovingian Dogs from the Boat-Graves at Vendel', in J.P. Lamm and H.A. Nordstrom (eds), *Vendel Period Studies: Transactions of the Boat-Grave Symposium*, Stockholm: Statens Historiska Museum, pp. 169–76.
Patton, K. (2006) '"Caught with Ourselves in the Net of Life and Time"': Traditional Views of Animals in Religion', in P. Waldau and K. Patton (eds), *A Communion of Subjects: Animals in Religion, Science, and Ethics*, New York: Columbia University Press, pp. 27–39.
—— (2000) '"He Who Sits in the Heavens Laughs": Recovering Animal Theology in the Abrahamic Traditions', *Harvard Theological Review* 93.4: 401–34.
Pliny (1970) *Naturalis historiae*, K.F.T. Mayhoff (ed.), Stuttgart: Teubner.
al-Qazwīnī, H. (1928) *The Zoological Section of the Nuzhatu-l-Qulūb of Ḥamdullāh al-Mustaufī al-Qazwīnī*, J. Stephenson (ed.), London: Royal Asiatic Society.
Qutayba, I. (1949) *The Natural History Section from a 9th Century 'Book of Useful Knowledge': The Uyûn al-akhbâr of Ibn Qutayba*, F.S. Bodenheimer (ed.), L. Kopf (trans.), Paris: Academie Internationale d'Histoire des Sciences.
Seiwert, H. (1984) 'Ethik in der chinesischen Kulturtradition', in P. Antes, H. Bechert, and W. Dupre (eds), *Ethik in nichtchristlichen Kulturen*, Stuttgart: Kolhammer, pp. 135–62.
Serpell, J. (1995) 'From Paragon to Pariah: Some Reflections on Human Attitudes to Dogs', in J. Serpell (ed.), *The Domestic Dog: Its Evolution, Behaviour, and Interactions with People*, Cambridge: Cambridge University Press, pp. 108–21.
—— (1985) 'Best Friend or Worst Enemy: Cross Cultural Variation in Attitudes to the Domestic Dog', in K. Lorenz (ed.), *The Human-Pet Relationship: Proceedings*, Vienna: Institute for Interdisciplinary Research on the Human-Pet Relationship, pp. 112–16.
Springer, M.T.C. (1931) *Nature Imagery in the Works of St. Ambrose*, Washington DC: Catholic University of America Press.
Thucydides (1981) *History*, H.R. Rawlings (ed.), Princeton, NJ: Princeton University Press.
Woods, B.A. (1959) *The Devil in Dog Form*, Berkeley: University of California Press.

9

The material culture of pet keeping

Katherine C. Grier

According to the American Pet Products Association, 68 per cent of American households contain a pet animal; 44 per cent have more than one (APPA 2013). The growing enthusiasm reflected in these numbers has resulted in a large and distinctive array of specialized consumer goods intended to enhance a distinctive human-animal bond. Commercial services – from dog-grooming businesses to small-animal veterinary practices; recreational spaces such as dog parks; and the memorial spaces of pet cemeteries – all rely on their own constellations of artifacts, from surgical equipment scaled for small bodies, to park dispensers containing special biodegradable pet-waste bags. Study of the material culture of pets and tracing its change over time is a useful method for parsing the evolving relationships between people and the animals they live with at home. The object-traces of human involvement with pets, past and present, reveal the evolving contours of routine practices, the cultural assumptions that underlie them, and the complex, deep feelings that pet owners have about their animals.

This chapter considers the material culture of pet keeping through a series of brief biographies for artifacts associated with dogs: an antique collar; a refrigerator dish intended to contain dog food; a *carte-de-visite* pet photograph; and a plastic squeaky toy. It closes with a preliminary consideration of a distinctive public material culture that has been neglected by scholars of animal-human interaction: that of the animal shelter and its equipment, particularly the catch-pole used by animal-control officers.

The term 'material culture' originated with the British ethnologist and archaeologist Augustus Henry Lane-Fox Pitt Rivers (1847–1900), who posited that the study of artifacts provides crucial insights into the culture, institutions, and daily practices of individuals and groups of people. Almost a century later, American archaeologist James Deetz defined material culture as 'that sector of the physical environment that we modify through culturally determined behavior', embracing 'all artifacts, from the simplest, such as a common pin, to the most complex, such as an interplanetary space vehicle' (Deetz 1977: 24–5). His broad definition, which even embraced the bodies of domestic animals as being culturally shaped, suggested that the most ordinary or ephemeral artifacts could be particularly diagnostic of a moment in time or a way of life. My research on animal-human interaction recuperates the material culture associated with pet keeping, and I continue to use artifacts as a medium for reconstructing daily practices and the shifting constellations of ideas, assumptions, and feelings that underlie them (Grier 2006).

An extraordinary array of things relating to pet keeping does survive, but examples are rarely found in museum collections. Rather, they belong to private collectors who have assiduously built sizable assemblages, sometimes relating to their professions, as I found in the case of a small-animal veterinarian whose practice looked like a museum of proprietary medicines for animals, and a man who raised cage birds for a living and possessed a large collection of antique cages and their accessories. I too became a collector of both artifacts and imprints to support my research, and the objects illustrated in this chapter, along with a print advertisement for dog food discussed below, come from my own, more modest holdings.

A definition of the 'pet' animal will help to frame biographies of exemplary artifacts associated with pet keeping:

1 The pet animal is chosen by people, who take responsibility (however imperfect or inadequate their efforts may be) for its well-being and intervene routinely in its life course.
2 The pet animal is kept in close proximity to the people who care for it, including residing in and around a dwelling.
3 The pet animal is not required to 'earn' its living except insofar as it contributes to the emotional economy of private life.
4 The pet animal is kept as a companion or the object of emotional ties.
5 The pet animal may be kept at least in part or wholly for the pleasure its beauty provides. This may be either physical appearance or some other characteristic, as in the case of birdsong.
6 The pet animal may be a source of leisure activity, which may include development of substantial expertise, as in the case of caring for reptiles or parrots.
7 The pet animal may be a source or a signal of social status. These animals may be exotic, 'purebred' (created from a closed genetic pool of similar individuals), or have other special characteristics, as when their status is associated with the potential danger associated with ownership.
8 The status of the pet animal is contingent. An animal can be raised into the status of 'pet' or lose that status through abandonment or other failures of care.
9 Not every circumstance of pet keeping meets all these characteristics or conditions, although items 1 and 2 are fundamental to this animal category.

This list of characteristics and conditions should be considered a 'lived definition' of the category; some of its elements are seen in the performance of the relationship, a performance that relies on artifacts to articulate its characteristics, rather than in verbal or textual discussions.

All types of animal-human interaction are shaped by distinctive constellations of artifacts, from horse tack to milking machines. The material culture of pet keeping consists of artifacts that articulate aspects of the definition offered above. This chapter introduces three categories: control, accommodation, and display; care and cultivation; and the performance of emotions. Many pet-keeping artifacts occupy more than one category, and some shift categories over time, as when utilitarian objects become mementoes associated with deceased pets. While I regard the bodies of pets as bio-cultural artifacts, shaped by various human interventions, space does not permit discussion of this category; for the present, it must suffice to note its existence.

Control, accommodation, and display: Frank's collar

The first, and perhaps the oldest, category of the material culture of pet keeping is artifacts of control, accommodation, and display. This includes collars, leashes, and chains; doghouses, kennels, cages, and hutches; and containers for fish and other aquatic creatures. Domestic spaces shaped to keep animals from leaving the household precincts, as in the case of fenced back gardens or the special bumped-out windows that some pet owners now install to serve indoor cats, may also be included in this group. These objects can be strictly utilitarian or even improvisational; for example, wooden barrels, turned on their sides and braced to keep them from rolling, were commonly used as doghouses in the nineteenth century. Birdcages, on the other hand, were often designed to be beautiful objects in their own right, framing and enhancing the beauty of the occupants and contributing to interior décor.

Leashes, chains, and collars for dogs may be the oldest artifacts of this category for pets. At their functional minimum, collars control the movements of dogs by controlling their heads, circling the throat in a way that makes holding onto the dog easy and allowing pressure to be put on the dog's airway. The leash or chain allows the dog to move but keeps it close by its human handler or limits its ability to leave the precincts of home. Images of what seem to be leather or metal collars and leashes or chains on watchdogs, hunting hounds, and small companion dogs appear on Egyptian wall paintings, Greek pots, and on Roman mosaics and ceramics (Lazenby 1947).

While the collars of pet dogs in the past did give owners something to hold on to, they had other functions. They both identified the dog as 'owned' and suggested the role of the animal. Collars for hunting or guard dogs were once covered with studs or spikes to protect the animal's throat from attack, while collars for pet dogs were often a form of fashionable dress. Let's focus on one particular pet-dog collar, which probably dates from the 1880s or 1890s (Figure 9.1). The collar features a name plate that has been engraved: 'FRANK/Jos Wood/704 N. 3rd Street/Phila.' This is an address in the Northern Liberties neighborhood of the city, which combined both commercial and residential blocks. My research has not yet been able to match this name to this particular address, so the identity of Frank's owner is still unknown.

Frank's collar was clearly worn for some time. When it was new, it was bright nickel-plate, but the plating has worn away in spots, exposing the stamped brass of the broad links. There is no maker's mark, but collars like this were sold in what were called 'fancy goods' stores, where people bought gifts and other trinkets. By the 1890s, modern pet stores selling a range of supplies and equipment could be found in US cities, and these emporiums also sold fancy collars.

In the eighteenth century, well-to-do dog owners in England often adorned their pets with wide, flat brass collars, and some were imported to the United States. Their manufacture was a small part of the brass trade. Although they were sometimes lined with leather or felt, these one-piece collars can't have been terribly comfortable for the dogs who wore them, but they certainly and clearly signaled ownership. They were often inscribed with the name of

Figure 9.1 Dog collar, 1880–90. Nickel-plated brass. American, maker unknown. Author's collection

the owners. but more rarely with the name of the dog, which made them easily transferable from one pet to another. (The most famous example is the inscribed brass collar given to Prince Frederick by Alexander Pope in 1736, identifying the dog, who was also a gift, as 'his Highness' dog at Kew.') The collars were valuable enough that advertisements for lost dogs sometimes called for the return of the collar in any case, and there is evidence of collars being reused, in one case turning the collar inscribed with an owner's name inside out and reshaping it (Ciarlo *et al.* 2011).

In the eighteenth and nineteenth centuries, most dog collars were made of leather, which was much less expensive than brass, and cobblers and harness-makers produced them as a sideline. By the 1870s, however, the business of pet supplies and equipment in America was developing its own array of manufacturers and channels of distribution. Pet-dog collars became varied and stylish, as in the case of Frank's collar where the links are stamped in the 'aesthetic' style. In 1890, the Medford Fancy Goods Company in New York City, which billed itself as 'the Only Exclusive Manufacturer of Dog Collars and General Dog Furnishings', devoted most of a large, colorful trade catalog to its varied lines of pet collars, even including special pug collars ornamented with jingle bells and a celluloid novelty shaped like a shirt collar (Medford Fancy Goods Company 1890).

Frank's collar was heavy, yet designed for more comfort than its eighteenth-century one-piece predecessors since the flat links conformed to the dog's neck. The links were actually a variation of the box or Venetian chain style seen in jewelry of the era, supporting the idea that Frank's chain was also a form of dress. Unlike eighteenth-century collars, this collar has a metal ring intended for a tie or leash, but the fact that the owner put the name on the collar suggests that Frank either was out and about on his own or that there was concern that Frank might escape the confines of home. Philadelphia had professionalized animal control by then, and the collar prevented Frank from being labeled a 'stray'. Frank wasn't very large, however; the collar is only about 11 inches around. This suggests another characteristic of many pet dogs, especially in cities; they were often smaller than dogs kept for sport or as workers.

Frank's collar was somewhat adjustable but what held it on his neck was not a buckle but a small lock, now missing. The use of small keyed locks on collars was common in the past, and in the US some collars retained small hasps to accommodate keyed locks until the mid-twentieth century. In the eighteenth century, the lock protected against opportunistic theft of a valuable piece of metal. However, another reason to lock a dog into a collar was to discourage theft of the animal itself. By the nineteenth century, urban dog thieves gathered up pet dogs and sold them to working-class dog dealers, who were one of the less savory forerunners of the pet store. Frank's collar was locked but there is no evidence that it had a license tag. While American cities began to require that owners purchase dog licenses as early as the 1820s, metal licensing tags were not part of the material culture of collars until the 1890s when licensing efforts began in earnest.

Frank's fashionable collar continued with the practice of using collars not only to identify the dog but also to suggest the prosperity of his owner. In this case, Frank had a human name – a sporty nickname, in fact – and was the stylish four-footed representative of the family. He was also treasured long after he passed away. Frank's owner or owners chose to save his collar, and it was saved after they were gone, too. Unfortunately, the person from whom I bought the collar did not recall much about the estate sale where he had purchased it, but it apparently did not leave family hands until the early 2000s. The thousands of dog collars that do survive from the nineteenth and early twentieth centuries, sometimes with license tags still attached, suggest that keeping a collar was an important way to memorialize a deceased pet, more tangible than family photographs and more evocative of routines such as walks, perhaps even still containing traces from the dog's body in the form of wear spots, hairs, or perhaps even a faint doggy odor.

My own family offers a more recent example of this practice. My father kept the collars from several beloved dogs in the top drawer of his dresser; after his death a decade ago, my mother still keeps them today. They do not, however, look like Frank's collar. By the 1920s, these kinds of stylish collars were gone from pet shops and catalogs, replaced by simple flat or round leather or choke-chain collars. The next great era of fashion-conscious dog collars was not until the 1950s and 1960s, when companies created them to 'dress' the small poodles, chihuahuas, and other small dogs as part of a fad among pet owners of the era. Today, urban dog boutiques and online stores encourage dog owners (again, the owners of small dogs seem to take the lead on this) to purchase collar wardrobes for their dogs: a washable nylon collar for everyday, holiday collars ornamented with jingle bells, and bejeweled collars for other social occasions.

Care and cultivation: the ideal dog food refrigerator dish

The definition of the pet animal above emphasizes the centrality of routines of care and cultivation; artifacts from brushes to special soaps to proprietary medicines support practices such as brushing, bathing, treating disease, and feeding. Care may be done expertly or ineptly; in fact, failures of care are a primary motivation behind the activities of animal-welfare laws and agencies. The two words also suggest both old practices of husbandry and the newer interests of pet owners in shaping the behavior of pets through training, play, and environmental enrichment.

Until the mid-nineteenth century, pet owners faced challenges in the daily care of their favorites that are wholly unfamiliar today. Pet ownership was only partially commercialized, and the foods and medicines that pets needed had to be produced by family members or servants. Until the 1840s, for example, urban bird keepers had to gather their own seeds, ant eggs, and other foods to be able to keep cage birds. In that decade, 'bird stores' began to appear in larger East Coast cities, and the store stock included packaged foods and other supplies, including the first proprietary medicines. Between the 1840s and the 1920s the array of special foods, nutritional supplements, grooming equipment, and supplies, as well as over-the-counter medicines, medicated soaps, and insect powders, gradually expanded to serve aquarium fish, dogs and cats, small mammals, and even some reptiles. Many of these products were intended to be consumed, so their presence is documented by empty containers (which survive in surprising numbers) and labels, instructional and advice booklets published by the companies that made the products, advertising in a variety of forms, and 'premiums', which were objects that companies offered as incentives to loyal customers.

Here I offer an artifact biography that considers a round glass dish with a lid, made in the late 1930s, intended to be used for preserving leftover wet dog food. A 'refrigerator dish', it was offered as a premium in 1938 by Wilson & Co., Chicago meat packers and makers of Ideal Dog Food (Figure 9.2). A popular item in its day (examples appear with some regularity at flea markets and online auction sites), the Ideal refrigerator dish suggests the profound changes in foodways for both people and animals taking place in middle-class American households between the 1930s and the 1950s. Advertising associated with this dish also suggests some of the fundamental tensions in the commercial pet-food business, including the issue of discomfort on the part of American consumers as they pondered both industrialized food for themselves and the cleanliness and health of food that they themselves were unwilling to eat but were being asked to present to four-footed family members.

Figure 9.2 Ideal Dog Food refrigerator dish, *c.* 1938. Glass. Hazel-Atlas Glass Company, West Virginia, USA. Author's collection

Commercial dog food was available for sale in the United States by the 1870s, introduced by Spratt's Patent Ltd., a British company that brought its dry 'dog cakes' to the American Centennial Exposition in Philadelphia in 1876. Ordinary dog owners were slow to add it to their household shopping lists. First, it was expensive, and Spratt's sold its products mainly to well-to-do sportsmen and dog breeders. Second, anyone marketing commercial dog food had to overcome longstanding habits of household thrift. American housewives were used to cooking up their own dog food, called 'dog stew', a mixture of table scraps and starches along with scrap meats either left over from their own processing of large cuts of meat or purchased from butcher shops.

By the early twentieth century, commercial dog food began to make significant inroads into some parts of the household market. Sometimes the purchase of dog food may have reflected new concerns about healthy eating in general, as the new professional field of home economics began to publicize changing understandings of nutrition, including the discovery of 'vitamins'. Meeting these concerns, some makers of commercial health cereals for human beings produced dry dog food; for example, Potter and Wrightington, a Massachusetts health food company, began to market Old Grist Mill Dog and Puppy Bread around 1905.

Canned dog food appeared for sale for the first time in the early twentieth century. Meat had been canned for human consumption for decades, at first largely for military use. Over time, meats such as 'corned' beef began to appear on family dinner tables. By the 1910s, canned meats were big business for large-scale packers in Chicago and other cities; canning was a way to use cattle, termed 'scrubs' or 'canners', whose meat was too lean and tough for fresh sales. Corned beef and over sixty other canned meat 'dainties', from potted chicken to Vienna sausages, kept giant canning machines running, each capable of filling and sealing 22,000 one-pound cans per day. Only about one-half of a cow carcass was suitable for human food, however; and packers were creative in pursuit of avenues for using the rest of the animal.

The first canned dog food was actually processed horsemeat, cooked by small companies who had access to broken-down horses; given the American distaste for horsemeat in the human diet, dog food processing provided a potentially more lucrative use of the carcasses than did other rendering. By the 1920s, the major Chicago meat packers began to get into the canned dog food business because it too was a way to make money from ground bone, meat scraps, and 'tankage' (a liquid slurry of meat particles and water) that could not be sold for human consumption. By the 1930s, all the major packers had also gone into the dog food business. While scores of small packers also canned pet food, the large packers were aggressive marketers who regularly bought full- and half-page advertising space in popular mass-market magazines.

Although it seems counterintuitive given the deprivations many Americans experienced during the Great Depression, by the mid-1930s canned dog food had become a commonplace in city grocery and pet stores. The product offered convenience at a time when middle-class American women were increasingly interested in streamlining their household routines and had gotten into the habit of purchasing a wide variety of canned foods. Further, urban households now had different habits of purchasing and consuming meat. Housewives were much less likely to keep or process large cuts of meat or to have on hand the quantities of organ meats and scraps that once fed the family dog. Given these two changes, canned food for dogs seemed an easy choice, especially after grocery stores began to carry it (Grier 2008).

This was the same decade when American families, despite the Depression, also embraced the modern electric refrigerator. As the technology of electric-powered household refrigeration was perfected, prices dropped and installment-buying plans made the refrigerator widely available, at least to middle-class households – the same households that provided a market for dog food (Nickles 2002). When American housewives got a new refrigerator, it often came equipped with a matching set of dishes intended for storage of food leftovers. Made from ceramic or Pyrex™ glass that could be heated in the oven, refrigerator dishes not only decorated the interior of the new appliance; they were also intended to be attractive enough to serve food directly.

Introduced in the 1930s, Ideal Dog Food quickly became a success. Wilson & Co. was notable for the creativity of its marketing efforts for Ideal, which it advertised as a 'seven-course meal in a can'. A salesman's scrapbook, 'Presenting a Modern Plan to Stimulate Demand for Ideal Dog Food', called dog food 'The Big Little Giant' of the American canning industry: 'The public will spend over seventy million dollars for it in 1937' (Wilson & Co. 1936–7). The scrapbook outlined a campaign to increase demand in New York City and Long Island, including painted building walls, newspaper advertisements, and a promotion in movie theaters featuring lobby displays, trailers, live performances by Snoozer 'The Ideal Dog', and giveaways of movie tickets.

By 1938, the company had added another technique to its marketing efforts, a premium scheme that offered 'Beautiful Gifts' in exchange for Ideal Dog Food labels. The selections in the small gift catalog suggested the middle-class character of Ideal's customers; it included full sets of tableware by a well-known American pottery, towels, and small appliances such as toasters. The easiest premium to achieve, however, was the 'sanitary, convenient, practical'

Ideal refrigerator dish, which only took twelve labels and 10 cents for postage. The catalog even included a tear-off order form specifically for the refrigerator dish. Although the dish could be used for family leftovers, the intended use was the 'unfed portion of an opened can of Ideal Dog Food' (Wilson & Co. 1938). Manufactured by the Hazel-Atlas Glass Company, one of the largest manufacturers of glass tablewares and bottles, and marked with its 'HA' logo, the lidded container was decorated with the red 'Ideal' logo and Wilson & Co.'s trademark. A little over six inches in diameter, the transparent dish embodied a tension in attitudes toward commercial canned dog food: it simultaneously segregated leftover Ideal Dog Food from other leftovers in a marked dish while also rendering it visible within the white, enameled sanctum of the new electric refrigerator.

The Ideal refrigerator dish was offered as a premium for at least ten years, evolving into a square container by the time it was featured in an advertisement that ran in the *Ladies Home Journal* in 1948. It featured a drawing of a smiling woman placing an Ideal glass refrigerator dish with leftover dog food into her spotless fridge, and the text seems to have been created in response to perceived consumer concerns about the quality of canned dog food and to continuing scruples about storing it next to human food: 'Ideal is so clean and wholesome I keep it in my refrigerator.' The text assured readers that 'it's quite natural to keep a fine clean U.S. Gov't Inspected food like Ideal in the refrigerator. This is quality food. [. . .] You can tell by the pleasant, meaty aroma.' To deal with a rising threshold of disgust among middle-class consumers, the makers of canned dog food had to suggest a level of palatability to which dog owners could relate. The advertisement's list of the seven principal ingredients of Ideal reflected both an emerging science of nutrition for pet animals and the relative simplicity of recipes for all kinds of canned goods before the emergence of the modern science of food additives: meat by-products, bone, wheat and barley, wheat germ, carrots, fortified cod liver oil, and soy grits. A tiny cat's head also appeared at the bottom of the ad; many dog food companies, including Wilson & Co., still advertised that their products were equally healthy for cats. (Specialized wet cat food did not really succeed commercially until the 1950s, reflecting a lag in research on the dietary needs of felines.)

Oddly, the family dog was visually absent from this advertisement, but the text suggested that he too was entitled to special care on the part of the family food shopper, suggested by the use of the special refrigerator dish. This was also the era when American pottery companies began to make and sell brightly colored, decorative dog bowls from which pets would eat that canned dog food within the kitchen. Wilson and Co.'s Ideal Dog Food refrigerator dish and the related magazine advertisement suggest how the daily practices of pet care were being reshaped by both producers and by changing household consumption patterns. Thus practices of dog feeding reflected ongoing negotiations of the definitions of both food for people and food for pets. Specialized food containers like the Ideal Dog Food refrigerator dish made both the connections and the distinctions between the two evident. The rise of dog food distanced dogs from the family dinner table at the same time that it suggested that their dietary needs required special attention from experts, rather than the family cook.

Performance of emotions: pet photos and pet toys

Many of the surviving vintage or antique objects associated with pet keeping, as in the case of the dog collar discussed above, are preserved as mementoes, physical traces of feelings. Artifacts in this category fall into two groups. The first is consumer goods that pet owners purchase as expressions of their feelings; they may wish to give their pets pleasure, for example, by providing food 'treats' or toys, or by making pets participants in human special occasions such as the celebration of holidays, including Halloween or Christmas. Many of these goods 'describe' the role of the pet animals as being childlike, which is not to suggest that pet owners actually confuse their animals with human children. The second grouping consists of images of the pet animal, ranging from painted portraits to photographs in all their historical formats and now including video and digital imagery. Carefully preserved and displayed photographs of deceased pets are common memory-traces in American homes, but the way such images do their emotional work is somewhat different from preserved artifacts. As in the case of the collars preserved by my father, artifact mementoes are largely private or are shared with a few family members or close friends. Photographs, on the other hand, are meant to be shared – and shared they were, and are.

Because pet photography has been collected and reproduced in a number of books, this discussion focuses on how images were used and shared rather than visual analysis. While it is impossible to estimate how often pets were photographed in the second half of the nineteenth century, the survival of thousands of studio portraits suggests just how powerful the drive to commemorate these animal-human bonds could be. Owners brought dogs, a surprising number of cats, canaries in cages, and even beloved chickens to photographers' studios. Beginning in the 1880s, as amateur and casual photography became common, widespread documentation of pets at home now recorded routine daily practices such as feeding and play.

As a material performance of emotional bonds, pet photography gathers meanings through a series of actions, including the process of creating the photograph, its display, and the sharing of the photograph with family members and friends, often in albums. (Unfortunately, finding old pet photos in their original album settings is rare, since antique dealers remove them for sale as individual pictures.) Sharing photographs often took place in person, as part of social intercourse, and this is evinced in inscriptions. The image here, a *carte-de-visite* (cdv) of a small dog, is an example of a studio photograph in a format that was specifically created for sharing (Figure 9.3). Produced in quantity with a special camera lens, the cdv is literally a 'visiting' or calling card, the same size as engraved calling cards. Like calling cards, cdvs were meant to be left in card baskets and collected by recipients as a social nicety that now provides documentation of networks of social ties.

On its back, this cdv bears two blocks of text: one is the printed identification of the photographer's studio, A.P. Chillman of Philadelphia, while the other is handwritten: 'Miss Topsy's compliments to Mrs. A.P. Broadbent Jan 17 [18]78.' The inscription makes the dog into a member of a social network: it is she who leaves the card to Mrs. Broadbent, and the image on the front is a portrait of the dog that presents her as an individual without a visible owner. Studio images of pets – mostly dogs but a striking number of cats – without owners are less common than those of owner and animal together, but their numbers are sizable enough to suggest that many owners regarded their pet animals as unique individuals.

By the early 1900s, pet photos were also shared widely through the US mail in a new format, the photographic postcard. Both amateur and studio photographers developed their images onto sensitized postcard blanks and mailed them to friends or family members. The messages on these cards often include comments on the animals depicted, sharing names and other news; as in Miss Topsy's *carte-de-visite*, sometimes the pet is the 'author' of the text, embellishing the representation of the animal as a social being. Today, the act of sharing pet photographs, along with millions of homemade videos, has been altered by the rapid development of new media; in recent years, family photographs in general are given materiality only through the devices that allow their patterns of electrons to be shared, commonly on public websites.

Figure 9.3 *Carte-de-visite,* albumen print on cardboard mount, 1878. P.E. Chillman, No. 18 South Street, Philadelphia, Pennsylvania. Author's collection

Unlike photographs, the three-dimensional artifacts associated with the performance of feeling in pet keeping remain largely unstudied. Commercial pet toys, consumer goods that were specifically intended to make pets happy, are selected by owners to facilitate play between themselves and their pet, as in the case of toys made for games of 'tug' or 'fetch', or to enhance play behavior by the animal on its own, as in the varied cage toys now created for parrots. Although toys are only a small percentage of expenditures among pet owners, the timing of their purchase (they are often birthday and Christmas gifts) and the forms they take indicate the emotional position of many pet animals within families. Further, interaction with pets through toys and play reveals a widely shared vernacular theory of animal minds, present long before science rejected mechanistic interpretations of animal cognition. For example, in my experience pet owners know that their animals have preferences in toys, repeatedly turning to a bedraggled favorite, and they recognize when their pets – dogs in particular – invite interactive play using toys.

For all this, commercial pet toys are relatively new on the scene. By the second quarter of the nineteenth century, some pet animals were regarded as companions and friends; others were not children per se but were clearly seen as childlike, with loving, innocent natures that mirrored new understandings of the nature of children. Period images – prints, pictures in books, and vignettes on ceramics – depict children frolicking with dogs or playing with bunnies or kittens, and diaries and letters occasionally report such play activities as teaching tricks or involving pets in pretend play. By the late nineteenth century, amateur photography allowed documentation of such play. However, pets did not have collections of toys that 'belonged' to them. People improvised with sticks, balls made of rolled rags, or bits of yarn or string; rubber balls intended for childhood sports were sometimes commandeered for play with dogs.

Toys created and sold specifically for pets appeared in the early twentieth century. One early marketer was the Dr. A.C. Daniels Company, a proprietary veterinary medicine company. It took advantage of the traditional association of cats and catnip, which was widely planted in America as a medicinal herb, and sold both loose catnip and a hollow ball that could be filled with the stuff. An illustration of an intoxicated cat on the back cover of the company's pamphlet *The Cat*, published around 1917, shows the ball to be the 'Exerciser for Felines' patented by Walter E. Smith of Salem, Massachusetts, in 1907 (Dr. A.C. Daniels Company 1917–24, back cover). In the patent application, Smith noted that his catnip ball would provide 'amusement' and 'much needed exercise' for 'well-fed, fat, and lazy' housecats; amusing pet owners was a secondary consideration.

Around the same time, some pet toys began to take forms that humans would find amusing. The catnip mouse, a staple of modern cat ownership, was the object of a patent in 1918 by Evelyn M. Ludlam of Waltham, Massachusetts. It is possible that she was claiming a toy that had already been improvised by many others from sewing-basket odds and ends. In the 1920s, catalogs of dog supplies from the urban sporting-goods stores Abercrombie & Fitch of New York and Von Lengerke & Antoine of Chicago featured rubber balls containing bells, 'meowing' rubber chew toys shaped like the heads of cats and other animals, squeaking rubber rats and pork chops, and burlap 'rag dolls' containing 'squeaker boxes'.

While the actual shapes of dog and cat toys have no meaning for pets, they carry humorous associations for people and are shaped by a set of conventions. Well-fed housecats don't need to chase mice but we find their play with faux mice entertaining. Dogs may chase real cats and would love to get their paws on the pork chops planned for dinner; rubber substitutes materialized jokes, but only the owners got the punch lines. The same principles apply to similar toys marketed to very young human babies – and, by the late 1950s, the similarities between toys for babies and toys for pet dogs were quite striking. Hollow latex or vinyl toys, almost always containing a squeaker, were widely sold in the shapes of cartoonish little people, animals, and objects for the enjoyment of both human babies and pets, and it is sometimes difficult to figure out for whom a particular item was intended (Figure 9.4).

The Stevens 'Squeaker' dog toy, a cartoon dog wearing a Santa Claus suit, is labeled as 'durable, washable' and having 'harmless colors'; it could as easily be identified as a toy for a small child but it retains its original packaging. The *c.* 1960 catalog of Exclusive Pet Furnishings of Clearfield, Pennsylvania, devotes three pages to these toys, and the array of squeaking Santas, mice, bunnies, dogs, and vegetables with human faces is little different from rubber toys sold for babies. In American pet stores today, flexible plastic or rubber squeaky toys remain, but owners now seem to prefer comical plush animals for their pets, particularly for dogs, just as parents and relatives do for small children. Some American pets now have so many toys that pet stores sell personalized toy boxes to contain the clutter.

Figure 9.4 'Squeaker' Dog Toy, W.B. Stevens Company. Painted latex rubber in original packaging, 1966. United States. Author's collection

Containment, control, and display in public: the artifact-world of animal control

Despite the network of services, institutions, and municipal regulations that surround pet owners, the status of the pet animal is still regarded largely as a private matter, and it can be highly contingent. Pets can lose their identity by straying, or by being abandoned or relinquished by owners. The practice of community 'animal control', which cost an estimated \$2.4 billion in the US in 2007, is a result of the messy realities of widespread pet ownership (Rowan 2007). It too has spawned an array of artifacts and even a specific architecture. This material culture is largely invisible to the general public, although animal-control practices, along with the array of live traps, special vehicles, and other equipment that supports the work, have become more familiar through reality television shows.

In the eighteenth and nineteenth centuries, American cities were full of wandering animals, including owned cats and dogs that were free to come and go from their owners' households. As early as the 1820s, New York City tried to control loose city dogs by sending out a cart with men armed with clubs and crude wooden catch-poles with rope nooses, the two tools of early animal control. Early 'dog pounds' seem to have been located in precincts such as the neighborhoods of livery stables or near riverfronts. The location near water was because one common method for euthanizing appears to have been drowning; another was clubbing to death. It was a nasty business, run by poor and underpaid men, members of the urban underclasses.

When US animal welfare groups began to organize following the Civil War, they soon began to take on community animal control, arguing that stray animals deserved an opportunity to be reunited with owners, to be adopted out, or to experience a painless death. In 1869, the Women's Branch of the Pennsylvania Society for the Prevention of Cruelty to Animals decided to build what they called a 'shelter' for stray dogs and successfully lobbied the mayor of Philadelphia to take over animal control in the entire city. Their new brick building included a carbon monoxide killing chamber for adult animals (kittens and puppies were killed with an overdose of chloroform). It is unknown who designed this apparatus, but it represented a revolution in thought about the treatment of dogs and cats without owners. The entire enterprise was so interesting to the trans-Atlantic animal welfare community that the Women's Branch published diagrams of the facility and explanations of how it worked (Pennsylvania Association for the Prevention of Cruelty to Animals 1881).

The story of the evolution of animal-shelter facilities remains to be told, but several developments can be noted here. First is the development in the twentieth century of specialized equipment used in animal shelter operations, from vehicles designed for safe and climate-controlled animal transport to items like the SnugglePuppie™, a flat plush figure that can hold a bottle and provides heat and an electronic heartbeat for orphan animals (described by its makers as 'an award-winning virtual mom').

Some tools of the trade have a long history. Interviews with animal-control field officers reveal that the catch-pole (also called control pole) is still the mainstay of animal control equipment, especially for work with dogs, but it has become a sophisticated tool intended to protect animals along with controlling their movements. The magazine *Animal Sheltering*, a publication of the Humane Society of the United States, even offers training advice for proper use of the tool in *Animal Sheltering* (1996). The makers of one brand describe their poles as featuring 'non-slip grips, automatic locking mechanism, spring loaded noose, protective bite guard, quick noose release knob, vinyl coated noose, and a 360 degree kink resistant swivel head' (Tomahawk Live Trap Company 2013). Variants on the basic catch-pole reflect the changing conditions of animal-control work, including 'snake tongs' and syringe poles that permit a tranquilizer to be administered to dangerous or agitated animals.

The annual *Shelter Pages*, published by the Shelter and Community Services division of the Humane Society of the United States, collects the complex assemblage of shelter material culture into a single publication. Perusal of its pages shows how much this material culture is never intended for the public to see, and is related both in form and concept to animal laboratories and veterinary hospitals, likewise driven by overarching concerns for sanitation and segregation. Kennels, examination tables, and bathing tubs are made of stainless steel, polyvinyl chloride (PVC)–coated metal, and plastic that can withstand hot-water-and-disinfectant cleaning. However, the *Shelter Pages* directory also includes listings for architectural firms that specialize in shelter design, which suggests the second change in animal control facilities and equipment: the increasing importance of a new, friendly, public face for shelters. While this development has not been studied, it seems clear that animal shelters – or at least well-to-do, privately operated shelters – have increasingly created carefully designed public spaces that are intended to invoke the look of an imaginary pet store crossed with a nursery school. Their pleasant lobbies are decorated with pictures of past adoptees, children's art and inspiring posters, and 'store windows' of adoptable cats and kittens. The new shelters also feature individual rooms where potential adopters can meet their future pet in a private setting that facilitates match-making.

Conclusion: why study the material culture of pet keeping?

The material culture of pet keeping in America is part of a larger system of text, talk, and object through which the attitudes of Americans toward the animals they care for as companions, ornaments, sources of leisure, or emblems of social status are articulated and performed. The first part of this chapter focused on artifact biographies relating to the ownership of pet dogs. These suggest the ways that animals are brought into human society through a set of practices that express much more than the simple fact of ownership. Rather, Frank's fancy collar, Miss Topsy's *carte-de-visite*, the Ideal Dog Food refrigerator dish, and the Santa Claus squeaky toy all suggest ways that ownership involves a constant play of metaphors describing pet dogs and placing them in relation to the human members of the household. Sometimes these objects move pets into a closer relation to people; at other times they suggest difference. A similar process of generating metaphors and playing with their meanings through the routines of daily life undoubtedly occurs with other species kept as pets. More close study of the artifacts of pet keeping will suggest just how this dynamic process of relationship-building between humans and animals in the context of the private household works.

At the same time, my brief discussion of the relatively unstudied material culture of animal control and the animal shelter suggests that a different set of public meanings and metaphors are also in play. The world of the shelter demonstrates how the identity of the pet is contingent. The environment of the shelter blends equipment intended for enforcing control and care in a setting that recalls both the laboratory and the hospital behind the scenes, while its public spaces are intended to restore the lost identities of adoptable animals – or provide new ones. This chapter focuses on the United States, but I close by suggesting that research on the artifacts associated with pet keeping in other countries and cultures will offer equally compelling stories, along with opportunities for more nuanced comparative research on animal-human relationships.

References

Abercrombie & Fitch (*c.* 1920–30) *For Dogs*, New York.

Animal Pet Products Association (2013) 'Pet Industry Market Size & Ownership Statistics', available online at www.americanpetproducts.org/press_industrytrends.asp, accessed 20 March 2013.

Ciarlo, N.C., De Roas, H., Elkin, D., and Dunning, P. (2011) 'Evidence of the Use and Reuse of a Dog Collar from the Sloop of War HMS *Swift* (1770), Puerto Desado (Argentina)', *Technical Briefs in Historical Archaeology* 6: 20–7.

Collins, J. (1924) *The Story of Canned Food*, New York: E.P. Dutton.

Crisset, F. (1971) *The Story of Food*, Chicago: Rand McNally.

Deetz, J. (1977) *In Small Things Forgotten: An Archaeology of North American Life*, New York: Doubleday.

Dr. A.C. Daniels Company (*c.* 1917–24) *The Cat*, Boston.

Grier, K.C. (2006) *Pets in America: A History*, Chapel Hill: University of North Carolina Press.

—— (2008) 'Provisioning Man's Best Friend: The Early Years of the Dog Food Industry, 1870–1940', in R. Horowitz and W. Belasco (eds), *Food Chains: From Farmyard to Shopping Cart*, Philadelphia: University of Pennsylvania Press.

Heart of the Earth Animal Equipment (2013) available online at www.animal-traps.com, accessed 10 April 2013.

Humane Society of the United States (1996) 'How to Use a Control Pole', *Animal Sheltering*, available online at www.animalsheltering.org/resources/magazine/sep_oct_1996/how-to-use-a-control-pole.pdf, accessed 1 May 2013.

Lazenby, F. (1947a) 'Greek and Roman Household Pets', *Classical Journal* 44.4: 245–52.

—— (1947b) 'Greek and Roman Household Pets', *Classical Journal* 44.5: 299–307.

Medford Fancy Goods Company (1890) *Dog Collars and General Dog Furnishings*, New York: by the company.

Nickles, S. (2002) 'Preserving Women: Refrigerator Design as Social Process in the 1930s', *Technology and Culture* 43.4: 693–727.

Pennsylvania Society for Prevention of Cruelty to Animals (1881) *Twelfth Annual Report of the Women's Branch of the Pennsylvania Society for Prevention of Cruelty to Animals, for the Year Ending December 31, 1880*, Philadelphia.

Rowan, A.N. (2009) 'Animal Sheltering Trends in the U.S.: A Historical Lesson from – and for – Animal Shelters', *Humane Society of the United States*, 11 March, available online at www.humanesociety.org/animal_community/resources/timelines/animal_sheltering_trends.html, accessed 24 April 2013.

Snuggle Pet Products (2013) available online at www.snugglepetproducts.com, accessed 1 March 2013.

Tomahawk Live Trap Company (2013) available online at www.livetrap.com, accessed 1 March 2013.

United States Patent 848,136 (1907) 'Exerciser for Felines', W.E. Smith, Salem, MA, 26 March, available online at www/google.com/patents, accessed 10 April 2013.

United States Patent 1,265,926 (1918) 'Catnip Mouse', E.M. Ludlam, Waltham, MA, 14 May, available online at www/google.com/patents, accessed 10 April 2013.

Von Kengerke & Antoine (*c.* 1920–30) *Catalog of Sporting Goods*, Chicago.

Wilson & Co. (1936–7) Salesman's Scrapbook, Chicago.

—— (1938) *Beautiful Gifts for Ideal Dog Food Labels*. Chicago.

10

The adored and the abhorrent

Nationalism and feral cats in England and Australia

Adrian Franklin

The possibility of feral animals as a type emerges from a taxonomic imaginary, one with origins in eighteenth- and nineteenth-century Europe, in which a relatively fluid, promiscuous, and interconnected natural order gained a greater ontological sense of separable and fixed categories. Through the social construction of distinct, unvarying categories of the natural order, anomalous and improper categories became possible and 'of interest'. Feral animals are anomalous because of their very nature as outsiders; they serve as warnings, highlight or stand for wider aspects of social anxiety and danger. They are often accounted for as active and disordering entities, whose presence as an external agent is all the more dangerous because it is unbounded by the natural order of things (rules, norms, legitimate categories) that pertain to the world. Such animals gain their potency as sources of danger because they threaten stable and proper categories of the social world. They give the appearance of a natural foundation, legitimacy, and solidity to what are always arbitrary and contestable social norms.

The source of distrust, where it exists, can be attributed to changing taxonomies and the rising power of science to change the ontological relationships between humanity and nature. As the environmental social anthropologist Roy Ellen put it:

> Effectively all landscapes with which humans routinely interact are therefore cultural: and our environment is every bit as much what is made socially as what is not. [. . .] How strange, then, that in another version of the biological imagination (that of classical evolutionary taxonomy) domesticated animals and non-endemics are, somehow, *not* the real thing. The complexities of biological reality, enhanced by the insights of modern ecology and genetics, make drawing the boundary between what is cultural and what is natural, almost impossible.
>
> *(1996: 14–15)*

As a label for animals, 'feral' is strange indeed. It belongs to the social practice of taxonomy and boundary making (and maintaining) far more than the natural practice of animals, yet it is justified on natural grounds and signifies the legislative power of science (Bauman 1989). Moreover, while the nature of nature is predicated on fluidity, evolution, instability, and turbulence, the concept of ferality is of interest because its force as a bad thing in nature stems exactly from the notion of a stable, unchanging, and settled nature – with legitimate community members having precedence over outsiders. Although feral animals are often 'wrong' where they are deemed to be a contaminate through the 'unnatural' work of humanity in the landscape, paradoxically they then call into practice yet more intervention from humanity, to the point where in places nature can no longer be trusted to manage itself independently and even more paradoxically renders all of nature effectively domesticated (though its social form of domestication might be called *nationalization*).

In this way, when used as a classificatory concept, feral animals masquerade in modern Western societies as *scientifically* constituted because they are constituted through the *natural* sciences, gaining their meaning, credibility, and significance from overarching meta-narratives such as ecology, environment, or ecosystem. Their presence can threaten to destabilize or destroy a proper natural order, demanding in some places that a course of purification or control be enacted. I will argue that, as in the case of so many other anomalous animals, this action is ritual, with the hope that social order and control be restored/maintained. However, feral animals are not always anomalous; their mobility and the cultural mixture they represent can also be seen as a social ideal, in those societies where these are valued.

The claim that feral animals are designated through purely objective, scientific procedures is flawed. 'Native animals', which they oppose in a taxonomic sense, are constituted by the social and political processes of nation-formation that simultaneously enrols some animals as social insiders and others as social outsiders. Nation-formation is an arbitrary and often contested division of global territory and gives rise to those powerful social sentiments, *nationalism*, that typically anchor social solidarity in new national territorializations as its natural estate. As Ernest Gellner (1983) reminds us, nation-formation precedes nationalism, not the other way around: nations are not given in the scheme of things, but are assembled through political will, often as assemblages of many unrelated parts (territories, ethnicities, cultures), often in tension, since they typically consolidate formerly hostile elements. They also group together very different ecologies along with animal and plant communities, and give them a unity that cannot be made on 'natural' grounds alone. It is from this simultaneous connection with social inclusion and social exclusion and its simultaneous mapping onto new natural territorializations that the sense of feral animals 'being in the wrong place' and 'out of control' is made.

Precisely because animal ferality has very powerful social and political foundations it is thus also highly variable in the way it is specified and acted upon in different nations. As a concept it defies a universally applicable natural-science foundation precisely because it is constituted socially, within very different and complex histories of nation and empire. Feral animals are created mostly by scientists coming under the influence of different and even opposed national interests and configurations. This instability, variability, social composition, and impact on environmental policy make it an important subject for human-animal studies. In a section below I will illustrate these points in a comparative analysis of the place of feral animals in the UK and Australia. While not exhausting the range of possible iterations of its social construction, these case studies do illustrate two very important dimensions, as between non-colonial and colonial (not to mention post-colonial) settings. Both types of modern societies have introduced animals from outside but their resulting designation, status, and treatment is symptomatic of their respective social structure, rather than the behaviour of the animals in their respective places.

The word 'feral'

The social history and temporalities of words, as Raymond Williams (1983) repeatedly showed, reveal a great deal about their conceptual relevance, their representational traction, and their instability. It is significant that the word 'feral' was coined in the early seventeenth century when, in England at least, people began to sense a gulf opening up between the 'taskscapes' of the medieval world, when man and the natural world were *inextricably intertwined*, and the modern world, when humanity became increasingly concentrated in the largely human worlds of the city, although this separation was nowhere near as complete as was often supposed (Ingold 1993; see also Thomas 1983; Franklin 2002).

The modernizing world was intellectually antithetical to promiscuous mixing and the interpenetration of nature and culture. It was based on a taxonomic imaginary, distilling difference from contamination, enacting purifications, separations, and change where nature ultimately became something separable from humanity in *both* conceptual and lived terms (Latour 1993; Ritvo 2012). Its conceptual separation made possible its use as a new *aesthetic* object, as a new moral object, as a socially differentiating object representing profound social change and, critically, as a *national* object.

Wildness or ferality then became a defining, absolute measure of this separate, wild nature. As Harriet Ritvo (2012) shows, taxonomists began to distinguish between the domestic and the wild, and consequently to remove domestic animals from inclusion in a newly emerging preference for wild, *native* animals. The earliest sense of feral as a synonym for wild did not refer to domestic animals that had escaped domestication and established populations outside human control and/or beyond human cultivated lands. In England it was a word belonging to the expanding interest in, and aestheticization of, nature as 'countryside', a lived world of countryside retreats and of returning to live in and experience the country during the seventeenth century. Wildness and the feral was a conceptual space that opened up with the decline of England as a rural culture.

As Trevelyan (1946) pointed out, this had to do with the extraordinary spatial reversal in the human life world of England. In 1700 that human world was based solidly (75 per cent) in the countryside whereas a mere 15 per cent of the population lived in towns. In 1700 towns had all the allure and pulling power of a cultural centre of gravity, creativity, wealth, trade, and vibrancy. While cities were an extraordinarily *vital* space and perhaps the most exalted form of life on earth, places like rural England were widely considered to be brutish and ugly. The earliest use of the term 'feral' in the early seventeenth century referenced this sense of savagery, but we can detect a change and a softening by the mid-century, where it identified the untamed, the uncultivated, and the wild in a more positive way. In the mid-seventeenth century, for example, 'wildernesses' were deliberately planted and designed as garden features, and their popularity spread rapidly so that even towns, through their numerous parks and gardens, shared in the aesthetic of the feral. So did the menagerie, as yet another 'room' in significant high-status gardens. The movement of exotic wild animals into these princely collections was a very prominent beginning to an accelerated age of animal transfers that culminated in the acclimatization societies of the nineteenth century.

By 1800, 25 per cent of the population had shifted to the town as trade and industry grew rapidly, and by 1851 the majority had become town dwellers. As cities grew distinct from the countryside around them, their presence grew ever more pronounced, but not in a positive way. By 1800 many writers had begun the fashion of making unfavourable comparisons between town and country, about their coal-fouled air, their filthy streets, their overcrowded and ugly poor, their despicable immoralities and debauchery.

The arrival of ferality as an aestheticized concept was also related to shifts in the distribution of power that accompanied demographic change. A richer, more powerful landowning class consolidated its grip on the English countryside and, as they emptied their estates of their tenants and appropriated the common lands, they created a palpable sense of emptiness that attached itself to these former human lifeworlds. Romantically dubbed 'wild', they were nonetheless crafted and included in the highly socialized spaces of sport and leisure. This was epitomized by the arrival of foxhunting in the eighteenth century, a sport predicated on the power of the local landowning class to roam across country unfettered by property boundaries or historic rights (Franklin 1996a). Ferality in this sense was especially associated with animals because it was associated with a force, agency, or push that could be felt and admired by humanity.

So although the wild or feral opened up a space beyond the human world, almost as soon as it emerged it became sacralized, obtaining special powers of purity, belonging, morality, and propriety precisely because it was unsullied or flawed at the hands of humanity. In turn, such a space was appropriated by social elites for its therapeutic and redemptive properties, where immersion in its airs, waters, woodland, and hills would purify, cleanse, and renew its pilgrim visitors. Despite its undisputable co-evolution with humanity, its sacralization also involved associating it with a putative primeval era, before mankind, from where it could be seen as the wellspring of particular human groups, destined to become nations. The wild and the feral therefore was the perfect religious object for nationalism in Durkhemian terms, an object without which it might be difficult to conceive of large, abstract social concepts such as peoples, folks, or nations.

As a term largely used for non-native domesticated or captive animals that have established wild populations in new places, feral was first used in the mid-nineteenth century. The availability of two indistinguishable terms for wild was useful for a new category of animal that was being created through the convergence of acclimatization movements and the spread of nationalism, particularly in newly colonized territories, but also in Europe where reverse flows were substantial and widespread (Rolls 1969; Anderson 1983; Gellner 1983; Gruffud 2011). From then on, 'feral' wild was reserved to set against an increasingly aestheticized and sentimentalized notion of 'native' wild, but this in itself did not determine animals' popularity, status, or meaning.

The sacred and despised: the social constructions of feral cats

On 16 September 2012, I googled the term 'feral cat', first with the addition of 'UK' and then with the addition of 'Australia'. The domestic cat (*Felis catus*) is a non-native species to both countries, but as an animal designated feral they are treated in inexplicably different ways in the two locations. In the British case their ferality is a cause for concern, prompting almost universal action for their welfare and care. In Australia, it is the opposite: it prompts action for their extermination and control. In Australia the feral cat is largely disliked because of its assumed adverse impact on native species. A similar case might just as easily be made in Britain, but isn't.

I am going to suggest that the cat has become a useful anomaly to those who want to uphold a state of anxiety about belonging and not belonging in Australia. The feral cat has been a very useful animal for those who seek to maintain the borders around Australian society against unwanted migrants and asylum seekers, or, at least, those who seek to define and purify a sense of that boundary. As anomalous, dangerous animals, feral cats look perilously like metaphors for the universally unwanted asylum seekers – creatures that cross borders of their own volition, becoming independent, outsider figures capable of threatening a properly Australian social order. As 'domestics', such creatures are supposed to live among us but can, and do, abandon us once 'here' in order to establish independent communities. They threaten to fragment that fragile and barely formed reality: 'Australia'. So they are despised and hated, and have become subject to 'species cleansing' (Franklin 2011).

In order to work as an animal that is good to think with, the cat has to slide in and out across the boundaries of nature and culture to offer a sense of certainty, grounded in the veracity of nature, for what is actually a fragile, contestable, and arbitrary social entity. The feral may gather these anxieties together and offer a means for their articulation that could not be expressed in other (human, political) ways.

The total UK cat population is in excess of 10 million; about 1 million of those are feral cats. In Australia the feral population is estimated to be between 14 and 18 million but given its greater land area the population of feral cats in the UK is actually substantially denser than it is in Australia (4.1 per square kilometre as compared with between 1.8 and 2.4 per square kilometre). On the surface then, their potential impact on native animals would seem more substantial but there is practically no scientific effort put into their eradication and control nor demand for it, despite very substantial public support for all native animal categories at risk. By contrast, in Australia anti-cat sentiments are widespread and substantial, from the political leadership through to most levels of civil society (Smith 1999). Indeed, there is evidence which supports the idea that feral cats are widely considered to be 'unAustralian' (Franklin 2011). Between the British and the Australian usages then, much might be learned about the vilification of cats in Australia and the political work it performs there.

Cats and Britain

The cat was probably introduced as a high-status animal to mainland northern Europe by the Romans some 2500 years ago, and the species' earliest presence as a domestic animal in Britain was during the second half of the first millennium BC (Davis 1987: 182). Although the cat has lived in the UK for over 2000 years there are good reasons why it might not be tolerated. Contemporary British culture is extremely focused on the care and welfare of its native animals, and there are few other places where birds in general have become such objects of aesthetic veneration and national pride. Many species are threatened and, although their decline is often the result of anthropogenic disturbance of their habitats, the large numbers of feral cats could provoke zealous calls for control if not extermination. This would seem to be the case particularly in respect of native birds, one of the main types of animal that provokes 'concern' about feral cats in Australia.

And yet, it is difficult to find any evidence that Australians share the same degree of enthusiasm (or obsession) for birds as the British. In Britain there is one bird species for every 376 square kilometres compared with one for every 6348 square kilometres in Australia. Two-thirds of all British households claim to feed birds, and a recent estimate of the number of active birdwatchers put it at 3 million (Unwin 2005). Over 1 million Britons are now members of the Royal Society for the Protection of Birds (RSPB), paying a minimum of £36 per year for the privilege. It is the twelfth-largest charity in the UK.

Britain's many small mammals such as shrews, moles, dormice, and harvest mice also invoke great concern, and they too are potentially threatened by feral cats. According to the report by Woods *et al.* (2003), mammals made up 69 per cent of the animals cats killed, birds 24 per cent, amphibians 4 per cent, and reptiles 1 per cent. More than 44 species of wild bird, 20 species of wild mammal, 3 species of reptile, and 3 species of amphibian were recorded. Summing up the research on cat predation in the UK, Woods *et al.* (2003) argue that cats could have a serious impact on native animals where their population densities are high.

Their research shows that only 10 per cent of domestic cats surveyed do not bring dead animals back to their homes, suggesting that a very high proportion of householders are aware of cat predation on native species. Nonetheless, as the authors acknowledge, anxiety about feral cat welfare is far more significant than anxieties about cat predation on native species (Woods *et al.* 2003: 176). It is extremely difficult to find evidence for significant levels of condemnation from either science or civil society. According to the RSPB, there is no necessary relationship between high levels of predation and species decline:

> Although large numbers of birds are killed, there is no scientific evidence that predation by cats in gardens is having any impact on bird populations UK-wide. This may be surprising, but many millions of birds die naturally every year, mainly through starvation, disease, or other forms of predation. There is evidence that cats tend to take weak or sickly birds.
> *(2012)*

It seems that the science of feral animals is being transmitted to the general public through bird welfare organizations so that the cat is judged on its actions and impact rather than its feared potential.

It is instructive that the first ten hits on a Google search using the key words 'feral cats UK' showed that welfare *for* feral cats was indeed the dominant theme (and repeated in the case of the subsequent ten). Eight out of the first ten hits were sites promoting welfare and care for feral cats. These were mostly sophisticated organizations capable of intervention and action across the country, and although most were charitable organizations and made calls for donations, it was clear that they had significant funding and were therefore well supported by a popular social base. These organizations were mostly those *specifically* set up to provide/promote welfare and care for feral cats, such as Feral Cat Welfare (a national organization to promote the welfare and well-being of feral cats), the RSPCA, Animal Trust, Catchat (helps cats move out of shelters and into homes), Cat Action Trust (an anti-kill welfare/help organization for feral cats), Wrexham Council (advice on how to help feral cats), and Messybeasts (also an advocate of no-kill methods of welfare and control through the use of trap-neuter-release schemes). Only two did not have welfare as a primary aim. The Scottish Wildcat Association is concerned that full-blood wild cats are dwindling owing to interbreeding with domestic and feral cats, and they aim to create a 7000-square-mile reserve for them free from all except *Felis sylvestris grampia*. The Game and Wildlife Protection League provides advice on controlling feral cats for its landowning members who sell shooting to paying clients. Critically, it reminds visitors to its site that feral cats are a protected species, who enjoy exactly the same welfare legislation provisions as those kept as pets. This is not the case in Australia; rather, it is the opposite.

Feral cats in Australia

Cats were introduced into Australia shortly after first settlement in 1788 when they accompanied the slow expansion of farming along the coasts and deep into the rich farming areas of the interior. Their role was to contain small mammal damage, largely from introduced mice and rats, and no doubt many left the confines of farms and settled areas when mouse, rat, and rabbit numbers reached plague proportions. These spontaneous, informal introductions were augmented by planned releases outside of the settled areas. Cats were released in the goldfields to control mice in the mid- and late nineteenth century and they were widely released to control plagues of rabbits and native rats in the nineteenth and twentieth centuries (Dickman 1996: 1). In contemporary Australia there are now approximately 3 million domestic cats and estimates for the feral cat population vary between 14 million (New South Wales Government 2011) and 18 million (McLeod 2004).

The first ten items revealed from a Google search using the term 'feral cats Australia' were all concerned with three main themes: their negative impact on Australia, their control, and their eradication. Negative impacts were multiple: environmental (against native animals, biodiversity), social (for example, against indigenous people's cultural values), health (as possible transmitters of diseases such as rabies), and agricultural (in the transmission of diseases affecting livestock). Indeed these themes characterized the first 100 items found. The top 20 sites consisted of approximately 50 per cent governmental organizations, 20 per cent research organization sites, 20 per cent media sites, and 10 per cent information sites. Further down the list were many voluntary organizations specializing in environmental work, hunting, farming, and pest control. In the top 50 sites there were only 3 that were not negative.

This suggests that the case against feral cats is being made by the most powerful and influential organizations in Australia. It is very difficult to find any other view being expressed, and since scientists and scientific organizations are endorsing the governmental case, they tend to be the ones most cited by the most powerful media organizations, which in turn find their way into educational and other sources of information on the environment.

On 24 October 2002, ABC TV broadcast the documentary film *Ten Million Wildcats* (Wild Visuals/Discovery 2000). According to Tim Flannery – a previously consistent supporter of eradication and former Environmentalist of the Year – and other biologists/ecologists featured in the film, environmentalists have wrongly vilified the feral cat in Australia; as it turns out, its diet consists mainly of reptiles, which exist in such numbers that their future is assured, and introduced mammals (mice, rats, and rabbits). According to Flannery, only on a few islands and coastal places has the cat made a serious impact on native wildlife, and, at the other extreme, the island of Tasmania has not lost a single native animal due to cats despite their presence there for over 200 years. A Tasmanian study concluded that 'there is little evidence that cats in mainland Tasmania are having a significant negative impact on the native fauna' (Schwartz 1995: 59). And for the mainland, as a whole, Flannery argued that 'there is no scientific evidence to say that feral cats were solely responsible for the loss of any native species', and indeed there are studies that demonstrate this very clearly (Jones and Coman 1982: 537–47).

But *belief* in the serious environmental damage done by cats seems completely unimpeded by the lack of firm scientific evidence for it. Data neither impedes scientific inference nor political demands for eradication policies. So, for example, in his definitive *Overview of the Impacts of Feral Cats on Native Fauna*, Chris Dickman argues that

> acceptable evidence for impact would be any demonstration that cats have caused a decline of 25 per cent or more in the population abundance or geographical distribution of any native species. *Unfortunately, unambiguous evidence of this kind does not exist.*
>
> *(1996: 15, emphasis added)*

Why does he add the 'unfortunately'? Nor is it clear why the compelling test case of Tasmania does not count as 'unambiguous'. It may well be that the cat has in fact done more damage than science has managed to document, but, even if that is the case, it is hard to understand why such a strong response is made without *more* compelling evidence.

According to Nick Smith,

> This eco-nationalist vision of a feral-free Australia is shared by many, including Alan Newsome, formerly of the Commonwealth Scientific and Industrial Research Organisation's (CSIRO) Division of Wildlife and Ecology [. . .] [I]n order to regenerate native biota, he advocates the removal of all feral species.
>
> *(2011: 5)*

He also gives examples of

> the way environmentalist narratives utilise the opposition between feral and native nature as a site of projective identification. In order to be imaginable, ideas of nature must be represented and given cultural form in narrative (amongst others). Native nature, (or what Australians colloquially refer to as the 'bush'), is often depicted as instrumental in shaping the national character.
>
> *(2011: 6)*

All designated serious pests have been declared Key Threatening Processes and this designation in turn requires the writing of a Threat Abatement Plan (TAP). A TAP exists for the feral cat, yet the plan states, 'Convincing evidence that feral cats exert a significant effect on native wildlife on the mainland, or in Tasmania, is scarce. [. . .] There is no evidence of feral cats causing extinctions in mainland Australia or Tasmania' (Department of the Environment and Heritage 2013).

On governmental and scientific websites (and among other sources influenced by them), the feral cat is invariably shown in emotive photos, with a colourful native bird kill rather than its more typical prey, again the introduced mouse, rabbit, or rat. It is almost always implicated in native animal extinctions. It is even considered a 'hypothetical' problem as a vector of zoonoses, should rabies be accidentally introduced to Australia. So the cat appears to have acquired a negative image for reasons other than its actual behaviour and impact on the Australian environment. Since its 'problematic' origins cannot be traced to incontrovertible scientific evidence that justifies such an extensive policy instrument, it is conceivable it originates from other social, cultural, and political sources, and there is growing evidence for this.

Nick Smith (1999) cites numerous examples of Australian politicians making use of anti-cat sentiment in Australia. A Liberal MP, Richard Evans, made a call in 1996 for the eradication of all feral cats by 2020 and the introduction of native pet species. Here is a relatively common example of the simple preference for 'properly Australian' animals and open hostility to alien species. In 2004, Trish Worth, another Liberal MP, linked the need for quarantine controls on cats to controls on asylum seekers (Seven News Network 2004; Franklin 2006). This is an example of animals serving to reinforce the preferred notion of Australia as a 'closed nation' or something requiring protection from outsiders, 'a land where others are no longer welcome' (Smith 2000). Australia is a small population occupying a huge land mass: this generates a nervous tension, the idea that it cannot enact proper surveillance and it might be overrun. This anxiety is very clearly established in both of the major parties with leaders competing very publicly to be seen as the toughest and least open to asylum seekers

Anti-cat sentiments of this kind were very clearly expressed in the 1998 Constitutional Convention, a preliminary national conversation in the lead-up to the referendum for an Australian republic. In this, Janet Holmes à Court's speech placed anti-cat emotions alongside the most sentimental of national attachments:

> We need the smell of eucalyptus in this and the feel of red dust. We need to have the feel of swimming in the sea and all those things that make us feel so passionate about this country and love it so much – eating beef and no feral cats.
>
> *(quoted in Smith 2000: 120)*

According to Philip Smith and Tim Philips, 'What has emerged clearly over the past 20 years is growing symbolic potency of the "*Un-Australian*" in the vocabulary of public life in contemporary Australian society' (2001: 325). It is not hard to find the idea of Un-Australian being attached to feral animals. For example, Smith gives the example of *Charlie Carp*, a liquid fertilizer made from the introduced European carp:

> The leaflet that accompanied my bottle reads: 'There's only one good European carp and that's a dead European carp.' Carnivorous carp are described as the enemy of Australian fish, 'the mighty Murray cod', yellow belly and catfish. They eat 'all that is good.' Carp are 'UN-Australian.' The leaflet also notes that the continued use of carp fertilizer guarantees the ongoing eradication of alien carp. The redemptive tone of this strategy is marked on the pamphlet by an image of a carp suspended by a noose through its gills, headed by the phrase: 'A good fish dead.'
>
> *(2011: 12)*

To be properly Australian one should not only want to protect native species but also, as migrants, one would relinquish any sentiments or sympathies for alien animals from other countries of origin. According to Professor Rob Morrison, for example, who is also Chair of the Anti-Rabbit Research Foundation:

> These un-Australian sympathies [for rabbits, cats, etc.] seem to be rooted in our European background and Americanised culture. But [. . .] native species deserve our support not for arbitrary reasons but for empirical reasons, independent of human benefits.
>
> *(Morrison 1996)*

This quote is surely instructive: it suggests that those with European backgrounds (most are British) are understandably sympathetic 'to their own' animals (ironically though the rabbit and the cat were both introduced species to Britain), yet Australians support their own native wildlife not for such sentiments but on the hard empirical evidence of their status as threatened by introduced animals. Nick Smith disagrees: 'for many conservationists (and people who would not think of themselves as such), getting rid of feral biota (and reintroducing native ones) is a way of making the country *and themselves* more Australian' (Smith 2011: 7).

In Tasmania, where native animals have lived alongside cats for 200 years remarkably well and without loss – where it might actually be expected that anti-cat sentiments were softened by experience – in fact the opposite seems to apply. In 2003, Tasmanian Greens Member of the Hobart Assembly, Nick McKim, advised that

> the government is to take urgent action to control feral cats and flagged the introduction of a two-step strategy to firstly introduce a system of cat registration and de-sexing to control recruitment from the domestic to feral population, and secondly an intensive eradication programme for feral cats.
>
> *(Office of the Tasmanian Greens 2003)*

Nationalism and animals in Britain and Australia

> Introduced aliens were often appreciated or deplored in the same terms that were applied to human migrants.
>
> Harriet Ritvo, 'Going Forth and Multiplying'

Different societies such as Australia and Britain illustrate how processes and timescales of nationhood can affect the life chances and social signification of feral animals. Britain's long national history, characterized by 'unifications' and 'additions', generated a generally eclectic and cosmopolitan sense of national identity. Introduced and unmonitored wild populations established themselves both before and after initial nation-formation in the tenth century under King Athelstan, and they were assimilated into the native animal community alongside the long list of migrants, traders, and invading humans who brought them (Davis 1987; Wood 2000). The dormouse and the brown hare were introduced during the Roman period, for example, while the rabbit and fallow deer were introduced by the Normans. Romans, Normans, and all of their introduced animals are all naturalized British.

Later animal introductions came from Britain's colonial territories and were in many ways prized as trophies and treasures. This largely positive attitude towards the exotic continues to influence their designation and treatment, often to the detriment of British nature. After all, as Ritvo argues, 'There is a difference between the imposition of the European biota on the rest of the world and the transfer of exotic animals to the homeland' (2012: 8). The American grey squirrel is held responsible for the decline of beech woodland in southern England; it is widely considered a nuisance and a pest; and it is (contentiously) blamed for the decline of the red squirrel. But little is actually done about it. It has not entered popular culture as a completely vilified animal. For example, aesthetically it is widely appreciated and enjoyed. As the Young Persons's Trust for the Environment (YPTE) puts it, 'the grey squirrel is a pretty, appealing and entertaining little animal' (YPTE 2013). Even the Forestry Commission, who have identified it as a major pest to tree growers, recognize that it is nonetheless a naturalized species:

> But is eradication a feasible, let alone desirable, option? The Working Group on the Government Review of Non-Native Species Policy concluded that the grey squirrel will remain, for the foreseeable future part of the fauna of England. Grey Squirrels are enjoyed by many as part of our wildlife.
>
> *(quoted in YPTE 2013)*

From the late nineteenth century on, the numbers of zoos and private menagerie collections proliferated, and, inevitably, many animals escaped. The Australian red-necked wallaby, sika deer, grey squirrel, and Reeves muntjac deer all formed viable colonies in the early twentieth-century UK with very little social comment and certainly not a concerted effort to remove or vilify them. They have become attractive to seek out and experience; all are popular with children and if anything they have blurred the distinction between the wildlife park and the countryside (Crawford 1999).

Coypu escaped fur farms in the 1930s and mink followed in the 1950s. Ring-necked parakeets escaped aviaries and established several very popular colonies in the late nineteenth century. The new colonies have been given adoptive names: the Kingston Parakeet, the Twickenham Parakeet, and the London Parakeet.

Recently, the UK has been gripped by big cat fever, a constant series of sightings and debates about the veracity or otherwise of breeding populations of exotic big cats: lynx, leopard, panther, puma, and others. It seems that there is some truth in these claims, not least because individual animals have been killed, found dead, or caught on infrared deer-census cameras. There are also highly credible stories of releases of captive animals following new legislation in 1976 that raised the cost of keeping them in captivity. Their presence is generally seen as exciting, welcome, and fascinating (Buller 2009). Such animals could pose a very significant danger to many of Britain's larger native animals, but there is practically no anxiety expressed about this, rather the opposite. Buller argues that their popularity expresses a yearning for the overly tame British countryside to become properly wild again (2009). Hurn's (2009) work on credible sightings of leopards in Wales is also a story of their being welcomed into the landscape.

The history of the social signification of introduced feral animals in Britain offers a way to understand attitudes to feral cats. There is only a very weak differentiation between categories of native and introduced animals; introduced animals that can live as wild animals tend to be freely naturalized and adopted; introduced animals are routinely tolerated and mostly held up as special and attractive. Critically, introduced animals enjoy the same rights and protection extended to domestic animals. These relate to Britain's relatively early transition into nationhood, its nation-formation arising from the unification of opposing elements, and its subsequent history of adding yet further elements through colonization. For a long time the British have expressed clear connections to, admiration of, and a longstanding desire to recreate Rome's eclectic and inclusive form of 'civilization'. Subsequently Britain has a tendency towards the toleration of difference, an interest in difference and its social articulation, and a strong tradition of civil liberties and freedom. In many ways British attitudes to feral cats therefore embody these values. Their liberty and freedom, their being at large in the country among native animals does not offend the British sense of identity and social solidarity because the cats' standing is more determined by their individual rights as animals. They are thus seen instead as 'at risk' themselves because they cannot enjoy the same access to care that they would enjoy as domestic cats, and the feeling is that they have typically been abandoned rather than chosen to live the much harder life as a vagrant animal in an unforgiving climate. They therefore evoke pity and charity.

Nationalism and animals in Australia

In Australia, the opposite case can be demonstrated. In 1788 Australia was settled by Britain through a series of colonial extensions of her expanding empire. By the mid-nineteenth century, a free settler society was established but their association with Australian nature, particularly its animals, was *unsettling*. There were four reasons why. First, the animals were anomalous to the zoological taxonomy of the world known to the British in the late eighteenth century. They were not just distinguishable from all known categories; they seemed preposterously different and of a different order from them. Hence they acquired the notions of 'failed creation' or an 'inferior nature'. Second, they appeared to lack utilitarian value, with none suitable for domestication, farming, or hunting. Third, they fell short on aesthetic grounds and particularly the birds for their absence of pleasing song. And fourth, they were encountered in such unprecedented abundance that in most areas suitable for farming they rendered agriculture almost impossible, and certainly extremely stressful.

The first two reasons produced a generally low regard for native animals while the latter two prompted very new forms of relationships with wildlife. The settlers simultaneously initiated very drastic attacks on indigenous species while calling for their replacement by British and other superior foreign species. Between 1852 and 1894 every colony (now states) formed its own Acclimatization Society and in all cases they were the most prestigious institutions comprising the governing class, landowners, scientists, and the educated classes.

In 1853 acclimatization was still a new idea, the brainchild of French anatomist Isidore Geoffroy Saint-Hilaire, and an idea that was simple enough: 'The prospect was nothing less than to people our fields, our forests and our rivers with new guests; to increase and vary our food resources, and create other economical or additional products' (quoted in Franklin 2011:

76). Their subsequent acclimatizations were on an epic scale and momentous (Rolls 1969). From the introduction of around 130 animal species, at least 81 (25 mammals, 20 birds, 4 reptiles, 31 fish, and 1 amphibian) have established wild populations (Hart and Bomford 2006; Australian Wildlife Conservancy 2012). This is only exceeded by New Zealand, where 87 were successfully acclimatized (32 mammals, 33 birds, 18 fish, 3 amphibians, and 1 reptile). Some, such as the rabbit, very nearly destroyed Australian agriculture in many areas.

As it proceeded alongside other intensifying developments in agriculture, forestry, and mining, the scale of change and loss of familiar nature saddened an emerging generation of Australian-born artists, writers, and academics who became increasingly vocal and supportive of indigenous animals and their habitats. Acclimatization Societies gradually lost their appeal: in Victoria they stopped importing new animals in 1870; in New South Wales they turned to building zoological collections; and by 1890 the Tasmanian Society had switched their energies to protecting indigenous species.

In 1901 Australia became a nation. Up until then the island continent was a series of separate colonies of the UK. Much of their independence and autonomy was preserved in the new federal model, and separation from Britain was the defining feature of the birth of an Australian nation. However, it did create an entity, social and territorial, that was called Australia for the first time, and, through citizenship, it created a sense of belonging to the Australian landscape, whereas before, for many, loyalty had been to a land 11,000 miles away. Notwithstanding an increasing sense of nativism from the Australian-born, it was not immediately apparent to anyone what Australia might and should be, and something of an identity crisis characterized a time when they divested themselves of the certainty of Britishness and mantled themselves instead with the uncertainty of an unfolding Australianness.

The act of forming an Australian territory also created a new category of animal that had not existed before. Until Australia came into focus, there were no Australian animals, only animals indigenous to the localities in which they were found, and on an island continent as large as Australia these localities were many, varied, and separated by very significant distances. The act of nation-formation and the creation of a single territorial entity created a category of animals, *native* animals, that were united by virtue of their inclusion within its boundaries. They became simultaneously 'Australian native animals', but the nature of their political creation as a category was always masked by their apparently 'natural' designation or differentiae. This is how the foundational basis of many nations seems to slide from acts of political will and construction of arbitrary territorial amalgamations to the possession of natural antecedence and integrity. The case made against the cat, that it does not belong and did not evolve in Australia, is coherent enough. However, the same can be said of most native animals too, since most were not pan-Australian species but very localized. The science of feral animals recognizes its largely social basis when it admits that 'there are few socially acceptable control techniques to reduce native animal damage' (Hart and Bomford 2006: 9).

The timeline for Australian natives has been set, tellingly, at 1788, when Europeans first settled Australia. It is of course an arbitrary line, which demonstrates the force of nationalism in the taxonomic construction of ferality. The dingo, for example, was introduced from Asia a long time before then and is thus designated native. The large number of introduced species that were rapidly introduced by 'the British' immediately before Federation fared less well.

Equally, some introduced feral animals such as the brown trout and various deer species are strongly associated with Australian social elites and despite being very environmentally invasive, as in the case of the trout, and damaging, as in the case of the deer, they seem to escape environmental persecution (Franklin 2011). Again, it seems that ecology is used inconsistently, a smokescreen for self-evidently social processes.

Tellingly perhaps, among most Aboriginal groups the same antipathy to introduced animals is lacking (Rose 1995; Trigger 2008). During much of the colonial and post-colonial period Aboriginal people were effectively displaced from almost all areas of Australia, and in fact were not 'Australians' either before Federation or after (they were not admitted as citizens until 1967). Hence it is perfectly logical that the concept of native animals and thus feral animals was meaningless to them. More meaningful to them was the concept of *country*, the nutritious land they occupied and looked after. It seems likely that feral species appeared in these gradually over such long periods of time that they became accepted, welcome, and hunted species. The cat is hunted and considered a special food among some groups, in others the cat has its own *Dreaming*; in other words, has become sacred (Trigger 2008).

Conclusion

It is possible that almost any of the introduced feral animals could have become Australia's *bête noire*, so why was the feral cat singled out for special treatment, meaning largely hated, even more than the cane toad? I think there are three main reasons.

First, there is something about the feral cat's 'social character', as apprehended by Australians, that makes it seem *treacherous*. As a recipient and repository of familial love and affection, the cat becomes a domestic insider. Yet that apparent social bond of valued interdependence is illusory because cats can and do choose to leave in order to live independently of humans. They are therefore apt creatures to represent a very feared social figure: a person whose loyalties are uncertain and unproven, one who might be let in but not join in; one who enacts division and fragmentation; and one who therefore questions and challenges the idea of a beleaguered Australian culture, with its monolithic values always potentially undermined by the arrival of others.

Second, the feral cat may be singled out because of its unique physical apprehension. Of all the introduced animals capable of living wild (with the exception of the cane toad), the cat is singular in that it lives and hunts within and around human settlements, particularly because of the omnipresence of mice and rats. Feral cats are very visible and noted predators, not so obviously of nocturnal pest creatures as of native birds, who are active during daylight hours and present in elevated numbers owing to the human habit of planting and watering extensive areas of native flowering plants. In the absence of native carnivores, banished by humans from their towns and cities, the cat stands out as the main killer of the natural jewels of city gardens and parks.

Third, the cat may also be the victim of vaguely remembered but longstanding historical beliefs about their association with dark, magical, and pagan forces (see Holmberg in this volume).

The national myth of Australia began with a small population of mostly British origin occupying a vast landmass close to heavily populated Asian countries. They feared being culturally annihilated through migration and social fragmentation. Maintaining Australianness was a survival strategy and upholding it through purification, filiation, and closure became a very significant guiding principle. Britain was the opposite, a dense population occupying a small island that had established a stable sense of nation and place in the world. Its success had been built on hybridization, alliance, and civil liberty, and these also became generalized guiding principles. Between the two, Australia and Britain illustrate the range of social processes of designation, treatment, and impact of feral animals in contemporary societies, and show how they relate strongly to historical, social, and cultural features of their respective societies.

References

Anderson, B. (1983) *Imagined Communities*, London: Verso.
Arvanitakis, J. (2006) 'Border Protection between Australia and UnAustralia: (Or, Why I Am an Internally Displaced Person)', *Cultural Studies Association of Australasia Conference Proceedings*, Canberra: University of Canberra.
Australian Wildlife Conservancy (2012) 'Feral Animal Control', available online at www.australian-wildlife.org/Conservation-Programs/Feral-Animal-Control.aspx.
Bauman, Z. (1989) *Legislators and Interpreters: On Modernity, Post-Modernity and Intellectuals*, Cambridge: Polity.
Bomford, M. and O'Brian, P. (1995). 'Eradication or Control for Vertebrate Pests?', *Wildlife Society Bulletin* 23: 249–55.
Buller, H. (2004) 'Where the Wild Things Are: The Evolving Iconography of Rural Fauna', *Journal of Rural Studies* 20: 131–41.
—— (2009) 'The Woozle's Return: Angels, Monsters and Imaginary Spaces', workshop paper, School of Geography, University of Exeter.
Crawford, P. (1999) *Living Britain*, London: BBC.
Davis, S.J.M. (1987) *The Archaeology of Animals*, London: Batsford.
Department of Environment and Heritage (2013) *National Threat Abatement*, 17 July, available online at www.deh.gov.au/biodiversity/threatened/tap/index.html, accessed 30 July 2013.
Dickman, C.R. (1996) *Overview of the Impacts of Feral Cats on Native Fauna*, Sydney: University of Sydney for Australian Nature Conservation Agency.
Douglas, M. (1966) *Purity and Danger*, London: Routledge and Kegan Paul.
Ellen, R. (1996) 'Introduction', in R. Ellen and K. Fukui (eds), *Redefining Nature*, Oxford: Berg.
Franklin, A.S. (1996a) 'On Fox Hunting and Angling: Norbert Elias and the Sportisation Process', *Journal of Historical Sociology* 9.4: 432–56.
—— (1996b) 'Australian Hunting and Angling Sports and the Changing Nature of Human-Animal Relations in Australia', *Australian and New Zealand Journal of Sociology* 32.3: 39–56.
—— (2002) *Nature and Social Theory*, London: Sage.
—— (2006) *Animal Nation*, Sydney: University of New South Wales Press.
—— (2011) 'An Improper Nature? "Species Cleansing" in Australia', in B. Carter and N. Charles (eds), *Human and Other Animals: Critical Perspectives*, London: Palgrave Macmillan, pp. 195–216.
Gellner, E. (1983) *Nations and Nationalism*, Oxford: Basil Blackwell.
Gruffud, P. (2011) 'On the Prowl with the Possum Posse: Nature and Nation in Aotearoa', in B. Carter and N. Charles (eds), *Human and Other Animals: Critical Perspectives*, London: Palgrave Macmillan, pp. 217–35.
Hart Q. and Bomford, M. (2006) 'Australia's Pest Animals: New Approaches to Old Problems – Revised Edition', *Science for Decision Makers*, Bureau of Rural Sciences, p. 5.
Hurn, S. (2009) 'Here Be Dragons? No, Big Cats! Predator Symbolism in Rural West Wales', *Anthropology Today* 25.1: 6–11.
Ingold, T. (1993) 'The Temporality of the Landscape', *World Archaeology* 25.2: 152–74.
Jones, E. and Coman, B.J. (1982) 'Ecology of the Feral Cat, *Felis catus (L.)*, in South-Eastern Australia', *Australian Wildlife Research* 8: 537–47.
Latour, B. (1993) *We Have Never Been Modern*, New York: Harvester.
Low, T. (1996) 'Cats: Scoundrels or Scapegoats?', *Nature Australia*, Autumn.
—— (1999) *Feral Future*, Ringwood: Penguin.
McLeod, R. (2004) *Counting the Cost: Impact of Invasive Animals in Australia, 2004*, Cooperative Research, Centre for Pest Animal Control, Canberra, available online at www.pestanimal.crc.org.au/info/Mcleod.pdf, accessed December 2006.
Morrison, R. (1996) 'Ockham's Razor', *ABC Radio National*, 17 November.
New South Wales Government (2011) 'Feral Cats', available online at www.environment.nsw.gov.au/pestsweeds/FeralCats.htm, accessed November 2013.
Office of the Tasmanian Greens (2003) 'Press Release', 20 August.
Ritvo, H. (2012) 'Going Forth and Multiplying: Animal Acclimatization and Invasion', *Environmental History* 17: 404–14.
Rolls, E.C. (1969) *They All Ran Wild*, Sydney: Angus and Robertson.

Rose, B. (1995) 'Land Management Issues: Attitudes and Perceptions amongst Aboriginal Peoples of Central Australia', Central Land Council Cross Cultural Land Management Project.
RSPB (2012) 'Are Cats Causing Bird Declines?', available online at www.rspb.org.uk/advice/gardening/unwantedvisitors/cats/birddeclines.asp, accessed 1 December 2012.
RSPCA (2012) 'Feral Cats', available online at www.rspca-halifaxandhuddersfield.org.uk/feral_cats, accessed 25 November 2012.
Schwartz, E. (1995) 'Habitat Use in a Population of Mainland Tasmanian Feral Cats, *Felis catus*', Grad. Dipl. Science thesis, Zoology Department, University of Tasmania.
Seven News Network (Operations) Ltd (2004) 'Greens, Dems Slam Worth's Comments', 31 August, available online at http://seven.com.au/news/topstories/115446, accessed 12 March 2008.
Smith, N. (1999) 'The Howl and the Pussy: Feral Cats and Wild Dogs in the Australian Imagination', *Australian Journal of Anthropology* 10.3: 288–305.
—— (2000) *Nature, Native and Nation in the Australian Imaginary*, PhD thesis, School of Sociology, La Trobe University.
—— (2011) 'Blood and Soil: Nature, Native and Nation in the Australian Imaginary', *Journal of Australian Studies* 35.1: 1–18.
Smith, P. and Philips, T. (2001) 'Popular Understandings of "UnAustralian": An Investigation of the Un-national', *Journal of Sociology* 37.4: 323–39.
Thomas, K. (1983) *Man and the Natural World: Changing Attitudes in England 1500–1800*, London: Allen Lane.
Trevelyan, G.M. (1946) *English Social History*, 3rd edition, London: Longmans Green & Co.
Trigger, D.S. (2008) 'Indigeneity, Ferality, and What "Belongs" in the Australian Bush: Aboriginal Responses to "Introduced" Animals and Plants in a Settler-Descendant Society', *Journal of the Royal Anthropological Institute* 14: 628–46.
Unwin, B. (2005) 'Out of Hiding: How Britain Has Become a Nation of Twitchers', *The Independent*, 21 February, pp. 23–4.
Wild Visuals/Discovery (2000) *Ten Million Wildcats*, Sydney: ABC.
Williams, R. (1983) *Keywords*, London: Fontana Paperbacks.
Wood, M. (2000) *In Search of England*, Harmondsworth: Penguin.
Woods, M., McDonald, R., and Harris, S. (2003) 'Predation of Wildlife by Domestic Cats in Great Britain', *Mammal Review* 33.2: 174–88.
Young Persons's Trust for the Environment (2013) 'Squirrel (Grey)', available online at www.ypte.org.uk/animal/squirrel-grey-/176, accessed 10 June 2013.

11

Animal conceptions in animism and conservation

Their rootedness in distinct *longue durée* notions of life and death

Istvan Praet

Investigating the nexus between conservationism and animism

Understanding the interface between modern practices of wildlife conservation and (supposedly) traditional conceptions of the relationship between humans and animals is one of the great, pressing problems of contemporary anthropology. How does the environmentalism of those who strive to preserve vulnerable species accord with the animism of the indigenous peoples in whose territories those species often find refuge? Do the former view animals in roughly the same way as the latter? Some would argue that both perspectives are wholly – or at least to a significant extent – incompatible. Classically, social anthropologists have tended to stress specificities, oddities, and unexpected aspects: Bronislaw Malinowski described the close link between fruit bats and flying witches in the cosmology of Trobriand seafarers, Mary Douglas focused on the special role of pangolins in Lele rituals, and so forth. Others believe that one should not exaggerate the difference between 'the West' and 'the rest'. In this view, overplaying the pertinence of indigenous quirks and surprising twists of *la pensée sauvage* is unwarranted when it comes to environmental issues. Many conservationists have thus tended to assume that protecting wildlife and defending nature is a common goal they share with the ancestral inhabitants of the various areas designated for protection. Cultural subtleties may exist and must be taken into account, but they are not all *that* relevant at the end of the day. Be that as it may, the evidence gathered by ethno-biologists and cultural ecologists lends a great deal of support to this idea of a mutual compatibility between animism and conservationism.

From Roy A. Rappaport's (1968) seminal study of the relation between ancestor rituals and the ecology of pigs in highland Papua New Guinea to contemporary work on Huna Tlingit people's sustainable harvesting of gull eggs on Alaska's Pacific coast (Hunn *et al.* 2003), the occasional congruence between age-old animistic practices and the aims of environmentalism has been documented in considerable detail. What is more, investigations in the Amazon and other tropical rainforests have indicated that indigenous horticultural techniques may not merely preserve local levels of biodiversity but actually augment them (Balée 1994; Posey and Balick 2006). In recent years, however, the picture of an inherent harmony between green politics and animism has been adjusted in several ways. Anthropologists working on human-wildlife conflicts have questioned the self-evidence of the putative alliance between conservationists and indigenous people (e.g. Orlove and Brush 1996; Knight 2000; Borgerhoff Mulder and Coppolillo 2005). In specific circumstances, lofty ideals of animal protection clash with the need to defend one's crops, and both sides may turn out to be adversaries rather than partners. When local people do not share in the benefits of eco-tourism and other environmental projects or are excluded from fortress-like nature sanctuaries, they may reveal themselves as cunning poachers rather than responsible stewards, at least in the eyes of the conservationists. In Central Africa, for example, the role of indigenous hunters in the bush-meat trade and its impact on the decline of great ape populations is frequently referred to (Lewis 2005). The recognition that cases such as this are far from exceptional has led to a widely endorsed critique of received notions such as the 'noble savage' or the 'ecological Indian' (e.g. Conklin and Graham 1995).

At first glance, this situation is somewhat confusing: on the one hand, animism and conservationism seem to be perfectly compatible, but on the other hand, both frameworks appear to be incommensurable to a significant degree. So what is it? Part of the problem is perhaps that I have posed the question in overly general terms so far. In the abstract, 'animism' and 'conservationism' are fairly hollow terms and their usefulness is a matter of empirical enquiry rather than something that can be determined a priori on philosophical grounds. As David Anderson and Eeva Berglund (2004) have shown, specific conservationist practices and local understandings of environmentalism need to be approached ethnographically just as much as the more conventional topic of indigenous worldviews (see also Milton 1993). Establishing how to combine both kinds of ethnography adequately is an important task still lying ahead of us (Descola 2011). In this chapter I contribute to that aim by looking at one very particular aspect, namely the way in which animal conceptions are embedded in distinct *longue durée* notions of life and death. Drawing on my own fieldwork, I start with a vignette that describes how indigenous and conservationist understandings of animals meet 'on the ground'.

Spearfishers, raconteurs, and purported guardians of biodiversity

Spearfishing is a central activity of the Chachi, the riverine inhabitants of the Rio Cayapas area in Esmeraldas, Ecuador. Men, women, and children usually fish from canoes using a considerable variety of specialized spears and harpoons. Occasionally, however, they visit the headwaters of one of the river Cayapas' many tributaries. Using a canoe becomes impracticable in this case; if one fishes far upstream, one must wade through shallow stretches and, where the current becomes too strong, hop from rock to rock. The intrepid Florentino, a widely respected elder from the hamlet of Estero Vicente, had perfected this technique into a form of art. One afternoon in April 2004, he appeared in the house of his son-in-law (and my host) Alfonso with a rather impressive catch of assorted freshwater fish. 'You are a veritable woodpecker!' the latter exclaimed. As an aspiring anthropologist, my curiosity was immediately raised. Predictably, I wondered whether I had just witnessed something akin to one of Amazonian anthropology's most notorious controversies, namely the claim that 'the Bororo are parrots' or, more precisely, that the Bororo Indians of Central Brazil – famously studied by Claude Lévi-Strauss – conceive of themselves as red macaws. But my host made clear that saying 'you are a woodpecker' simply indicates that a fisherman has been exceptionally lucky. It is a commonly used expression throughout the Chachi territories. To be sure, Alfonso did not envisage his father-in-law as a woodpecker.

When I probed further, both men responded somewhat hesitatingly by telling a few stories about how Chachi hunters died in the past and how their respective ghosts acquired specific animal forms. This reluctance surprised me at first, as I had been on good terms with them for some time, but I later realized it was the appropriate thing to do. Chachi people generally avoid talking about the dead, even when they are not directly related (Barrett 1925; Praet 2006; Taylor 1993). In one story, a woman goes fishing together with her husband. While she perceived him as a human being, the husband was actually a woodpecker. At some point, he went ahead of his wife and started pecking a tree trunk. The wood splinters and leaves that fell into a nearby rivulet transformed into fish. When his wife caught up she was pleasantly surprised, for they could immediately return home with a great catch. Yet, when she wanted to put the fish on her roasting rack, they revealed themselves as merely splinters and leaves. Here she realized who he really was. As soon as the true identity of the husband was revealed, he shape-changed into a woodpecker and flew off.

In another story, a woman had a husband of small stature who continuously drank rum even though he needed very little to get drunk. She remarked that he behaved like a hummingbird, because these tiny birds drink nectar without pause and tend to fly from flower to flower in what is sometimes described as a 'drunken' way (uniquely among birds, they are able to hover in one place and some varieties can fly backwards and even upside down). As soon as she had said this, the husband realized he had been found out; he instantly metamorphosed into his bird-shape and flew away.

These sudden transformations are equivalent to saying that the men died, because death is always conceived of as metamorphosis in Chachi animism (Praet 2005), just as in many other indigenous societies of lowland South America (e.g. Vilaça 2006). One can easily find similar stories for most other kinds of birds and indeed most other kinds of animals in the Chocó rainforest and the adjacent western Andean cloud forest. A common thread running through most of them is the link between wild animals and the dead; to all intents and purposes, woodpeckers and all the rest manifest themselves as animate representatives of death or tangible ghosts. Remarkably, pets and tamed animals are never associated with the dead. For now, this puzzling link between untamed animals of the forest and the domain of death will have to remain in suspense, but I will return to it in due course.

A few weeks after Florentino's memorable fishing trip, the men and women of Estero Vicente were attending a meeting organized by an American conservation organization that has been operating in the Rio Cayapas area since the early 1990s. The aim was to discuss issues concerning the Cotacachi-Cayapas nature reserve, which stretches from the Andean *sierra* of Imbabura to the coastal plains of Esmeraldas. Internationally recognized as a global 'biodiversity hotspot', the reserve is incredibly varied and includes tropical lowlands, fog forests, *páramo* highlands, and even a snow-capped volcano (the eponymous Cotacachi). The Chachi ancestral territories border the westernmost side of the reserve, where there are particular problems of poaching and illegal logging. That day, the three delegated conservationists seemed to be in some kind of didactic mood. In Spanish (which most Chachi understand), they lectured about the importance of biodiversity and the dangers of deforestation. As far as I could tell, the attendees sat through the event in a way that took the middle ground between patient benevolence and boredom; clearly, most of them had heard the conservationists' rather paternalistic discourse a couple of times before. They were too polite to say it loudly, but some of my acquaintances remarked that the highlight of the day – and the main reason they were there – was the free meal and soda drinks (generously offered by the environmental NGO working under the auspices of the American conservation organization).

There was, however, another moment of considerable excitement: when the delegates distributed a richly illustrated brochure documenting the unique variety of animal species in the reserve, various younger men and even a few experienced hunters engaged in a spirited discussion on miscellaneous topics:

> 'How long is the tongue of tree-dwelling anteaters? Surely, it is more than half a metre; those biologists must be wrong.'
>
> 'Are pumas yellow, as the brochure suggests, or are they red, as various hunters assert and the vernacular name (*llukela*, literally 'red wildcat') suggests?'
>
> 'Has anyone ever seen that spectacled bear depicted here? No, they live high in the mountains! But perhaps they descend to the forest from time to time? Who knows . . .'
>
> 'Would that iridescent green lizard that biologists refer to as a plumed basilisk be the same as the thing we Chachi call *paapiñi*, the underground dragon? No, they are too small to be dragons. Maybe the gringos missed the bigger ones when they were doing research for the brochure?'

The agitation went on for about an hour and the conservationists seized this moment of enthusiasm to bring home their own message. Cleverly, they first praised the Chachi for what they described as their age-old, respectful attitude towards the natural environment. 'You, indigenous people, always had an ecological way of thinking,' one of the delegates underlined. 'You are the original conservationists!' But that laudatory tone quickly changed into something more threatening as they went on to warn about the irreversible loss in global biodiversity.

They explained about human-induced extinction and mentioned a few *causes célèbres*: the elephant birds of Madagascar, the Tasmanian tiger, and the famous flightless dodo of Mauritius. It was impressed upon the Chachi attendees that this could also happen in their home area. Ultimately, they were assured, the massive extinction of species could lead to a collapse of the entire planet. Therefore, it was argued, protecting vulnerable animal species and refraining from 'illegal logging' and 'poaching' were so important (the thorny question of what makes a particular subsistence activity illegitimate was conveniently avoided on the occasion). At the time, it was my impression that this rather apocalyptic message was received with a mixture of bemusement and vague disquiet. Yet, what down-to-earth spearfishers and first-class raconteurs like Florentino and Alfonso really thought about this whole business of animal protection, biodiversity, and worldwide extinction remains difficult to gauge. How far is their way of viewing animals compatible with the narratives of protection and stewardship promoted by contemporary environmentalism? To what extent are they truly worried about an impending ecological disaster triggered by their propensity to over-hunt and over-fish?

Obviously, the piece of ethnography I have just presented is merely one instantiation of a broader comparative problem that many anthropologists have pondered. But before I compare animistic and conservationist animal conceptions in a more systematic manner, it is appropriate to offer a brief sketch of the wider background.

Historical antecedents

One key insight anthropologists and other social scientists have gained in recent years is that many of the people whom their predecessors used to portray as primitive and isolated actually have long histories of involvement in long-distance trade and global commercial networks (Wolf 1982; McNeill and McNeill 2003; Elvin 2004). Since time immemorial, Southeast Asian forest dwellers have supplied tiger whiskers, dried geckoes, and many other ingredients used in traditional Chinese medicine; Siberian fur trappers have participated in the export of sable skins to Europe for centuries; and in the savannahs of East Africa generations of local expert hunters have provided the rhinoceros horns used to craft the traditional daggers that used to be a status symbol in some parts of the Arabian peninsula. The ornamented ostrich eggs, narwhal tusks, and bison horns (formerly conceived of as dragon eggs, unicorn horns, and griffin claws), the stuffed crocodiles, the ivory chess pieces, and the feathers of peacocks and birds of paradise that are conserved in the palaces, castles, churches, and cathedrals of the Old World are a testimony to the often underestimated antiquity and the surprising pervasiveness of global networks of commerce and exchange (Daston and Park 1998).

What is more, not only dead specimens or valuable body parts were transported from the farthest reaches of the known world but also live animals. Medieval emperors, princes, and dukes had a passion for the wondrous and the exotic; the game parks, zoological gardens, and aviaries that frequently formed part of their courts were also emblems of their extraordinary power and influence (Hahn 2003; Nijsten 2004). In the thirteenth century, the English kings kept lions and leopards in the Tower of London. The royal menagerie also included a polar bear held on a leash (occasionally, it was allowed to swim and catch fish in the River Thames) and an African elephant acquired during one of the crusades (it died fairly soon after its arrival in Britain, allegedly after having drunk an overly generous amount of red wine). Such examples – anecdotal as they may be – are nevertheless salient, since they give us a basic sense of the long, intricate history of global trade in wildlife and animal body parts. The frequent mention of such exotica in archival sources and their relative abundance in treasure rooms and historical collections corroborate – albeit indirectly – the now commonplace assertion that no society, however primitive or isolated, has ever operated in complete autarchy.

So-called indigenous peoples, animistic hunters, and local suppliers have dealt with merchants and other powerful outsiders much more frequently than previously presumed, whether voluntarily, by coercion, or in some intermediate way. In such a long-term perspective, it becomes clear that the latter-day confrontation between conservationism and animism is not the historically unique moment it is sometimes presumed to be in present-focused anthropological writings but rather something that has been quite recurrent – albeit in slightly different manifestations – throughout the ages. The encounter between outsiders' and native conceptions of wildlife and the interweaving of localized and more metropolitan trade webs involving animals is a longstanding feature of indigenous histories rather than a specific phenomenon of the past century (Fausto and Heckenberger 2007). But does this apparently persistent pattern of mutual contact and regular interaction imply that both kinds of worldview were brought into alignment long ago? Some have indeed suggested that anthropologists tend to exaggerate the contrast between animism and Western frames of thought such as conservationism, either because of an inappropriate penchant for nostalgia or an outdated taste for the exotic (MacClancy 2002). This line of reasoning prefers to downplay the gap between 'us' and 'them', and highlights the levelling effects of world history.

While recognizing some of its concerns, this chapter nevertheless emphasizes difference over similarity. My principal argument is that the gap between animistic and conservationist conceptions of animals is much wider than generally acknowledged, since they are embedded in utterly different and largely invariant axioms about life and death. My aim in the remainder of this chapter is simply to illustrate that poorly understood yet generally neglected gap.

Why do conservationists care about animals?

'Could you please explain to us, what is a conservationist?' One day, the men of Estero Vicente asked me this impromptu question during one of their regular village meetings. My response, at the time, was perhaps not very coherent. I said something about the overbearing human influence on the Earth's natural resources and about the need to establish protected zones to safeguard animals. That is why they established the Cotacachi-Cayapas nature reserve, I pointed out. But my Chachi interlocutors wanted to know about their deeper interests: 'How are they making money with this? How can they pay the speedboats in which they visit us?' My explanation that they are funded by people all over the world who are concerned about the increasing vulnerability of many species did not entirely convince them.

'Why do all those people love those animals so much? And why the animals in *our* forest, which they have never seen – isn't that strange?' I conceded that conservationists are a strange tribe indeed. The villagers' question – 'why do they care about animals?' – was perfectly straightforward but, to my own surprise, I found it difficult to come up with a plausible answer. The best I could think of, under the circumstances, was a rather lame narrative about the duty to respect our fellow living beings: 'After all, aren't we all part of the same tree of life?' Thankfully, the meeting took place in the village school so I could point to a poster that depicted a Darwinian tree of life and refer them to the captions that explained – in the vernacular language – how humans, apes, and all other animals and plants are ultimately related. Still, judging from the facial expressions of my Chachi friends and acquaintances, my account was not entirely persuasive.

Later, I often thought about this episode and it occurred to me that I never really found a satisfactory answer to the problem the men of Estero Vicente had posed with such refreshing downrightness. What, in fact, is a conservationist? Of course, there is no dearth of descriptions and analyses of the work and the ideas of those involved in the global conservation movement. Yet such discussions always seem to be cast in a specific mould. The inherent values of 'biodiversity' as well as its benefits for humankind are consistently emphasized. The species richness of coral reefs, tropical forest canopies, and other 'hotspots' is celebrated both on aesthetic grounds and in terms of its economic and medicinal potential. The notion of a beautifully intricate yet fragile 'biosphere' (epitomized by the famous Blue Marble photograph taken by the astronauts of the Apollo 17 spacecraft) is also very central.

The Earth is viewed through an ecological lens, as a complex and delicate whole of interdependent ecosystems. A related conservationist hobbyhorse is the interrelatedness of biological species. Their insistence on the high percentage of DNA similarity between humans and chimpanzees is a classic of the genre. We humans, so the reasoning goes, are very similar to our fellow mammals and especially to the great apes; we just happened to take a slightly different evolutionary path. While many different traditions of conservationism do exist worldwide, each with its own accents and approaches, these basic ideas are a common element (see Scheffer 1991; Sale 1993; Lee and So 1999; Weiner 2002). What comes back time and again is a certain schizophrenic characteristic not entirely unlike what has been documented about millennial cults: one extols the diversity of the natural world but at the same time predicts its global collapse; one glorifies the grandeur of life but simultaneously fears its disappearance in an impending catastrophe. The overtone is often openly messianic: we, living beings, are doomed but we can save the planet if we behave in a morally appropriate way. The 'sixth extinction' – the first one induced by humans – may be well underway but the ultimate breakdown can still be averted (McNeill 2000).

The linchpin of this entire construction is a specific notion of life that has its roots in modern biology. Yet this historically contingent foundation has rarely been taken into account, not even in the most critical treatments of conservationism. In *Les mots et les choses*, Michel Foucault (1966) famously argued that the idea of life did not exist prior to the eighteenth century. The natural philosophers of the previous centuries studied animals and plants and all sorts of natural wonders and curiosities, of course, but these various lines of enquiry were not conceived of as constituting a single endeavour with an overarching goal. The unification into what is now known as 'the life sciences' only began to take place during the period generally known as the Enlightenment. Life, in that sense, is a relatively recent invention – it is not older than a couple of centuries.

Conservationists, however, seem to have embraced the idea as if it were an eternal, self-evident truth. In their worldview, everything that moves and breathes (or is made of cells or contains DNA, to use criteria that are currently more in vogue nowadays) is alive. Life is therefore extensive: it includes walruses, parrots, apple trees, goldfish, microbes, and everything else that pertains to what is customarily referred to as 'the tree of life'. In this extensive conception, the quality of 'life' is the greatest common denominator: 'we' are all living beings, connected through the vagaries of biological evolution. To protect vulnerable walruses and all the rest is to protect beings that have something in common with humans – we must care for them because they are our 'fellow living beings'. But is this deeply ingrained conviction of a basic commonality between all that is subsumed under the caption 'life on Earth' really an absolute certainty or is it just a widely shared and therefore inordinately influential convention?

In what follows, I make a case for the latter. To grasp the strangeness of the particular notion of life on which conservationists premise their outlook, it is useful to consider notions of life in societies that are usually designated as animistic. Elsewhere, I have documented the ubiquity of a *restricted* conception of life and an *expanded* conception of death among so-called indigenous peoples in the Americas, Asia, and Africa (Praet 2013a, 2013b, 2013c). Taking these complementary concepts into account is essential to understanding human-animal relations within many (if not most) forms of animism. In the aforementioned publications, I have illustrated this in some depth with reference to my own ethnographic fieldwork among the Chachi in Ecuador.

The central conclusion is that, while pets and companion animals are generally included within a restricted sphere of life, prey and other wild animals form part of an expanded and very tangible realm of death. To put it even more crudely: everything that is sufficiently tamed is alive, but all creatures modern observers would normally think of as 'wildlife' are – in fact – representatives of death. Within Chachi animism, it is indeed more adequate to speak of 'wild-dead' than of wildlife. In other words, simply 'being animate' does not guarantee that one will be included within life; it is perfectly possible to be animate and to pertain to the realm of death at the same time. This expanded conception of death – at least, it appears expanded from a contemporary Western perspective where death has all but disappeared from the material world and has been evacuated to the transcendental sphere – used to be more widespread than is commonly acknowledged.

In what follows I illustrate this by means of the particularly richly documented historic example of the Lakota Sioux warriors and hunters who, as we know from countless films and TV series, lived their nomadic existence on the Great Plains of North America until the early twentieth century. This choice is apposite because the Sioux figure prominently in Shepard Krech's widely debated book on what he termed 'the ecological Indian', in which he examines how far native people such as the Lakota possess an inherent environmental awareness. My aim here is not to illustrate how contemporary Lakota people have creatively fitted modern ecological ideas into their ancestral philosophies and have invented their own distinctive forms of environmental activism – Bornali Halder (2002) has already provided a fascinating account of this kind of innovative process. Instead, I focus on what is nowadays often considered a more old-fashioned anthropological preoccupation, namely that of animistic 'customs and traditions'. Whereas the current attitude of extreme care not to 'exoticize' indigenous people is laudable in certain respects, I think it is nevertheless important not to overdo it to the point of denying them any kind of originality. And Sioux notions of life are arguably original.

To be clear, I add that the aim of this exercise is not to criticize the conservation movement in itself. My intention here is not to prove that the notion of life it has inherited from modern biology is somehow deficient or counterproductive. Rather, I merely wish to reveal its peculiar character. If there is a critical point, then, it is more about conservationists' failure to recognize their own peculiarity as well as their curious and perhaps slightly annoying tendency to impute *their* extensive idea of life to everybody else. People such as the Chachi and the Sioux may have adopted that idea to some degree out of pity, benevolence, calculated self-interest, or perhaps genuine curiosity (their true motivations could only be determined by ethnographic enquiry), but that does not mean that it totally and definitively supersedes age-old animistic ideas. The latter, I suspect, may be more resilient than is commonly assumed.

Sioux animism in the nineteenth century

Before we focus on human-animal relations in Sioux animism it is useful to summarize some basic ethnographic observations. The Sioux used to refer to themselves as 'Lakota' or 'Dakota', which can be translated as 'the people' but also as 'all those who are alive', since – as will be documented in some detail in what now follows – their particular form of animism was characterized by a restricted conception of life (Hassrick 1964; Walker 1982; DeMallie 2009: 189). Raymond DeMallie has convincingly demonstrated that speaking the same language was the foremost criterion of inclusion; their common language, he states, was the Sioux's 'most important symbol of social identity' (2009: 190). Those who did not speak it were outside the bounds of Sioux kinship and did not qualify as 'people'; strictly speaking, they were not even considered to be 'alive'. They were associated with death and referred to as 'those related to as enemies'.

Moreover, vernacular notions of humanity and life were always grafted onto the pervading motifs of roundness and circularity; hence the well-known Siouan obsession with hoops and circles. 'Within the circle all were relatives', DeMallie writes, and he adds that this was true 'from the intimacy of a single family's tipi, to the camp circle, and finally to the broadest identity as Lakota or Dakota' (2009: 201). In the 1930s, the widely respected 'medicine man' Nicholas Black Elk famously phrased it thus: 'everything [a Lakota] does is in a circle [because the world] always works in circles and everything tries to be round' (Halder 2002: 111). That is why Sioux tipis are circular, for example, and why the mouth of a pipe should always be moved about in a circle before it is smoked. To understand Sioux animal conceptions, it is necessary to keep these crucial aspects of roundness and the common language at the back of one's mind.

Anthropological discussions of Sioux animism have often focused on the central notion of *wakan*. The term is conventionally translated as 'holy', 'sacred', or 'divine'. The Lakota qualified buffalos, elk, wolves, grizzly bears, eagles, and everything modern observers would label 'wildlife' as *wakan* (Walker 1991). Ethnographers have generally interpreted this as a prime example of a 'sacred ecology' or a 'spiritual view of nature'. Animals supposedly have a 'holy' character and are therefore treated with due respect; in this perspective, there is an inherent convergence between Lakota animism and the aims of contemporary wildlife conservation.

However, something that the advocates of this view have conveniently neglected is that many other things were considered *wakan* as well. Intriguingly, most if not all of these things had a connection to the realm of death and/or were used at funerary occasions: certain rocks, tobacco, sagebrush, sweat lodges, ghosts, the deceased, etc. Why is it that both wildlife *and* everything related to the dead were referred to as *wakan*? What is the logic behind this at first glance incongruous link between wildlife and the dead? To this day, the answer to this fairly straightforward question remains in suspense. Within the limits of this chapter I cannot offer an exhaustive answer, but by focusing on the specific case of raptor birds and their link with Siouan clowns known as *heyoka* I hope to provide some elements that lead towards a solution.

Heyoka clowns, eagles, and thunderbirds

Lakota people often talk about *wakinyan*, the 'thunderbird'; in many explanations, the clap of thunder is its voice and lightning is its glance. While they are usually hidden in the clouds, thunderbirds are occasionally spotted. Consider the following account of a Sioux man, which Bornali Halder recorded in 1998:

> [My wife and I] were coming back from a ceremony, about two-thirty, three o'clock in the morning, and it was kind of cloudy and lightning was flashing, and we were just coming down to a stop sign out here, coming from Rosebud onto the highway, and lightning flashed behind a cloud and all of a sudden I see this huge bird standing in the sky and I instantly knew what it was – that's the Thunderbird, that's the original Thunderbird. [. . .] [I]ts head was behind the clouds, and its feet were below the horizon, but I would say, based on the perspective that I was looking at it, it could be anywhere from two to five miles high. It was real high up in the sky, and its wings were hanging down like this. It was in the north, and its head, I could see by the feathers in its neck, its head was turned toward the west. So lightning flashed and I saw it, and then of course it went dark.
>
> *(2002: 103)*

In the nineteenth century, this kind of experience was relatively common and those who had it turned into *heyoka* or 'clowns'. Ordinary people shunned them as they were thought to bring bad luck. The Sioux's most notorious historical leaders such as Crazy Horse and Sitting Bull were all clowns at some point in their lives. In their daily activities clowns would maintain a close relationship with raptor birds such as hawks and eagles, which tend to soar on thermal updrafts in front of approaching thunderstorms and were also envisaged as minor incarnations of the thunderbird. In fact, everything indicates that the modern division between culture, the supernatural, and nature made little sense in Sioux animism: clowns, thunderbirds, and eagles were envisaged as different manifestations of the same thing.

Clowns would don ridiculous outfits: they walked around in rags, wore conical hats and false noses, or simulated baldness by putting a bladder on their head. A key characteristic is that they consistently spoke and acted in a contrary manner. Their speech was literally inverted: they pronounced their sentences backwards. If one recalls the importance Sioux people attributed to speaking a common language, one realizes the significance: clowns deliberately placed themselves outside the sphere of humanity and, arguably, outside the sphere of life. As the inverse of proper Lakota – that is, as non-humans – their behaviour is conspicuously unconventional. They would wear their clothes inside-out or sit the other way round on a horse. In times of scarcity, a clown would complain about how full he was. When it was extremely hot, he would shiver, put on gloves and wrap himself in a warm blanket.

Reversely, he would walk around naked when it was freezing. As the ethnographer Verne F. Ray has put it: 'the exact opposite of normal human behaviour is the ideal sought' (1945: 75). It is an instance of perspectivism, to use current anthropological parlance: ordinary Lakota and clowns have inverse points of view, just as the living and the dead have inverse points of view.

It is not surprising, then, that clowns had an explicit association with death. For one thing, they often painted their faces black, just as mourners do. They were expected to give away all their belongings, again just as mourners do. They also had a close relationship with cedars, the tree that usually marks somebody's grave. Not coincidentally, people would seek shelter under such cedar trees when there was a thunderstorm. All this indicates that clowns were envisaged as tangible incarnations of death – that is why people avoided them and that is why they always engaged in contrary behaviour (the conception of the dead as 'opposite beings' is of course widely documented in Amerindian ethnography, both in North America and in South America; e.g. Taylor 1993).

The so-called '*heyoka* ceremonies' or 'kettle dances' that were organized as spring approached and the first thunders arrived are a further illustration of this particular strand of perspectivism (Bray 2006: 44–5; Halder 2002: 104). On such occasions, clowns strangled, butchered, and boiled puppies in kettles. Eventually, they would plunge their arm into the boiling water and retrieve chunks of dog meat. They often provoked great hilarity by throwing the meat towards the onlookers. The clowns would also scoop up handfuls of boiling water and hurl it over each other's backs, all the while complaining that the water was too cold (they did not burn themselves because they would have rubbed their skin with a special ointment from a particular prairie wildflower). The precise aim of such ceremonies need not concern us here; what matters is that this inversion of cold and hot, as well as this uncharacteristically cruel behaviour (ordinary, living Lakota would never do such a thing to their pets) indicate that clowns are palpable representatives of an expanded realm of death that also includes the various birds and the thunderbirds with which they are associated.

In this context, eagles and hawks must be understood as 'wild-dead' rather than as 'wildlife'; they may be animate and tangible but they are dead nevertheless. To substantiate that point I must say a few words about Sioux warfare as it was practised until the early twentieth century.

The shape-changing techniques and temporary death of Siouan warriors and buffalo hunters

In fact, *heyoka* clowns were not just buffoons but also formidable warriors who could draw thunder, lightning, storm, and death upon their adversaries in a quite literal sense. Unpredictability of movement was their central quality, which they emulated from the thunderbird, the eagle, and various lesser flying creatures:

> The jagged line of lightning; the sudden fall and climb of a dragonfly; the tipped swoop of a flock of swallows ahead of thunder; the squeaking night flight of bats; a horse's sudden run or the awesome drop to the kill of [. . .] the red-tailed hawk – all were characterized by suddenness, unpredictability. [Lakotas recognized this erratic motion] as a property of the [Thunderbird and its] messengers.
>
> *(Bray 2006: 65)*

In times of combat clowns decorated the rumps of their horses in streaky lines and painted their own bodies with the motif of hailstones and lightning streaks.

They also adorned themselves with hawk or eagle feathers and blew eagle bone whistles. Such decorations and paraphernalia transformed them into veritable 'spotted eagles', that is, into concrete incarnations of the thunderbird. This is in any case how Crazy Horse, perhaps the most famous Sioux warrior of all time, typically went into battle in the nineteenth century. Bray has described his technique perceptively:

> Crazy Horse worked hard to assimilate Thunder's message, fronting charges to act as the leading edge of an unpredictable lightning strike. Against the Crows, and later in such large-scale clashes with the Americans as the Battle of Rosebud, he led long oblique lines of charging horsemen that rippled across the terrain like lightning.
>
> *(2006: 65)*

Recall our earlier remark that Sioux notions of life are primarily based on ideas of roundness and circularity. The warriors'/clowns' jaggedness and their irregular, streaky movements are an indication of their (at least temporary) location outside the restricted sphere of life. Their shape-changing into raptor birds or other flying creatures and their metamorphosis into thunderbirds or thunder itself (all of which are manifestations of the same thing anyway) corroborate that interpretation: they are palpable incarnations of the dead.

This also implies that eagles, hawks, swallows, bats, dragonflies, and – potentially – everything that moves in irregular, unpredictable ways form part of the expanded realm of death that is such a characteristic feature of most forms of animism (Praet 2013a, 2013b, 2013c). It is not surprising, then, that eagles also played a prominent role when somebody died. Just like thunderbirds and their avatars, the recently deceased were deemed capricious and unpredictable (Halder 2002: 106–7). In fact, they are often said to become eagles that soar into the sky, thus reaching the land of the dead.

What is more, eagles and flying animals more generally were not exceptional in this respect: buffalos and other prairie animals were equally part of the expanded realm of death. Their iconic buffalo hunts, for example, were always preceded by various rites that were considered *wakan* (Halder 2002: 268). Among many other things, hunters would smoke tobacco (a substance associated with death) and medicine men would paint a buffalo skull red (the colour of death): during the expedition, the hunters and their ritual assistants thus assumed the same position as the dead. In a perspectival scheme this makes perfect sense; for the hunt to be successful the predators must temporarily adopt a similar point of view as their prey. The communal hunts of the nineteenth-century Sioux could indeed be studied as classic examples of perspectivism, whereby buffalos must be grasped as 'wild-dead' rather than as wildlife.

Conclusion

The global conservation movement and Amerindian animism are premised on rather different configurations of life and death, which determine to a significant extent how human-animal relations are conceived. From a historical perspective, these distinct configurations appear to be remarkably stable, since they recur time and again, throughout the world. The extensive notion of life characteristic of conservationism – the idea of a biosphere covering the entire surface of the planet, the idea of a tree of life of which all existing biological species are offshoots – goes back to the incipience of modern biology as a scientific discipline in the early eighteenth century. The restricted notion of life characteristic of Amerindian and, possibly, other forms of animism is in all likelihood much older. If my hypothesis that one can find it (or at least detect traces of it) in indigenous cosmologies on all continents is confirmed, it may even stretch back millennia. Be that as it may, it is surely legitimate to speak of distinctive *longue durée* conceptions of life.

Long ago, Robert Hertz ([1907] 2004) wrote that the modern cosmology banished death to the margins, that is, to the realm of the transcendental and/or the inert. Conservationism proves to be a textbook manifestation of that cosmology, given that it consistently puts everything that moves and breathes under the jurisdiction of life. Its pet topic of biodiversity exemplifies this tendency: only life can be diverse – to speak of 'the diversity of death' makes little sense for some reason. In line with the modern cosmology from which it sprouts, the conservation movement has in effect reduced death to a null quantity. One can have wildlife, but not 'wild-dead'. To be sure, there is nothing inherently problematic about such a view. What *is* perhaps problematic, however, is that conservationists tend to imagine that this peculiar outlook is so self-evident that it must be shared by all other people in the world. That is why they often attribute an age-old concern for biodiversity to indigenous people, for example.

My account of Sioux animism illustrates why such attributions are misguided. We have seen that, for the Lakota, life used to be restricted to everything 'within the circle': one's own family, one's kinsmen, one's neighbours, friends, and allies (i.e. all those speaking the same language), and, *à la limite*, one's pets and horses. Only 'the people' were actually alive, that is, all those who participated in the yearly sun-dance ceremony (where the Lakota put their tipis in a great circle). Whereas life was tantamount to roundness, death was equivalent to jaggedness. Sioux animism is thus characterized by an expanded realm of death that not only includes the deceased and their erratically behaving ghosts, but also thunder, clowns, warriors, and enemies (all of whom manifest themselves in erratic, non-circular ways), as well as eagles, hawks, bats, swallows, dragonflies, and all other flying creatures that display erratic, non-circular patterns of movement. In addition, it comprehends the buffalo herds, wolves, bears, and all other animals of the prairies and plains that were customarily designated as *wakan*. The Lakota did not envisage them as 'other living beings' (that is merely an odd habit of modern biology) but as non-living beings. If this interpretation is correct, *wakan* must be understood as 'animate but simultaneously dead' and not as 'divine' or 'supernatural', as it is usually translated. The popular idea that the Lakota maintain a 'spiritual relation with nature' and operate in a 'sacred ecology' turns out to be apocryphal. Attentive readers will have noticed that the same thing recurs in Chachi animism: the woodpeckers and hummingbirds mentioned in the introduction must also be grasped as animate representatives of an expanded realm of death.

If one would like to describe Sioux animism using contemporary conservationist terminology, one must say that it was typified by a strikingly limited biosphere (essentially the Lakota themselves and their domesticated animals) and an equally strikingly extensive 'thanatosphere' (comprising everything that conservationists would designate as wildlife). Yet these two spheres are not totally impermeable because life and death were not inherent properties but positional qualities. That becomes evident when one considers the shape-changing techniques of medicine men, buffalo hunters, warriors, and clowns – all of whom temporarily die and are thus capable to function in the normally off-limits domain of death. Moreover, this permeability also worked in the opposite direction. Within the scope of this chapter I could not illustrate this in any depth, but one can infer from the ethnographic literature that horses and women captured from enemies were structurally dead at first but were subsequently *made* alive – by taming them (the horses) and by teaching them the common Lakota language and thus turning them into allies (the women), which arguably also amounts to a form of taming. As has been documented in many other forms of animism, life is always conditional. For the Sioux of the nineteenth century it required roundness, speaking the same language, and pertaining to the (circular) camp. It could be suspended by behaving in a clownish manner – i.e. irregularly, unpredictably, 'jaggedly'.

Within conservationism, by contrast, life is always a given: one is born alive and one cannot do anything to get rid of that status temporarily. There are no specific conditions whatsoever: that is why millions of species can be subsumed under the caption of 'life'. This chapter, I hope, has given you a sense of the unexpected strangeness of such an extensive idea of life. Conservationist ideology – and its implicit view of animals in particular – is more exotic than is commonly recognized. To be sure, people such as the Sioux and the Chachi have creatively incorporated environmentalist ideas within their worldviews in the present day and age, but it would be mistaken to infer from this that their longstanding – and often radically different – animistic views are no longer relevant at all. When it comes to understanding animals, conservationism and animism each have their own originality – anthropologists who fail to take that distinctiveness into account do so at their own peril.

References

Anderson, D.G. and Berglund, E. (eds) (2004) *Ethnographies of Conservation: Environmentalism and the Distribution of Privilege*, Oxford: Berghahn.

Balée, W. (1994) *Footprints of the Forest: Ka'apor Ethnobotany – The Historical Ecology of Plant Utilization by an Amazonian People*, New York: Columbia University Press.

Barrett, S.A. (1925) *The Cayapa Indians of Ecuador*, New York: Museum of the American Indian – Heye Foundation.

Borgerhoff Mulder, M. and Coppolillo, P. (2005) *Conservation. Linking Ecology, Economics and Culture*, Princeton, NJ: Princeton University Press.

Bray, K.M. (2006) *Crazy Horse: A Lakota Life*, Norman: Oklahoma University Press.

Conklin, B.A. and Graham, L.R. (1995) 'The Shifting Middle Ground: Brazilian Indians and Eco-Politics', *American Anthropologist* 97.4: 695–710.

Daston, L. and Park, K. (1998) *Wonders and the Order of Nature, 1150–1750*, New York: Zone Books.

DeMallie, R.J. (2009) 'Community in Native America: Continuity and Change among the Sioux', *Journal de la société des américanistes* 95.1: 185–206.

Descola, P. (2011) *L'écologie des autres. L'anthropologie et la question de la nature*, Paris: Editions Quae.

Elvin, M. (2004) *The Retreat of the Elephants: An Environmental History of China*, New Haven, CT: Yale University Press.

Fausto, C. and Heckenberger, M. (eds) (2007) *Time and Memory in Indigenous Amazonia. Anthropological Perspectives*, Gainesville: University Press of California.

Foucault, M. (1966) *Les mots et les choses*, Paris: Gallimard.

Hahn, D. (2003) *The Tower Menagerie. The Amazing True Story of the Royal Collection of Wild Beasts*, London: Simon and Schuster.

Halder, B. (2002) 'Mitaku'oyasin: An Anthropological Exploration of Lakota Sioux Environmental Activism', unpublished doctoral thesis, University of Oxford.

Hassrick, R.B. (1964) *The Sioux. Life and Customs of a Warrior Society*, Norman: University of Oklahoma Press.

Hertz, R. ([1907] 2004) *Death and the Right Hand*, London: Routledge.

Hunn, E.S. *et al.* (2003) 'Huna Tlingit Traditional Environmental Knowledge, Conservation, and the Management of a "Wilderness" Park', *Current Anthropology* 44: S79–S103.

Knight, J. (ed.) (2000) *Natural Enemies: People-Wildlife Conflicts in Anthropological Perspective*, London: Routledge.

Krech, S. (1999) *The Ecological Indian: Myth and History*, New York: W.W. Norton and Company.

Lee, Y.S. and So, A.Y. (1999) *Asia's Environmental Movements: Comparative Perspectives*, Armonk, NY: M.E. Sharpe.

Lewis, J.D. (2005) 'Whose Forest is it Anyway? Mbendjele Yaka Pygmies, the Ndoki Forest and the Wider World', in T. Widlok and W.G. Tadesse (eds), *Property and Equality*, vol. 2, Oxford: Berghahn, pp. 56–78.

MacClancy, J. (2002) *Exotic No More: Anthropology on the Front Lines*, Chicago, IL: University of Chicago Press.
McNeill, J.R. (2000) *Something New under the Sun: An Environmental History of the Twentieth Century*, New York: W.W. Norton.
McNeill, J.R. and McNeill, W.H. (2003) *The Human Web: A Bird's-Eye View of World History*, New York: W.W. Norton.
Milton, K. (1993) *Environmentalism: The View from Anthropology*, London: Routledge.
Nijsten, G. (2004) *In the Shadow of Burgundy: The Court of Guelders in the Late Middle Ages*, Cambridge: Cambridge University Press.
Orlove, B.S. and Brush, S.B. (1996), 'Anthropology and the Conservation of Biodiversity', *Annual Review of Anthropology* 25: 329–52.
Posey, D.A. and Balick, M.J. (eds) (2006) *Human Impacts on Amazonia: The Role of Traditional Ecological Knowledge in Conservation and Development*, New York: Columbia University Press.
Praet, I. (2005) 'People into Ghosts: Chachi Death Rituals as Shape-Shifting', *Tipití: The Journal of the Society of Lowland South America* 3.2: 131–46.
—— (2006) 'Courage and Fear: An Inquiry into Chachi Shape-Shifting', unpublished doctoral thesis, University of Oxford.
—— (2013a) 'Humanity and Life as the Perpetual Maintenance of Specific Efforts: A Reappraisal of Animism', in T. Ingold and G. Pálsson (eds), *Biosocial Becomings: Integrating Biological and Social Anthropology*, Cambridge: Cambridge University Press.
—— (2013b) 'The Positional Quality of Life and Death. A Theory of Human-Animal Relations in Animism', *Anthrozoös* 26.3: 341–55.
—— (2013c) *Animism and the Question of Life*, London and New York: Routledge.
Rappaport, R.A. (1968) *Pigs for the Ancestors*, New Haven, CT: Yale University Press.
Ray, V.F. (1945) 'The Contrary Behavior Pattern in American Indian Ceremonialism', *Southwestern Journal of Anthropology* 1: 75–113.
Sale, K. (1993) *The Green Revolution: The American Environmental Movement, 1962–1999*, New York: Hill and Wang.
Scheffer, V.B. (1991) *The Shaping of Environmentalism in America*, Seattle: University of Washington Press.
Taylor, A.-C. (1993) 'Remembering to Forget: Identity, Memory and Mourning among the Jivaro', *Man* 28: 653–78.
Vilaça, A. (2006) *Quem somos nós: os Wari' encontram os brancos*, Rio de Janeiro: Editora UFRJ.
Walker, J.R. (1982) *Lakota Society*, Lincoln: University of Nebraska Press.
—— (1991) *Lakota Belief and Ritual*, Lincoln: University of Nebraska Press.
Weiner, D.R. (2002) *A Little Corner of Freedom: Russian Nature Protection from Stalin to Gorbachev*, Berkeley: University of California Press.
Wolf, E.R. (1982) *Europe and the People without History*, Berkeley: University of California Press.

12
The emptiness of the wild

Philip Armstrong and Annie Potts

I

In January 1770 James Cook's *Endeavour*, the first European vessel to make extended landfall in New Zealand, lay at anchor in Queen Charlotte Sound at the northern tip of the South Island. Joseph Banks, the ship's naturalist (later president of the Royal Society), recorded his impression of a land full of wild vitality:

> This morn I was awakd by the singing of the birds ashore from whence we are distant not a quarter of a mile, the numbers of them were certainly very great who seemd to strain their throats with emulation perhaps; their voices were certainly the most melodious wild musick I have ever heard, almost imitating small bells but with the most tuneable silver sound imaginable.
>
> *(Banks 1962: 455–6)*

Anyone who has ever heard the korimako or New Zealand bellbird (*Anthornis melanura*) will attest to the accuracy of this description. Yet Banks' perception is also shaped by his culture's understanding of wild nature. He locates wildness in the 'very great' numbers of the birds, in the volume of their song (sufficient to wake him across a quarter-mile of water), and in the intensity of their passion (his remark that they 'strain their throats' suggests Banks has observed the way bellbirds puff their feathers, hunch their wings, and tense their whole bodies when they sing). Mingled with this paradigm – which associates wildness with numerical profusion, sensory extravagance, and bodily exuberance – Banks includes another, which imagines nature as a 'melodious [. . .] musick', amenable to 'emulation', to the kind of artistic and mechanical manipulation represented by 'small bells' with their 'tuneable silver sound'. In this way, Banks' response encapsulates the two versions of wildness that were predominant at the end of the eighteenth century. In the first, which would come to be embraced and celebrated by Romanticism, wild nature was the locus of an exorbitant and overwhelming intensity; in the second, which was consistent with Enlightenment rationalism, it was an instrument to be brought into harmony, a mechanism requiring calibration.

These two paradigms, and the management of the relationship between them, have indeed been central to the project of modernity throughout the two and a half centuries since Banks recorded the comments above. New Zealand provides an exemplary case history for observing these wild transactions for two reasons. The first is that the country was colonized by Europe at the very historical moment at which the relationship between Romantic and Enlightenment approaches to wild nature became a definitive cultural force. The second is that the endemic animals of New Zealand presented a biotic community exceptionally amenable to both cultural idealization and material manipulation. Unlike Australia, the islands of New Zealand were not home to any venomous species dangerous to humans, nor to any endemic predator larger than a small raptor. The only endemic terrestrial mammals were a couple of inoffensive bat species, and although a truly formidable avian predator had once occupied the top of the food chain – Haast's eagle (*Harpagornis moorei*) possessed a wingspan of eight and a half feet and had claws the size of a tiger's – it had been extinct for centuries by the time Europeans arrived (Tennyson and Martinson 2006: 62). The conjunction of these two factors – a newly mechanistic European attitude to wild nature and a relatively diffident zoological population – made New Zealand the perfect instrument for the Enlightenment's production of harmonious melody out of nature's wild music.

The human-animal studies approach exemplified here, and followed in the remainder of this chapter, is a twofold one. It involves analysis of the rhetorical and figurative modes by which human-animal relations are represented (in this case, the musical metaphor deployed by Joseph Banks), and the connection of this analysis to consideration of the material practices by which humans and animals actually interact (which in this context include natural historical observation, collection, and documentation, followed by widespread ecosystemic modification in the service of agricultural settlement). This approach has been summarized by Chris Philo and Chris Wilbert who, in the introduction to their paradigm-setting volume *Animal Spaces, Beastly Places*, assert the need for human-animal studies 'to give credence to the practices that are folded into the making of representations, and – at the core of the matter – to ask how animals themselves may figure in these practices' (2000: 5). The identification of these (sometimes fine-grained and subtle) interconnections between the representational and the material treatment of animals acts as a powerful analytical tool for revealing what might be called the 'animal unconscious' of a culture's dominant meanings, forms, and practices. After all, human-animal representations and relationships are everywhere around us, all the time – to such an extent that they most often form a taken-for-granted and invisible substrate of our everyday lives. By taking a commonly and often thoughtlessly used notion ('wildness') and paying attention to how it works both representationally and practically within a particular local cultural history (that of Aotearoa New Zealand), this chapter seeks to reveal ideological assumptions and functions that usually go unnoticed and unchallenged within a particular domain of human-animal relations.

II

As the last substantial habitable landmass on the planet to be settled by *Homo sapiens*, the islands in the New Zealand archipelago have been subject to the most rapidly extensive anthropogenic environmental change in recorded history. Of course, this process did not begin with the arrival of the *Endeavour* in the eighteenth century, nor was it initiated by Europeans. In their six or more centuries of inhabitation prior to the arrival of Europeans, the various populations of Māori are thought to have burnt a large proportion – perhaps as much as half – of New Zealand's original forest cover. During that time 30 or 40 native bird species became extinct and the New Zealand fur seal population was severely depleted (Anderson

2002: 20; Andrews 2009: 167–8, 219). Such a pace of ecological change was swift enough, but the first century of European settlement brought a further acceleration.

The country's wild fauna and flora suffered a radical depletion, as virtually all accessible land was transformed into a farm-supply and population-overflow unit for the rapidly industrializing British Empire. In the first few decades of the nineteenth century the remaining endemic fur seals were hunted almost to extinction for fur and oil, and the southern right whales for baleen and oil, to clothe the citizens, lubricate the factories, and illuminate the streets of northern hemisphere cities. By the end of that century, half the remaining native forest had been cut for timber or burnt off (along with the protective fern and scrub cover in areas already deforested by Māori) to make space for pastureland. Eighty-five per cent of New Zealand's wetlands were drained for the same purpose. 'Virtually no terrain, other than the higher mountains, was left untouched by agriculture': today 60 per cent of the total landmass of New Zealand is taken up by farms or production forests (Andrews 2009: 140–1, 210–15, 292). Consequently, from the mid-nineteenth to the mid-twentieth century, another 30 or 40 bird species vanished (Tennyson and Martinson 2006: 9). This ecosystemic emptying-out was accompanied by an equally vigorous re-stocking. European settlers flooded the country with new species: 'an astonishing 28,000 plant species', of which the number now established in the wild equals two-thirds of the number of indigenous species; 34 terrestrial mammal species and the same number of bird species; 20 freshwater fish, 3 frogs, 1 reptile, and some 2200 invertebrate species (Andrews 2009: 218–19).

Conservationist voices began to be raised in the last decades of the nineteenth century, and the government set up a few national parks and pest-free island sanctuaries for birdlife in the 1880s and 1890s. The Native Bird Protection Society was formed in 1923 (Nathan 2009). Yet Acclimatization Societies around the country were still hard at work well into the twentieth century, establishing introduced species for recreational, commercial, or nostalgic reasons, and laws were still in place to protect introduced species rather than endemic ones (Park 1999: 192). It wasn't until 1962 that a government first recognized conservation as an issue of significance to the voting public and so established the Nature Conservation Council. Over the next few decades, conservationist ideology grew to become a defining feature of New Zealand political and social life.

One of the turning points in the emergence of conservationism was the sensational rediscovery of a surviving wild population of an endemic bird species that had long been thought extinct. In 1948, Geoffrey Orbell came upon live takahē (*Notornis mantelli*) in an isolated part of Fiordland, in the remote south-west of the South Island. Nor was this the first resurrection of these ungainly, flightless, somewhat dodo-like birds. They initially became known to science during the 1840s when a few bones were accidentally acquired by naturalist Walter Mantell. From the outset the species was assumed extinct – until Mantell purchased the skin of a bird recently killed and eaten by sealers. It was like a resurrection: 'a bird once thought extinct was now authoritatively declared a living species' (Andrews 1986: 147). Despite many attempts to locate more specimens of recently living takahē only three were acquired: one shot by whalers in 1851, the others caught by dogs in 1879 and 1898 (Andrews 1986: 141–9). In the ensuing decades, when no more evidence of live takahē was forthcoming, the scientific establishment concluded that the species was, this time, truly extinct. It was not until 50 years had passed that 'a shrewd, well-informed amateur in the Victorian tradition', Geoffrey Orbell, discovered fresh tracks of the takahē and then found the living birds themselves, in the Murchison Mountains in Fiordland (Young 2004: 140).

Orbell's rediscovery of the takahē provided a highly significant foundational myth for New Zealand conservationism, just when it needed one. 'It inspired what might be called the

Lazarus approach [that] New Zealand scientists have, of necessity, become so expert in: bringing back endangered species from the brink of extinction' (Young 2004: 140). Captive breeding programmes of takahē, pāteke (brown teal, *Anas chloritis*) and whio (blue duck, *Hymenolaimus malacorhynchos*) were all undertaken successfully during the 1960s; today the most high-profile efforts are with kiwi (*Apterix australis*) and kākāpō (*Strigops habroptilus*). The most miraculous 'Lazarus' achievement of all was the resurrection of the Chatham Islands black robin (*Petroicus traversi*) after the species had declined to the smallest possible viable number – a single breeding pair. Don Merton pioneered an approach that employed parents from other species to incubate eggs from the last remaining fertile female, nicknamed 'Old Blue', who became a national hero for her efforts on behalf of her species (Young 2004: 191).

These stories of species saved from extinction function for New Zealanders as contemporary, secular myths of redemption – as shown by their retelling in the form of uplifting moral fables for children. For example, Mary Taylor's *Old Blue: The Rarest Bird in the World* – which won New Zealand's national book awards in 1994 for the best children's non-fiction title – presented as a 'story of hard work and miracles, and of people who cared enough to help save a little black bird on a group of lonely south-sea islands', allows Don Merton himself to draw out the wider moral of the tale: '[i]f the rarest bird in the world can be rescued, then given human determination and effort, no species need become extinct' (quoted in Taylor 1993).

The resurrection of the takahē, black robin, and others, and the ongoing efforts to rescue from extinction other severely threatened species such as the kiwi and kākāpō, manifest an intense and widespread cultural desire to redress the ecological wrongs of the past. At the same time these endeavours repeat in updated form the paradox that Joseph Banks encapsulated two centuries earlier: a Romantic idealization of wild nature, combined with an Enlightenment dedication to its manipulation and recalibration.

In a recent anthology entitled *Wild at Heart: The Possibility of Wilderness in New Zealand*, award-winning natural-history broadcaster Alison Ballance demonstrates the durability of this twofold approach when she describes her journey into wilderness in the company of 'kiwi hunter Jane'. Describing their arrival in the Murchison Mountains – the same remote region in which Orbell rediscovered the takahē – Ballance writes that she feels 'a great sense of being in the "wild", of truly being a visitor to another world. This sense became stronger as Jane and I began to drop down the steep mountainside. [. . .] At one point I heard her call back to me, "I feel I might be the first person to have ever set foot here"' (Ballance 2011: 85). Ballance recounts this experience to exemplify the 'three cornerstones' that she considers foundational to the 'best, purest wilderness experience': first, the sense of 'a place that is in some way physically inspiring' and that 'gives a sense of being untouched by human hand'; second, 'a natural ecosystem that has resilience, which is populated by creatures that are meant to be there, and that works, more or less, as it has done for thousands of years'; and third, the evocation of 'wonder, discovery, joy, fulfilment, and that sense of possibly being the first person to experience it' (Ballance 2011: 87). The paradox arises, of course, when we are reminded that the reason these two women are there is to change the harness attaching a radio transmitter to one of Jane's study kiwis – part of the increasingly desperate national conservation effort to save New Zealand's totem bird from extinction. Ballance holds the 'wild' kiwi while Jane readjusts the bird's electronics. 'I had never seen a kiwi that close, let alone held one', she writes, adding that 'it was with a great sense of privilege that I grasped its incredibly muscular legs in one hand and gently cradled its head with the other, while Jane went about her task' (Ballance 2011: 85).

Ballance is fully aware of the contradiction central to her appreciation of the 'wild'; she emphasizes it as she describes the rest of her journey with Jane, which involves checking the regularly placed wooden stoat traps designed to protect the kiwi and takahē from introduced predator species, and when she concludes that '[w]e have upset the balance of nature so much that we can no longer stand back and assume that large tracts of forest are a safe haven for wildlife' (Ballance 2011: 86). According to this view, New Zealand's wild animality is now so encompassed and pervaded by human influence that it is only by means of heavy-handed intervention that a sense of untouched wildness can be recreated: humans have an obligation to empty the ecosystem of introduced wild species and replenish it with endemics.

Of course this belief relies also on the assumption that human environmental managers know what is best for wild nature – that they possess, in regard to their constant tuning of nature's melodious wild music, a kind of biological perfect pitch. Certainly, the instruments at their disposal have come a long way since Banks' tuneable silver bells. A recent item on the TV news described the development, by the Wildlife Management and Conservation Centre at Lincoln University, of touchpad technology designed 'to enable species recognition within traps'. Dr Helen Blackie described the concept as follows: 'You've got a small plate in these systems; an animal runs over the system; it's able to say "hey that's a rat, that's a mouse, that's a kiwi or a weta", and so we're able to adapt these systems to act accordingly' (Pugh 2011).[1] Such endeavours typify the ironies of a political ecology that, in the words of Bruno Latour, claim 'to protect nature and shelter it from [hu]mankind', even as they actually involve expanding humans' intrusion into nature 'in a finer, more intimate fashion and with a still more invasive scientific apparatus' (Latour 2004: 20).

The urban-accessible equivalents of Ballance's encounter with the wild in the remote mountains of Fiordland are provided by wildlife sanctuaries and nature preserves. Driven by the same secular myth, that of the resurrection of lost ecologies and species, sanctuaries are prelapsarian gardens embodying one of the most powerful cultural narratives in New Zealand's contemporary national self-fashioning. As such they offer a very potent emotional experience for visitors. One example is the island of Tiritiri Matangi in Auckland's Hauraki Gulf, which can be reached in 40 minutes by catamaran ferry from Auckland City. Disembarking, visitors make their way from the jetty and climb hills clad in regenerating native bush. Pathways and boardwalks lead through stands of regenerating native trees: māhoe (whiteywood, *Melicytus ramiflorus*), pōhutukawa (*Metrosideros excelsans*), pūriri (*Vitex lucens*), and many others, clearly labelled with their Māori, Latin, and English names. Birdsong is everywhere: the chimes of the korimako and tūī (parson bird, *Prosthemadera novaeseelandiae*), the chitter of pīwakawaka (fantail, *Rhipidura placabilis*), the chatter of kākāriki (red-crowned parakeet, *Cyanoramphus novaezelandiae*), the penetrating cheep of tīeke (saddleback, *Philesturnus carunculatus*). Here, then, are many of the rarest and most elusive of New Zealand's decimated endemics, audibly abundant, visibly thriving. Even the unskilled birdwatcher can spot 20 different species within a couple of hours.

The real enchantment of this small island, though, lies in the birds' fearlessness. Even the most retiring of species – and New Zealand's forest birds tend to be very cryptic and timid – go about their business unafraid, flying or hopping to within a metre of their rapt human observers. To the bird-loving visitor who knows the litany of New Zealand's extinct and endangered native species, these encounters evoke an Eden-like past where humans and the avian natives of this land are united in benevolent harmony and mutual interest – a melodious wild musick indeed. Reaching the summit at the island's southern tip, the visitor also experiences the high point of this enchantment. At the foot of a storybook lighthouse stands a visitors' centre in which volunteers from the Friends of Tiritiri Matangi serve hot and cold drinks. But no matter how thirsty they are, visitors stop in their tracks and reach for their cameras before entering the café. Because there, waiting for them on the lawn surrounding the visitors' centre, are three or four takahē. Unmistakable in their turquoise plumage, the big birds pick their way among the picnic tables and chairs, checking the grass for crumbs, going from one group of excited visitors to the next. So it is that, by virtue of a 40-minute voyage on a jet-propelled catamaran and a 15-minute walk up a manicured pathway, any visitor to Tiritiri Matangi can re-enact for him-or herself the most famous 'return from extinction' story of New Zealand's history.

At the same time, and very deliberately and explicitly, New Zealand wildlife sanctuaries are designed to tell another story. On the voyage to the island, during the welcome talk by the Department of Conservation Guide upon landfall, and at every stage via information boards and displays, the visitor is repeatedly instructed about the primary means by which this sanctuary was created and is maintained: pest control. As Alison Ballance discovered even in the 'wilderness' of the Murchison Mountains, the heroism of New Zealand's 'Lazarus conservation' derives more often from extermination of wild animals than from anything else.

This story is even more visible and vocal in Zealandia, the mainland sanctuary in the nation's capital city, Wellington. Upheld as the epitome of conservation in action, 'recognized worldwide as a benchmark in urban ecological restoration' (*Karori Sanctuary* 2007), the sanctuary consists of 225 hectares of regenerating native bush and wetlands protected by an 8.6 km predator-deterring fence. The area inside the fence is populated by native birds, fish, insects, and reptiles. Zealandia professes a '500-year vision to restore [the] valley as closely as possible to its pre-human state' (*Zealandia* 2010). Upon entry visitors are plunged not into the serenity of an environment devoid of human interference, but into a technologically no-holds-barred warfare against introduced species. The route from the gatehouse to the sanctuary is lined with a sequence of signs and contraptions educating the visitor about possums, stoats, rats, and wild cats, and the various ingenious means by which they may be killed. The odour of a decaying mammal, killed in a trap and now on display, may follow the visitor for several metres along the path. The smell and sight of death is not only apparent at the entrance of Zealandia; every five to ten metres along the pathways possum bait stations have been set. These are highly visible and while warning notices are given as to the purpose and potential danger to humans of these traps, many a visitor cannot help but peek to see if an animal has been caught. For international guests this must seem a strange phenomenon – a kind of contemporary version of the Garden of Eden after the Fall, from which Adam and Eve and their descendants were barred access by an angel with a flaming sword.

As well as demonstrating – indeed embracing – the paradox of an intact wild dependent on the most pervasive, carceral, and lethal of human interventions, sanctuaries like Tiritiri Matangi and Zealandia exemplify the 'trouble with wilderness' famously outlined by William Cronon in his essay of that title (1995). For Cronon, both environmental studies and environmental politics are badly served by their continuing investment in a particular notion of 'wilderness' as the purest, most authentic, and most valuable form of 'Nature'. Cronon's essay traces the development of this concept of wilderness, which he regards as both artificial and tendentious, via biblical, Romantic, and touristic traditions, and then identifies some ways in which 'wilderness thinking' can inhibit environmental thought and action.

Representing the apotheosis of nature as that which is free from and untainted by the human stain, this notion of wilderness reinforces, and even celebrates, modern humans' alienation from the rest of the world. At the same time, seeming to provide a vacation – literally in the case of wilderness tourism – from the irremediable 'culturedness' of modern and especially urban existence, wilderness allows us to 'evade responsibility for the lives we actually lead' (Cronon 1995: 11): being 'green' can mean holidaying for a couple of weeks in a wilderness reserve but ignoring possible improvements to the urban or suburban environments in which we spend the other 50 weeks of our year. In New Zealand, it can mean enjoying a visit to Zealandia and admiring the ingenuity and vigilance of our heroic pest-controllers, while failing to question the everyday habits most detrimental to our wildlife and wild ecologies: for example, the production of meat and dairy and other agricultural products, which squeezes areas of native biodiversity into the rockiest and most remote corners of the country (or into offshore islands or walled reservations), while it contaminates waterways and contributes the largest proportion of the nation's greenhouse gas emissions.

III

Self-evidently, New Zealand's wildlife sanctuaries perform not only an ecological but also a cultural function. As well as providing safe places for the regeneration of endemic species, they are designed to express a dominant cultural rhetoric about what wildness means in this country. The severity of the distinction between those species given sanctuary within the fence and those banished to the other side of it, or killed and displayed in the entranceway, establishes for the visitor a categorical opposition between properly wild animals and improperly wild ones. In order to retain the motivating Romantic ideal of the term *wild*, however, a synonym with different connotations is required for those species whose wildness makes them 'pests'. The term *feral* performs this function. In New Zealand, then, feral animals are those whose wildness carries a taint of degradation: foreignness, invasiveness, over-fertility, lapsed domestication, ill-conceived or inadvertent introduction. The echoes of xenophobic and racist prejudice against human immigrants might seem clear enough – but for the ironic fact that all the vilified feral species in New Zealand were released here, often intentionally, by Pākehā (white) settlers. Indeed, as is so often the case with antipathy to human immigrants, the vilification of feral species performs a necessary redirection of blame, in this case away from the anthropogenic causes of depleted biodiversity and towards animal ones.[2]

Nowhere is this effect more apparent than in New Zealanders' attitude to the brushtail possum (*Trichosurus vulpecula*). This species was first successfully introduced from Australia to New Zealand in 1858. While some were imported to New Zealand as 'cuddly, if thorny, pets' (Druett 1983: 188), the main reason was to establish a profitable possum fur industry in the new colony (Warburton *et al.* 2000: 251–61). This venture was wholeheartedly supported by successive governments: hundreds of separate liberations were made by government agencies and Acclimatization Societies between 1895 and the early 1920s. The Auckland Acclimatisation Society argued during this time that 'we shall be doing a great service to the country in stocking these large areas with this valuable and harmless animal' (cited in Druett 1983: 188). For a period in the early 1900s possums even enjoyed protected status under the Animals' Protection Act (McDowall 1994). Released possums flourished in both the native and exotic forests of New Zealand, feeding on young leaves and shoots of trees and reproducing exponentially in the absence of their natural predators (Clout and Ericksen 2000: 1–9). It wasn't until 1922, following much debate and after concerted pressure on the government from farmers, that the Department of Internal Affairs declined further requests to release possums, although illegal releases continued into the 1940s (Druett 1983).

The turning of public, and official, opinion against the possum coincided with the rise of the 'Lazarus' conservationist ideology described above. In 1956 possums were classified as 'noxious animals', and in the same year the government initiated research on possum control. Following this change in status, anti-possum rhetoric intensified in New Zealand. This was only in part due to the general growth of environmentalist feelings from the 1960s onwards; it was more specifically the result of a concerted propaganda campaign mounted in the 1980s by government agencies (Druett 1983). One of the most pervasive results of this campaign was the widespread assimilation of the xenophobic attitude to feral animals mentioned above. Since the 1980s possums in New Zealand have been almost universally associated with notions of invasion and with military tropes related to defence and attack. The former Minister for the Environment, Simon Upton, states in his foreword to *The Brushtail Possum*, a landmark volume covering biological and social science perspectives on possums in New Zealand:

> 'Know thy enemy.' Sun Tzu's timeless injunction in the *Art of War* applies as well to New Zealanders facing the brushtail possum as it did to Chinese against invading hordes in the sixth century BC. The possum has been a spectacularly cunning and successful enemy.
>
> *(quoted in Montague 2000: ix)*

Department of Conservation spokesperson Herb Christophers uses the same register to defend the use of the controversial poison 1080 (sodium fluoroacetate) as a vital weapon against possums: 'It's critical to our fight to eradicate pests. There may be moral issues over the use of toxins to manage pests, but you don't suddenly stop fighting a war because someone doesn't like guns' (quoted in Marks 2006). The assimilation of both xenophobic and militaristic terminology into popular cultural representations of the feral wild can be observed in an episode of the highly popular local documentary series *Wild About New Zealand* (Bruce 2000). 'Kiwi bushman' Grant Latimer, who presents the show, is deer-hunting with a friend, Roger, through the Ruahine forests in the North Island of New Zealand:

> Trouble is [this forest] is also home to hundreds of thousands of possums who like to gather in one tree at a time and nibble it to death. I mean, *look at this* – a once mighty totara tree reduced to a dead skeleton by *POSSUMS*! You'd think possums would have more sense than to chew out their own food source, wouldn't you? But then possums have really small brains. And they're Australian, aren't they? (He laughs.) [. . .] They may look like cute little teddy bears, but they're *not*! They're pests! They've got teeth as sharp as razors and these claws can give you a nasty scratch. And they carry tuberculosis that can get into our cattle herds.

The next day the pair sets off deer-hunting and comes across a possum-control officer from the Department of Conservation setting poison baits. Despite the DoC officer's comment that possum numbers are low in that part of the Ruahine ranges, Latimer proceeds to tell the viewer:

> Hard to believe there are so many possums in the bush. There are *millions* in fact. Especially in the daytime – they're hiding in the dens. But if you know where to find them, they're there alright. We're just going to take a wee look over here.

Recalling the clichéd cooking show moment in which the chef declares 'and here's one I prepared earlier', Latimer walks a short distance to a part of the bush where a possum is conveniently found sniffing the leaves on the ground in the daylight (although of course possums are nocturnal). Grabbing the possum, who screams as he tries to escape, Latimer says: 'You can tell by their attitude that not all possums play dead when cornered. They'll have a crack at you if they can.' The possum is shown quietening down as he attempts to get a footing on the branches of a shrub; he then looks directly at Latimer, who states: 'I think I'll call this one Stu.' The scene shifts back to the hut in the bush where we witness Latimer cooking over an open fire; he tries some of the food from his pot: 'Hmmm and not a bad stew it is either.'

The scene-setting, prop-placement and scripting of this programme demonstrates the ongoing need to define the kind of wildness represented by possums as feral – promiscuous, rapacious, foreign, diseased, and appallingly fertile – even in cases where possums might be displaying none of those characteristics.

At the same time it obscures a further contradiction, which is the never-admitted investment of New Zealanders in keeping possum numbers *high enough*. For just as possums were originally introduced to create a fur industry they now provide opportunities for big profits – and not just for the pest-control and poison industries. Possum carcasses harvested through hunting, trapping, or poisoning may be transformed into clothing, accessories, home apparel, and food (mainly pet food, although some possum meat is sent to Asian markets for human consumption). Clothing and accessories made from possum fur (or using other possum-derived materials) are hot fashion in New Zealand at present, and particularly sought after by the ecologically concerned middle class. Because of the status of possums as pests, possum fur can be marketed as a solution to the problem posed by live possums; those who choose to purchase and wear it are depicted as environmentally friendly and loyal to their country. In this way, the contradiction between the desire for eradication of the possum and desire to exploit its continued existence is papered over.

In its advertising campaign, the up-market national fashion label Untouched World™ quotes Cameron Silver, a world-renowned fashion reviewer, who describes the company's product line as 'the definitive modern luxury casual brand for the active and conscious globe-trotter' (*Untouched World™ Winter Catalogue* 2006: 1). The mission statement for this business asserts that 'Untouched World™ finds its whole philosophy summarized in a symbol representing a Māori Kite, the emblem of an ideal relationship between man [*sic*] and nature' (*Untouched World™ Winter Catalogue* 2006: 89). Of course, the ideal relationship envisaged here between humans and nature relies upon, and cannot help but perpetuate, a complex history of very non-ideal human modifications to nature. These modifications include the original introduction of possums to New Zealand and their ongoing slaughter using a range of problematic methods, from lethal traps to guns to 1080 and other poisons. Avoiding all mention of these issues, pervaded instead by terms such as 'serene', 'unique', and 'natural', and assisted by imagery of spectacular New Zealand beaches or alpine scenery, Untouched World™'s promotional material insists that the company is organic, ecological, and GE-free. This appeal to environmentally conscious consumers is coupled with the recycling of possum fur in combination with merino wool to produce a 'luxuriously soft and durable' knitwear collection unique to Untouched World™ known as Merinomink.

In an article appearing in the *Independent*, Auckland designer Teresa Angliss (engaged at the time in making a bedspread from 66 possum skins) explained her allegiance to the possum fur industry in New Zealand: 'I call it eco-fur [. . .] I wouldn't work with anything endangered, it would be against my conscience. But this is a national pest, so it's really appealing. I'm exploiting a commercial demand to help contain an environmental disaster' (Marks 2006). In the same article the reader is also informed that anywhere else in the world 'designers who work with fur earn the wrath of animal rights activists. But in New Zealand, they are considered national heroes. Indeed, environmentalists and wildlife campaigners' (Marks 2006). On the website of the *New Zealand National Geographic*, in a news article titled 'Making Possums Pay', we are told that 'possum is becoming the fur *de rigueur*, the designer fibre'; during L'Oreal New Zealand Fashion Week in 2002, 6 of the 50 featured designers used possum fur in their creations, either as trims or borders, or, in one case, for a full-length coat (Cary 2003). Here is the full measure of the potency of the diligently cultivated distinction between wildness and ferality: it can produce an ethical exemption that allows fashion-loving consumers not only to leave behind the taint of slaughtered seal pups and embrace fur garments again, but also to proclaim themselves environmentalists and patriots in the process.

IV

Untouched World's™ fabrics and designs, and its corporate self-fashioning, showcase the emergence of a powerfully marketable taste for certain kinds of wildness. This same taste regime can be seen at work in the gastronomic domain, where it is exemplified by the phenomenon of wildfoods. Nowhere in New Zealand is the lability and adaptability of the contemporary wild more in evidence than at the Wildfoods Festival, held annually in Hokitika, a small town on the South Island's rural West Coast.[3]

Every year since 1990, on the second Saturday of March, 10 to 20,000 locals and visitors have gathered in Hokitika's Cass Square to chew with varying enthusiasm on spit-roasted steaks of chamois, wild pig, venison, and tar; on 'moa' burgers (actually made of ostrich meat) and emu kebabs; on sheep's testicles, wasp caviar, muttonbird, whitebait, and dozens of other slices of 'the wild'. But the strongest flavour is that of a certain brand of local identity: as Westland District Mayor John Drylie notes in the festival brochure, 'this weekend reflects the logo of our district: "Westland: For the Wild at Heart"' (Hokitika Wildfoods Festival 2001a). Like the festival itself, this notion of wildness cleverly – albeit superficially and temporarily – papers over the longstanding antagonism between the two main subcultures that now constitute the West Coast population: an older rural and small-town working class of miners, farmers, timber workers, hunters, and fisher-people versus a newer population of nature-loving conservationists from other parts of the country or the world. The festival allows these two groups to meet amicably and celebrate the consumption of wildfoods – particularly the hunted meat of introduced animals whose wild populations are blamed for the destruction of local ecosystems. In this way the festival suggests that the wild can be served, and served up, in a single process.

Of course, the festival has national and global functions too, representing as it does a lucrative market-branding of the Coast as a destination for foodies and eco-travellers; it advances taste tourism as an environmentally safe alternative industry to both mining and forestry. A glance at the list of stallholders in the festival brochure demonstrates that the wild is being processed in different ways for this variety of audiences and markets. As mentioned above, the term 'wild' designates in the first place a regional identity proud of its opposition to urban over-cultivation and in the second place a taste for consuming those introduced animals whose wildness is believed to constitute the most critical threat to the local environment. Third, the notion of wildfoods implies a strong emphasis on the meats of game and marine animals – wild pig, venison, tahr, chamois, eel, muttonbird, whitebait – supposedly freshly hunted or caught 'in the wild', as opposed to meats derived from farmed stock or purchased via retailers. Fourth, and paradoxically, 'wild' is used to describe the non-traditional (in New Zealand) meats of species that are recognizably native to (or 'wild' in) other locations, which have become part of an increasingly globalized trade and have been imported for farming here over recent decades: emu, ostrich, and wallaby, for example. Fifth, and crucially to the branding and marketing of the festival, 'wild' refers to foods that constitute a radical challenge to conventional canons of taste (at least those of townies): wasp caviar, mountain oysters, huhu grubs, pickled punga, and even – most notoriously – bull's or horse's semen.[4]

As food anthropologists have argued, the distinction between 'everyday foods' and 'festive foods' helps to manage change within cultural diets: an event such as the Wildfoods Festival offers a transitional space within which consumers can experiment with emergent tastes and challenge the boundaries of received or dominant ones, foreshadowing wider-scale shifts in society at large (Lupton 1996: 127). Accordingly, those foodstuffs that are 'wild' in the sense of being adventurous or potentially disgusting, while they are not the items most likely to be incorporated into diets on a regular basis, mark this 'festive' function and anticipate larger cultural transformations. At the same time, the experience of wildness-as-adventurousness also holds together the other sometimes contradictory forms of wildness, and manages interplay between them.

One form of widespread taste transformation celebrated and enabled by the festival applies not just to wildfoods, but to any foods that can be associated with a more primitive and thus authentic past form of production. Hence, for example, the taste for 'peasant cuisine', whereby products once solely associated with a rural, provincial, and working-class way of life are processed, repackaged, and marketed to consumers who until relatively recently had no taste for them: in particular to the urban middle-class amateur food expert or 'foodie' (Bell and Valentine 1997). As part of this trend, the increasing appeal of wildfoods can be measured by their spread, since the Hokitika Festival began in 1990, out of the wilds and into suburban homes and inner-city restaurants. Since 1998 New Zealand's *Cuisine* magazine, designed to appeal to the elite food consumer, has promoted an urban 'Wildfoods Challenge' in which chefs from elite *haute cuisine* restaurants in Auckland, Wellington, and Christchurch compete to produce the best dish encompassing wildfood (defined as 'non-farm produced') (Crimp 2000: 113–20). In 2011 Television New Zealand broadcast eight episodes of *Monteith's Wildfoods Challenge* in which two chefs travelled around the country meeting restaurateurs who competed to produce a winning dish 'using wild foods sourced from within 100 kilometres of their restaurant' and thereby to reveal 'the inspiration that makes a dish what it is – the soil and the surroundings where it all began. Inspired by the land' (TVNZ 2011).

This same reprocessing of a supposedly primordial nature for consumption by a middle-class urban market characterizes media coverage of the Hokitika Festival itself. The stay-at-home urban foodie, for example, tuning into Radio New Zealand National's 'Saturday Morning' programme – whose main audience base comprises the city bourgeoisie – could vicariously sample the wild flavours on offer via an interview by John Campbell (in the studio in Auckland) and Wayne Mowat (broadcasting from Hokitika) with Colin Cutler, an Aucklander who arrived in Hokitika during the 1990s to open a game restaurant, seeking to 'turn the wildfood into a commercial business'. Their discussion concerned precisely those 'distinctions' which embody the habitus of the urban middle-class foodie. They first evaluated the 'difference in taste between eating wild boar and eating farmed pork' and then insisted that 'the flavours can differ from pig to pig, depending on what they've actually lived on in the wild'. Savouring their finely developed tastes for the wild in this way, these radio gourmets then laid claim to the more established aspects of the elite palate by discussing which wine would best complement the meat. Finally, current concerns with healthy eating were combined with a purported reconnection to primitive authenticity: Cutler asserted that in game meats 'the fat content is far far lower [than in farmed meats], and of course it's all organic, there's no human process there except for the catching of it' (Campbell 2001).

In fact, the wild meats at the festival hardly come unmediated from bush and stream onto the plate or the palate – and neither to those served in Cutler's restaurant or in the venues taking part in the Wildfoods Challenge. Indeed, wildfoods are no less subject than farmed meats to procedures of selection and no less liable to chemical and biological contamination. As one festival stallholder described, hunters supplying the festival have to be careful to take the animals from areas 'away from the TB areas and 1080' (Campbell 2001); and as the festival website pointed out, possum – a favourite in previous years' festivals – was in 2001 prohibited by the Ministry of Agriculture and Fisheries (MAF) due to the impossibility of guaranteeing that the animals would be unaffected by the Department of Conservation's saturation use of 1080 in the region (Hokitika Wildfoods Festival 2001b). Another stallholder described the processes by which his wild meats were obtained: the deer had been 'harvested' from a range of locations around the country (mostly in Marlborough, a region 400 kilometres or 250 miles to the north-east, but none on the West Coast itself); moreover, again under MAF regulations, the meat had to come through 'licensed premises', for example, retail outlets such as South Island Gourmet in Christchurch, New Zealand's second-largest city, located on the opposite coast (Campbell 2001). This is the same route followed by game meats on their way from the wild to the kitchens of urban restaurants.

The foodie fashion for wildfoods is far from unique to New Zealand. Britain's Hugh Fearnley-Whittingstall's popular books and TV series represent the UK's best-known instance, while in Australia, cook Sandra Harris testifies that '[w]hat used to be called "bush tucker" has become sophisticated native cuisine and suddenly it's all the rage' (quoted in Morton and Smith 1999: 156). In one sense, obviously, the passion for wildfoods belongs to the generalized Romantic nostalgia for a more immediate relation to nature that has emerged globally over the past few decades, as a reaction against technologization and urbanization, and the accompanying development of intensified farming practices. More specifically, though, wildfoods represent the most sanguinary aspect of a related cultural phenomenon, which has come to be called 'the new carnivorism'. This curiously hybrid taste regime draws on two rather different structures of feeling: on one hand, it partakes of the new ethical orthodoxy of the environmental, organic, and free range movements, while on the other, it joins the backlash against the perceived 'political correctness' of vegetarianism and animal rights. Characterized by the scornful dismissal of ethical objections to the killing of animals for food, new carnivorism instead celebrates, and indeed publicly enacts and broadcasts, the hunting, killing, butchering, cooking, and consumption of animals. As Jovian Parry points out, this new taste regime's supposedly refreshing matter-of-factness about the origins of meat is carefully constructed to showcase certain ideologies and tastes – in particular the organic, environmentalist, and humane sensibility of their adherents – by way of contrast with older-fashioned, 'invisible' forms of industrial meat production. In the process a kind of extra value is added to the meat, since consumers can 'demonstrate their sophisticated and discriminating tastes while simultaneously purporting to marry virtue with pleasure' (Parry 2009: 47).

Self-evidently, the taste for wildfoods caters to this increasingly visible and vocal disposition. Attendees at the Hokitika Festival are encouraged to eat whatever most resembles the *living* animal rather than the usual kinds of meat that have been euphemistically renamed to conceal their origins. The stalls are decorated with the heads, horns, skins, and assorted other offcuts of the dead animals whose meats are being purveyed; the remainder of the animal is right in front of the consumer, often in the form of a carcass on a spit. Yet as suggested above, this relish for reinstating the connection between the edible substance and the living animal involves some rather complex cultural processing, including the sourcing and preparation of supposedly wild local 'game' via an urban gourmet meats franchise, the careful filtering of products to avoid contamination from the kinds of pesticide with which the local 'wilderness' is saturated, and the introduction of exotic 'wild' commodities from overseas. As Parry remarks, the 'search for origin and authenticity' through which the new carnivorism seeks to add to the value of its meat 'is undercut by the selectivity and superficiality of the knowledge offered' (Parry 2009: 43). Seen in this way, the new carnivorism becomes 'a heady brew of certain ignorances and certain constructed knowledges' (Bell and Valentine 1997: 192).

V

Saved from extinction, served through eradication, served up in restaurants: the vicissitudes of wildness and many of its functions are vividly exemplified by its history in New Zealand following the arrival of Europeans. In particular, this history illustrates the way wildness becomes, under modernity, simultaneously the emptiest and the fullest of the classifications we apply to nonhuman animals. Where industrial capitalism depleted the wild in order to refill it according to its own requirements, environmentally inflected consumerism produces a wild that is both replete of meaning (the most sublime, spiritually intense, essential, and pure of natural domains) and utterly empty (a consumer brand whose meanings are always up for grabs). And so the contemporary wild is nothing and everything at once; it is no more and no less than a vividly imagined site in which to play out human-animal relations in all their ideological and material variability.

Notes

1 The name 'weta' is given to various species of endemic cricket-like insects, a number of which are endangered.
2 This section draws on Potts (2009).
3 This section draws on Armstrong and Potts (2004).
4 'Mountain oysters' is a rural euphemism for sheep's testicles; huhu grubs are the larval stage of the endemic longhorn huhu beetle (*Prionoplus reticularis*), a traditional Māori foodstuff; punga is the endemic silver fern (*Cyathea dealbata*).

References

Anderson, A. (2002) 'A Fragile Plenty: Pre-European Māori and the New Zealand Environment', in E. Pawson and T. Brooking (eds), *Environmental Histories of New Zealand*, Melbourne: Oxford University Press, pp. 19–34.

Andrews, J. (1986) *The Southern Ark: Zoological Discovery in New Zealand, 1769–1900*, Auckland: Century Hutchinson.

—— (2009) *No Other Home than This: A History of European New Zealanders*, Nelson: Craig Potton Publishing.

Armstrong, P. and Potts, A. (2004) 'Serving the Wild', in A. Smith and L. Wevers (eds), *On Display: New Essays in Cultural Studies*, Wellington: Victoria University Press, pp. 15–40.

Ballance, A. (2011) 'The Heart of the Matter', in M. Abbott and R. Reeve (eds), *Wild at Heart: The Possibility of Wilderness in New Zealand*, Dunedin: Otago University Press, pp. 81–8.

Banks, J. (1962) *The Endeavour Journal of Joseph Banks 1768–1771*, vol.1, J.C. Beaglehole (ed.), Sydney: Angus and Robertson Limited.

Bell, D. and Valentine, G. (1997) *Consuming Geographies: We Are Where We Eat*, London: Routledge.

Bruce, B. (2000) *Wild about New Zealand*, Auckland: Red Sky; New Zealand On Air.

Campbell, J. (2001) *Saturday Morning* (radio broadcast), Wellington: National Radio.

Cary, P. (2003) 'Making Possums Pay', *New Zealand National Geographic* 63, available online at www.nzgeographic.co.nz/issue-63/making-possums-pay, accessed 20 April 2012.

Clout, M. and Ericksen, K. (2000) 'Anatomy of a Disastrous Success: The Brushtail Possum as an Invasive Species', in T.L. Montague (ed.), *The Brushtail Possum: Biology, Impact and Management of an Introduced Marsupial*, Christchurch: Manaaki Whenua Press, pp. 1–9.

Crimp, D. (2000) *The Wildfoods Cookbook*, Auckland: Reed.

Cronon, W. (1995) 'The Trouble with Wilderness; or, Getting Back to the Wrong Nature', in W. Cronon (ed.), *Uncommon Ground: Rethinking the Human Place in Nature*, New York: W.W. Norton & Co., pp. 69–90.

Druett, J. (1983) *Exotic Intruders: The Introduction of Plants and Animals into New Zealand*, Auckland: Heinemann.

Hokitika Wildfoods Festival (2001a) *Festival Brochure*, Hokitika.
Hokitika Wildfoods Festival (2001b) website, www.westlanddc.govt.nz/wildfoods_festival.htm, accessed 25 July 2001.
Karori Sanctuary Public Brochure, 2007.
Latour, B. (2004) *Politics of Nature: How to Bring the Sciences into Democracy*, C. Porter (trans.), Cambridge, MA: Harvard University Press.
Lupton, D. (1996) *Food, the Body and the Self*, London: Sage.
Marks, K. (2006) 'Alien Invasion: How the Possum Became Public Enemy No. 1', *Independent*, 22 December.
McDowall, R.M. (1994) *Gamekeepers for the Nation: The Story of the New Zealand Acclimatisation Societies, 1861–1990*, Christchurch: Canterbury University Press.
Montague, T.L. (ed.) (2000) *The Brushtail Possum: Biology, Impact and Management of an Introduced Marsupial*, Christchurch: Manaaki Whenua Press.
Morton, J. and Smith, N. (1999) 'Planting Indigenous Species: A Subversion of Australian Eco-Nationalism', in K. Neumann, N. Thomas, and H. Ericksen (eds), *Quicksands: Foundational Histories in Australia and Aotearoa New Zealand*, Sydney: University of New South Wales Press, pp. 153–75.
Nathan, S. (2009) 'Conservation: A History. Voices in the Wilderness, 1769–1907', *Te Ara: The Encyclopedia of New Zealand*, updated 2 March 2009 available online at www.TeAra.govt.nz/en/conservation-a-history/3, accessed 19 April 2012.
Park, G. (1999) 'Going between Goddesses', in K. Neumann, N. Thomas, and H. Ericksen (eds), *Quicksands: Foundational Histories in Australia and Aotearoa New Zealand*, Sydney: University of New South Wales Press, pp. 176–97.
Parry, J. (2009) *The New Visibility of Slaughter in Popular Gastronomy*, unpublished MA thesis, Christchurch: University of Canterbury.
Philo, C. and Wilbert, C. (eds) (2000) *Animal Spaces, Beastly Places: New Geographies of Human-Animal Relations*, London and New York: Routledge.
Potts, A. (2009) 'Kiwis against Possums: A Critical Analysis of Anti-Possum Rhetoric in Aotearoa New Zealand', *Society and Animals* 17.1: 1–20.
Pugh, A. (2011), *One News*, Television New Zealand, 30 August.
Rigby, K. (2002) 'Ecocriticism', in J. Wolfreys (ed.), *Introducing Criticism at the Twenty-First Century*, Edinburgh: Edinburgh University Press, pp. 151–78.
Taylor, M. (1993) *Old Blue: The Rarest Bird in the World*, Auckland: Scholastic.
Tennyson, A. and Martinson, A. (2006) *Extinct Birds of New Zealand*, Wellington: Te Papa Press.
TVNZ (2011) *Monteith's Wildfoods Challenge*, TV series website, http://tvnz.co.nz/monteiths-wild-food-challenge/monteith-s-4424321, accessed April 30 2012.
University of Otago (2006) 'Move over Rover' (advertisement), *Next Magazine*, February, p. 204.
Untouched World™ Winter Catalogue, 2006.
Warburton, B., Tocher, G., and Allan, N. (2000) 'Possums as a Resource', in T.L. Montague (ed.), *The Brushtail Possum: Biology, Impact and Management of an Introduced Marsupial*, Christchurch: Manaaki Whenua Press, pp. 251–61.
Young, D. (2004) *Our Islands, Our Selves: A History of Conservation in New Zealand*, Dunedin: Otago University Press, Dunedin.
Zealandia Visitor Guide and Map, 2010.

13

Feral Attraction

Art, becoming, and erasure

Bryndís Snæbjörnsdóttir and Mark Wilson

Foreword

In the summer of 1999 we (Bryndís Snæbjörnsdóttir and Mark Wilson) undertook a nine-day hike in Hornstrandir, an uninhabited and remote coastal area in the far north of Iceland. It was July and at that time of year, in that region, there is 24-hour daylight. Remarkably, however, for virtually the entire hike we were submersed in a shroud of dense mist. Consequently, despite the general light, for over a week we were unable to see much beyond a few paces, either back from where we had walked or ahead in the direction we were going. At the time, paradoxically, this had been a heady experience, close to epiphanic in its effect. Where the physical activity of walking in 'wild' landscape for that length of time is normally associated with retinal reward, with 'views' to draw the eye into a distancing and objectifying relationship with the terrain and away from the immediacy of bodily locus, in this case, because of the mist, our attention was entirely held in an enforced myopia. Unable to draw upon the reassuring and conceptual certainties of a commanding view and so (dis)placed beyond the controlling apparatus of representation we were cast instead into the stumbling blindness of uncertainty, of indeterminacy, instinct, intuition, of saving our skin – in short, into the awkwardnesses of close terrain negotiation, survival, and most significant of all into the ontology of 'the moment'.

Though revelatory, it was so in ways we could not easily express. We discussed it as a form of cerebral locking-in, where the deprivation of seeing either forward or back left us in a state of temporal suspension. The terrain remained to be negotiated (we were driven with increasing anxiety by the imperative of an arranged rendezvous with a boat many miles south

Figure 13.1 Tálkni peninsula, Snæbjörnsdóttir/Wilson, video still(s), 2010

of our starting point), but this necessitated navigational means, which were suddenly and lastingly bereft of the faculty of vision. Like most people, we have experienced conditions of uncertainty and fear many times but this was altogether more all-consuming and immersive. Simultaneously and crucially, it must be said, it was also exhilarating.

The point of this is as a reference from which to suggest that there are other ways (involving the relinquishment of control) of experiencing and understanding the world beyond what is deliverable to us by means of language, semiotics, and whatever means we customarily deploy in order to control. The story touches on ideas relating to the familiar and unfamiliar in the landscape. It turns the attention to methods we might use when confronted with the unknown, in order to soothe and calm anxiety and to populate our perceptual world instead with representations stripped of threat. It is no exaggeration to see the fear that prompts us to protect ourselves as being a key driver behind the acquisition of knowledge. The need to bring everything into the realm of what is understood and 'known' has led us to cut ourselves adrift from things which otherwise would tax us. But the reductionism implicit in this process has without doubt left us impoverished in other ways. Our insulation from environments beyond our urban or agrarian control has robbed us in turn of the know-how of how to be, not just 'in' the world, but 'with' the world. In the context of this chapter, what we propose is that the attraction of a feral condition lies in contradictory feelings provoked in us, in a disruption of order and an escape from what is known, named, or contained. It turns things upside down and calls into question the otherwise indisputable. It speaks of the intentionality of 'things' and, like the arrival of sudden, heavy snow in the city, reminds us that things remain beyond human control. The condition that the feral state stirs in us, between uncertainty and exhilaration, or more practically between a sense of inconvenience or the opportunity to see things anew, is its compelling attraction.

Introduction

The project *Feral Attraction* explores the disconnection between empiricism and cultural determinacy, and considers the effects of cultural blindfolding in a context of environmental fatalism. The project focuses on the site of the Vestfjords, a remote area in the Northwest of Iceland, which became the theatre for the enactment of urban/rural ideological tensions and ultimately a frenetic and awkward resolution involving the herding and eventual eradication of a community of feral animals. The Vestfjords is an environment that during the twentieth century has been increasingly host to controversy surrounding the inexorable population drain from a rural to an urban way of life involving two gradual processes. One was the migration of people from farming regions into coastal towns (including Reykjavík) and the abandonment of farms in large parts of the Vestfjords. The second was the persistent out-migration from the region as a whole, including its coastal towns, both small and large. The herding narrative that follows in some way mirrors the management of remote farming

families and small communities, the continuing presence of which came to be considered from an administrative perspective to be an unsustainable drain on the wider national project. Our art research project focuses on the significance of imagery in the story and on the particular resonance of visual information in the accumulation and instrumentalization of knowledge as the events unfolded.

Feral Attraction examines the particular incident in which a flock of feral sheep, resident for several decades on the remote mountain peninsula of Tálkni in Iceland, was finally and with great difficulty rounded up in order to satisfy agricultural protocols and the legal subordination of farmed animals in Iceland. As recently as the 1920s, although not strictly considered good husbandry, it was not unusual in Iceland to keep sheep out and grazing through the winter months – a custom known in Icelandic as 'útigangur'. For a number of reasons, including the increased capability for haymaking through mechanization and the need to address widespread land degradation and soil erosion, during the twentieth century the practice fell increasingly out of favour. Anyone now who allows the sheep to overwinter in the mountains not only transgresses what is thought of as good practice, but indeed is in breach of the law itself. What began as a way of exercising more control over stock eventually came to affect perceptions of the animal itself and its relationship to its environment of over a thousand years. This was signified by a reduced estimation of its capacity to survive in its adoptive land and a concomitant increase therefore in the assumed responsibilities of its keepers.

As artists, our enquiry engages with environmental and relational discourse, and so a scrutiny of the representation of others and other species is central to our work. In an earlier art project, *nanoq: flat out and bluesome* (2001–6),[1] concerned with the killing and capture of polar bears by British expeditions over the past 200 years, we mapped a transition in the culturing of a 'wild' phenomenon. In *Feral Attraction* we follow in reverse the passage of the Tálkni sheep from farmed to feral beings, acknowledging their independent survival in a wild landscape. In respect of both transitions (polar bear or sheep) the association with man anticipates a fatality, veiled in a representational transformation. Amongst other intentions, our work critiques the still prevalent primacy of human interests and environmental exceptionalism together with the apparent impossibility of humankind to divorce itself from its solipsistic regard for self-survival, both practically and theoretically. Instead we lean towards a relational and ecological paradigm in which the species *Homo sapiens* is accepted as merely a player amongst a multitude of players.

Through Icelandic history the polar bear has been an occasional visitor to the island shores; folklore has generally recorded horrific accounts from these meetings.[2] Far from seeking to underestimate the danger of polar bears under these circumstances, we want to take a step back to reflect and to consider alternative and what we consider more measured and inquisitive approaches and behaviour towards the 'aberration' of unexpected arrivals and migrations in the landscape. In Svalbard, a territory in which the encounters between polar bear and man are frequent, legal constraints are in place to avoid polar bear deaths whenever possible. The right of the indigenous animal to this landscape, which the Spitzbergen human community has come also partly to occupy, is paramount, instilling and reflecting a different sense of respect and environmental order. Whilst on an artists' residency in Longyearbyen, Svalbard in 2010, the local radio reported a group of tourists in crisis: a polar bear had shown up around their camp and was not responding to air rifles, flares, and other customary measures used to scare bears away. But instead of tranquilizing the animal and airlifting it to a new location, as might normally be the case, the ombudsman/sysselman ordered that the tourists instead be relocated by helicopter.

In respect of these examples on Svalbard and Tálkni, the potency of the encounter between 'man' and 'animal' signals complexity in the perceived constitution of environmental order and protocols. Australian ecofeminist Val Plumwood (2000), in her reflections on her experience of being attacked and nearly killed by a crocodile in the East Alligator Lagoon in Kakadu, northern Australia, highlighted a significant ethical perspective by recording her determination that, contrary to the normal response following such attacks, the animal in question should not be hunted down and killed, believing herself to be an intruder into its territory.

In taxonomic and other human systems designed to underpin the human position in relation to other beings, the differences between a crocodile, a polar bear, and a sheep are clear. But what can be compared usefully is our approach and attitude here to any species testing the margins of what we consider to be 'our' territory. In our inability to adjust to the signals of environmental threat (a condition sustained by such anthropocentrism), it is and will remain impossible to distinguish or redraw our taxonomic biases of significance. Dust, plants, animals, minerals, biomass, particles, waves, oxygen, cold, and densities are oceanic in their combined effects and mutuality. In this light, importance may not be measured in their apparent individuality, nor indeed in their 'human significance', but in their infinitely complex behavioural associations and interactions.

Historical context

There is a history to 'feral' sheep being on Tálkni. The flock initially came from a farm called Lambeyri, whose owner, due to personal circumstances, is thought not to have been managing his sheep strictly in accordance with the law. In the early 1970s he stopped farming, leaving the farm to his family. They chose to keep the house and the land, but did not wish to keep farm animals. It is understood that by that time, the remaining sheep at the farm had already taken to the mountain. In accordance with Icelandic law, landowners of registered farmland have a duty to contribute to herding sheep generally in their district. In the Vestfjords this involves negotiating the many steep mountains characterizing this landscape. It is a dangerous job, requiring detailed knowledge of the area as well as substantial agility and fitness.

The first officially recorded and acknowledged existence of the sheep in Tálkni was in 1984, when there was an outbreak in Iceland of the neurodegenerative disease scrapie ('riða' in Icelandic). At the time it was thought that between 40 and 60 sheep were on the mountain. An argument concerning the presence, or otherwise, of the disease ensued between local farmers and the chief veterinary officer. The farmers wanted proof that the disease was in their area before agreeing to cull their sheep. The veterinary officer was unable to provide such evidence, but pointed to the unregulated sheep on Tálkni as being amongst those that might have crossed the district borders in question and so were possible carriers of the disease. In the end, the chief veterinary officer ordered these sheep to be culled.

To fulfil the task, he called on a special division of the Icelandic Coastal Guard known as the 'Viking Squad' to shoot the sheep on the mountain from a helicopter. This was something at which this special squad was not expert. To make matters worse, the expedition hit bad weather, resulting in it having to be aborted, leaving some sheep dead but many more badly wounded. A few days later, when the weather had calmed down, the local rescue team then went over to the mountain and found, as reported by Lilja Magnúsdóttir when we interviewed her in Reykjavík on 25 June 2010, '30 sheep either dead or dying'. In respect of these animals, the team had no choice but to finish what had been started. They also took photographs at the scene, which locals would reference in support of their opposition to further remote directives of the chief veterinary officer back in Reykjavík. This visual evidence prompted two consequences: it helped constitute an identity for the remaining flock of sheep as 'Tálknafé' (Tálkni-sheep) and it galvanized the identity of the local farmers, affirming the division between themselves and the authorities based in the nation's capital.

Twenty-five years later, in October 2009, news came that the flock of feral sheep on Tálkni had once more been targeted, but on this occasion they had been herded from the mountain and slaughtered. As the story of their capture unfolded through the media, it revealed a fascinating tale of human dominion. A group of the best herders ('smalar') of the area, together with their sheepdogs, had risked their lives in climbing the mountain to retrieve the sheep. The adventure was undertaken by order of the chief legislative officer in the area. The consequence was that from a flock of 25, 14 sheep were captured alive, 5 perished as they fell from the cliffs in their attempts to avoid capture, and 6 managed to escape. The 14 captive sheep were loaded on board a boat that had brought the herders to the mountain and taken to the nearby town of Patreksfjord, where they were immediately slaughtered. In addition to the ewes, there were four rams and, retrieved from the other side of the mountain, three yearling rams. The six remaining sheep, two rams, three ewes, and one ewe lamb, were recoverd a few months later in January 2010, thus ending for the time being at least the existence of sheep on Tálkni.

Art and relationality

This story caused considerable controversy in Iceland at the time and for some provided a new focus for environmental concerns. In order to find out more, we visited the Tálkni area in the summer of 2010, together with Dr Karl Benediktsson, Professor of Human Geography at the University of Iceland, and Unndór Jónsson, an independent artist and researcher, and recorded a series of interviews with people who had been connected to the events. We gathered images, documentation, and other material along the way. On location we interviewed Ásgeir Jónsson, Ásgeir Sveinsson, Þröstur Reynisson, Sveinn Eyjólfur Tryggvason, and Ragnar Jörundsson. On our return to Reykjavík, we interviewed Lilja Magnúsdóttir, another local inhabitant temporarily residing in Reykjavík, as well as previous chief veterinary officer Dr Sigurður Sigurðarson, and Dr Ólafur Dýrmundsson, whose specialism is the Icelandic sheep and who works for the Farmers Association of Iceland.

Ásgeir Jónsson is a member of the local council, who had also for some years been responsible for organizing the autumn roundups of sheep. His role was important, but complicated, in that for many years he had turned a blind eye to the sheep being on the mountain. Due to his official capacity, when the court order came, he was forced to take part in the clearance. He also had valuable information for this project because, due to a recent minor accident, he was stationed back in the boat to receive the animals during the herding, which meant that he had physical contact with each sheep captured.

Ásgeir Sveinsson, a sheep farmer, lives and farms with his brother and elderly father. At the time of interview he had a flock of 800 sheep. He has a reputation as an excellent 'fjármaður' (shepherd) and is the owner of exceptionally well-trained sheepdogs. Ásgeir's interview provided a detailed description of the sheep's unique behaviour and their unusual reaction to humans and dogs. He is very knowledgeable about sheep breeding, and is the proud keeper of the only remaining known descendant of the sheep from Tálkni. This ewe was the progeny of a sheep that had escaped to Tálkni, but which he had managed to herd back from the flock. His description of the characteristics of this animal was of further interest. He told us that in spring, when released, she heads to the top of the nearest local mountain and spends the summer there, apparently alone.

Þröstur Reynisson took part in the herding as an employee of the town council and the owner of a good dog, prepared to stand its ground. His role was to be at the foot of the mountain with his dog to stop the sheep escaping along the beach. He talked about the wariness of the flock, reasoning that from time to time the sheep had been shot at by locals, some for target practice and others for meat.

Sveinn Eyjólfur Tryggvason was recruited by the governor of the local council and had been put in charge of the herding operation. He selected the men who went on this trip. In the interview he discussed the different characteristics of this flock and why the dogs did not work as they might in other sheep-herding exercises. He did not consider it unusual that some sheep fell from the cliffs, as it is behaviourally characteristic of sheep when cornered on a mountain. He talked from the perspective of animal welfare and observed that the flock would have been much larger had it been kept in the right conditions. He mentioned that only 1 sheep of the 19 caught had been earmarked, as all farmed sheep are. This animal had been on Tálkni for four years, indicating that it had chosen to join the feral flock.

In our interview with Ragnar Jörundsson, the governor of the municipality of Vesturbyggð, to which the village of Patreksfjörður (Patreksfjord) belongs, he talked about the police involvement and how the local council established jurisdiction to clear the area of sheep. He discussed the division between the local people and city dwellers 'who don't know anything about sheep'. He also accused the media of reporting the incidents in a particularly frenzied way in order to stir up opposition, deeming their reportage to be misinformed. He talked about the responsibilities of the district council towards sheep that are unclaimed and therefore 'in need'. These sheep by default belong to the council. He said the council takes advice from the chief veterinary office and the Farmers Union and that both thought it best to clear the sheep of the area.

Lilja Magnúsdóttir was born and raised at the farm opposite Tálkni. She was part of the first serious attempt at gathering the Tálkni sheep, which took place in 1992. She is interested in the breeding of Icelandic sheep and described the physical appearance of the sheep both as livestock and as meat during and after the 1992 gathering. In our interview her description concerned the particular shape of the feet, observing them to have been higher and thicker than in normal sheep. She also described their bodies as being longer and more slender. As meat she described the location of fat as being in the muscles themselves and under the skin, whereas in the farmed Icelandic sheep the fat is around the abdomen. She proposed that the latter sheep would not have survived, as they would too easily have been caught up with and trapped in the heavy snow. Her theory concerns 'natural selection' as she calls it, saying that the sheep originating from the Lambeyri stock – the ones that were 'abandoned' as it were when the farm closed – were more suited to the landscape and the weather, and, that this was the reason the majority of the flock looked as it did, despite newer additions.

Ólafur Dýrmundsson's comments were from what he considers the perspective of animal welfare. He put forward various reasons why sheep are not able to take care of themselves. He pointed out that one out of every four sheep taken in 2009 to 2010 were from other farms around the area. Despite that, he acknowledged that the majority was of a colour no longer prominent in Icelandic sheep. He said that this had caused difficulties for the elite coast guard squad when attempting to shoot them from the helicopter, because the colour blends with the landscape, making the animals difficult to see. He went on to say that this colour is the dominant outcome when mixing with white, and that the presence of a flock of sheep in Tálkni would always encourage other sheep to join the group, it being such a difficult area to herd.

In this respect he was sympathetic with farmers at not being able to go after them. While he denied there being any such thing as 'feral sheep', he estimated that of 470,000 sheep in Iceland approximately 500 are not accounted for.

Jón Þórðarson was one of the owners of the surrounding land closest to Tálkni. He was against the herding of the sheep, wanting instead to keep them on the mountain. At the time of our visit he was living in the nearby town of Bíldudalur, and from there he runs a tourist and fishing business together with guesthouse and art gallery/residency. His idea was that they could have become a tourist attraction. He tried to stop the sheep being herded by declaring them to be on his land; however, by law, in order for this to be acknowledged, he would have had to build shelters for them on the site, and so his claim was dismissed.

From the beginning of our research in Iceland, the role of the image was of great importance. Images were crucial in cementing the identity of the sheep as a 'special flock' by their unique appearance, on the one hand, and by means of the television footage documenting their attempts to escape capture on this inhospitable mountain, on the other. This footage stood in contrast to and conflicted with an image of domesticated livestock destined for the slaughter and consumption normally associated with 'réttir', the autumnal roundup. In the imaginations of many who protested from near and far, these sheep instead were independent beings of note, deserving of their right to live out their lives. The imaged embodiment of the animal in an effectively nonhuman landscape seems to challenge the scope of human representation by means of a paradigm shift. In its apparent self-determination, the animal in question can be seen to have grown into 'its larger self' through its adoption of this landscape as a

Figure 13.2 Sheep on Tálkni, video still, 2009

permanent home beyond human accessibility and control. The (albeit perhaps reluctant) acceptance of this by locals in the surrounding area for so long eventually in itself became a bone of contention, prompting the central government office to make demands for the flock to be recovered.

To return to the comparison between the polar bear and the Tálkni sheep in the context of the Icelandic landscape, it has always been deemed necessary to kill the polar bears arriving in the country, because the Icelandic wilderness is considered not to be their natural environment. As a non-native species and a carnivore, the polar bear is considered a danger to other Icelandic beings and impossible to contain humanely and securely. The Tálkni sheep on the other hand did not threaten anyone or anything. The land they occupied was not managed, occupied, or indeed coveted by anyone else. In fact, the family who owned Tálkni was quite comfortable with the flock of sheep remaining there. This however raised the legal necessity to erect houses for their shelter and upkeep – itself an impossible task, considering the nature of the land and its limited accessibility. Paradoxically, this would in turn have undermined the independence of the flock and therefore defeated the purpose of any armistice.

There are different ways of interpreting what happened on the mountain the day the flock was herded, leaving five sheep dead and six still at liberty. There are many questions to be asked regarding animal consciousness – whether, for instance, that in the context of new circumstances, jumping from a cliff is indicative of the exercising of choice. There are the complicated distinctions to be made between what is seen as a 'noble' and an 'ignoble' killing – the affront to the many Icelanders who protested was triggered by what was seen to be a bungled and, as a consequence, inhumane exercise of shepherding. Was this perhaps an unconscious conflation of the idealized concepts of tidiness and seamless erasure?

When humans slaughter animals, their imposed departure is one of transformation, not normally regarded as one of eradication. In order to preserve a sense of vital continuity within Western, particularly Anglo-American culture, it has been a characteristic that insofar as we are eaters of animal flesh we focus on meat as opposed to the extinguishing of life that such consumption necessitates. This death is a by-product of our desire to eat, but its visibility has been discreetly minimized in deference to this more culturally palatable focus of attention. Everywhere in the story, and not least in our interviews, there are contradictory perspectives and conflicting ideals. There is the unquestioning belief by some of the need for adherence to existing legislation. There are environmental perspectives, those based on animal rights, and there are matters of professional and moral pride, including the desire of farmers to be seen to be 'taking care' of the animals in their charge.

Exposition and process

From an early stage in the examination of this story and in the research material we accumulated, we stumbled again and again on the claim that something odd had happened to the sheep during their time on Tálkni. In the media, in popular accounts, and in the interviews we conducted with those on the ground, there is repeated mention of an adaptation in the sheep's physiology. A natural, adaptive, even proto-evolutionary process had occurred, due apparently to their constant negotiation of this demanding mountainous terrain. This adaptive response to the topographic constitution of the landscape that ensured their relative isolation and insulation from humans for so long seems to have been most conspicuous in a lengthening of their legs. But in the absence of concrete evidence, how is this ever to be tested?

In art, there is often more significance in one identifiable and well-articulated detail than can be relayed in a wealth of information, particularly where such a detail exposes a flaw, a fluctuation, or break in the rhythm of cultural (and human-specific) affairs. Relationality is key to our artworks and projects. It is reflected in the research process by which we seek to gather information and evidence through contact with individuals and organizations concerned. These meetings are often recorded or documented through photography, video, or audio, and are often pivotal in influencing the structure and the development of an artwork. The biggest significance of this story is in the exposure of the insecurities of 'expert' culture. Those who felt they should have power felt their power usurped. In managing the evidence (the disposal of the flock and subsequently of the carcasses and bones), the community of experts involved reduced the physical signs by which the history of this event (this nomadic becoming) could be remembered or told. In the absence of relics and data, all fact and fiction is conflated, all borders between them are blurred and therefore subject, potentially, to wholesale dismissal as myth. But, by ascribing greater significance to materiality and ensuring its strategic absence, the perpetrators of this act perhaps underestimated the ripple effect of that removal; into every void, the imagination will pour its will or its questions. Without the hard evidence to provide a satisfactory backstop to such suspicion, the impertinence of the questions is always likely to exceed what facts alone might have tempered or quelled.

In this work, the value of the leg and its transformation pertains to its role in the extruded and extruding process of liberation: the sheep remained out and over time became better capable of being so by a process of adaptive response to the environment. The symbolic driver in representational terms is in a retrieved relic – even (by necessity) a faked relic, designed and made in order to give value to something observed but allowed neither to survive nor be measured and corroborated by instruments of science. For this artwork, the bone is extended in correspondence with the lengthening of the legs as was reported by some observers of the

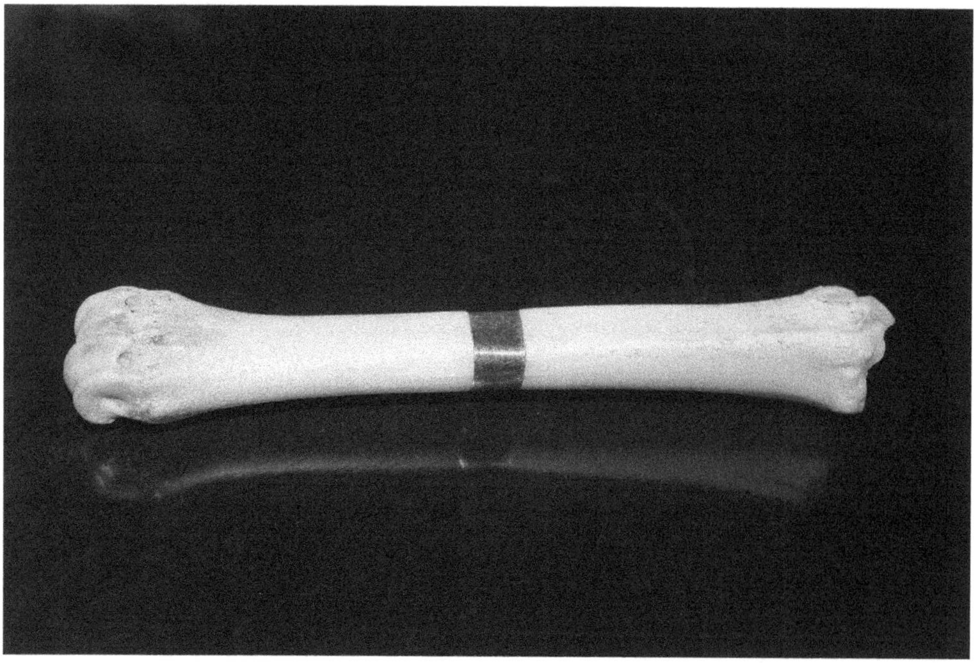

Figure 13.3 Relic, Snæbjörnsdóttir/Wilson, sheep bone and silver, 2012

sheep. The gap of significance is bridged by a hoop of silver, a material we accept culturally as being 'of value'. The human representation and symbolic conferral of importance may even be seen as a compromise here, in deference to the semiotics of a culture that often fails to recognize intrinsic value.

Here, silver gives presence to the missing, valuable, phenomenological, and symbolic 'effect'. It signifies the change that is intrinsic to a) a period of time; b) a specific location; c) the transitional condition of 'becoming'; and in addition both d) a theft; and e) a possible conspiracy where all evidence of the flock and its bodily remains were eradicated deliberately, before biologists (for example) had the chance to examine them. Broadly it is the embodiment of difference – in opinions and of the contested claim that the sheep adapted as a consequence of having become feral. As there was no opportunity for scientific study to be conducted on the flock before or after slaughter, we mark a space in which this extension hovers between being a memorial and a relational corollary of being feral in a mountainous landscape. With this intervention, we keep alive the story of a community of domestic animals, which despite climatic inclemency and the seeming impenetrability of this landscape survived without human care for three decades and indeed showed every sign that they might have continued to live there in perpetuity.

After-lives

The flawed nature of this enterprise, that is, the inefficient and messy nature of the herding of the Tálkni sheep, had a retrospective after-effect, calling into question the validity of the enterprise itself. The very representational tropes that ennoble human agrarian enterprise, for example, the promotion of efficiency in the management of land and animals of the kind implicit in historical paintings (Thomas Gainsborough's *Mr and Mrs Andrews*, Paul Potter's *Bull*, and innumerable seventeenth-century livestock paintings), throw the dubious nature of the reclamation of the herd into sharp relief. Whilst involving a starkly different kind of relationship to our 'landscape others', there are parallels also to be drawn (and they have been described above in this text) between the controversy and contradictions embodied in this episode and those prompted by what has become an intermittent but recurrent phenomenon in northern Iceland, again the arrival of 'stray' polar bears. Tidiness and order are threatened by the presence of the feral animal on the one hand and the appearance of the exhausted and dangerous alien wild animal on the other.

Rather than engaging with complexity and being open to the opportunities it may offer, the default position of local and national governmental authorities seems to be to excise the agent that would test its borders, thereby allowing the status quo to be maintained. The resultant human fault line seems to run between two ideologies: on the one hand, a national, establishment view, in which the integrity of Icelandic farming (and perhaps culture) is an imperative; on the other, a growing lobby of those whose interests can perhaps be said to be less locally rooted and who are able at this point to exercise little direct power, but whose collective voice increasingly coheres around environmental ideas extending far beyond nationhood.

In another art project from 2010 entitled *Uncertainty in the City*, we explored the idea of contested 'human' environments with specific relation to the presence of other species. The project hinged on an (albeit unwritten) assumption of neutral interspecific claim to territory, and we conducted interviews with hundreds of participants in relation to their encounters with animals within and around the margins of their home. Along with an invitation to retell their stories we invited them to consider ideas of ownership, colonization, and encroachment in this context.

Given the space to objectify their experiences, there was surprising openness towards questioning the rationality of their responses and to confronting the emotional inconsistencies within such experiences. The garden – a piece of land we suggested was a surrogate, albeit altered, tract of 'nature' – is a kind of cultural epidermis by which tolerances and affections for others moving through were tested and analysed. In this project and more widely we use aberrant exemplars to challenge accepted behavioural and cultural tropes. In the *Uncertainty* project, nonhuman encroachment on human systems was often, although not always, viewed as a negative occurrence. Typical cases were the presence of ants, for instance, around the door to a house; the margins of tolerance were drawn in relation to the proximity to the threshold or perceived infringement either towards or across that line. Urban foxes and seagulls are amongst the most consistently contested species and their presence is alternately construed as pleasant, desirable, or offensive, according to the experience and/or conditioning of the humans concerned. In the case of the Tálkni sheep, however, the migration away from the human, in giving further dimensionality to the phenomenon of human/nonhuman entanglements, reminds us that our presence is neither necessarily crucial nor desirable for most species, even ones we have domesticated and trained to be reliant upon us.

The feral flock was a thorn in the side of the agricultural community – not necessarily those in the local area in question, but more starkly and tellingly from a remote, central-administrative perspective. But, in the resistance of something, particularly an entity that is normally attributed with little self-determination, as artists we see something much more interesting, in that it breaks the mould of our expectations – it draws our attention. The expression and enactment of capabilities beyond what we are given to believe is expected forces us to re-examine our perception of that thing and our initial reasoning for arriving at such a perspective. Did we believe we had modified the behaviour and capacities of the domestic sheep to the extent that it had indeed become an unreconstructed model of our projected will upon it?

Just as we might enjoy the frisson of being lost when we believe that it is a temporary condition, so too do we find fascinating the idea that our constructed worldview is in some way destabilized by the will of another. In the same way therefore when such aberrant behaviour is suppressed, there is a sense that an injustice is done. Something that appeared to us to offer a new perspective, rather than being acknowledged and valued, preserved, and observed, is eliminated purely in order to restore the status quo and to reaffirm the behaviour-model. Such action is based on an anthro-utilitarian approach that sees adaption or evolution within systems we have engineered around other adopted organisms as running counter to our interests and therefore undesirable.

Simultaneously of course the phenomenon exposes the mythic projections we deploy in order to uphold our utilitarian requirements: if an organism is useful for this and that, then anything – any capacity, behaviour, or inclination that does not support that function-set – may, if noticed, be deemed undesirable and may be subject to extirpation. This thinking is the basis of intensive breeding programmes and the kind of genetic modification that gives us, for instance, hairless cats and seedless grapes. It is this single-mindedness that in modernity has caused us increasingly to consider things, places, and beings in isolation. This has been to the detriment of possible developments towards a more coherent and complex worldview, which might privilege, instead, an understanding and appreciation of ecologies and the acknowledgement of material interconnectivity.

There is a tension between what we hold culturally as being right and proper and what we observe as a bid by another agent to disrupt that order. At the heart of this case is something that may be dismissed by many to be trivial and inconsequential; for us, in ways resonant with those ideas proposed by Jane Bennett in her book *Vibrant Matter*, it serves as a vital pointer to expose how human systems suppress the inclinations and capabilities of 'things', seeing instead only what we have designated for them. We have a tendency to blind ourselves to the wills of those outside our systems whose actions do not correspond with, or seem at odds with, our own – who are simply not compliant in the human enterprise at hand. When the animal agent is one with which we technically coexist (a domestic animal), the oversight seems particularly acute. A lack of porosity is evident: a resistance to ideas or indicators of change; a reactionary dismissal of knowledge concerning environment and the adaptability of denizens to the shaping of existence by environment, or the capacity of discrete environments to model not only new biological permutations but to spawn new behavioural possibilities as a consequence of introductions or migration; a failure on our part still to acknowledge that a condition of 'becoming' is actually the norm in nature, and that stability and material independence are illusionary.

Acknowledgements

We thank Professor Karl Benediktson for his assistance and support of this research, and Unndór Jónsson for transcription of interviews and research assistance.

Notes

1 *nanoq: flat out and bluesome* traced stuffed polar bear specimens in Great Britain from their current locations in museums and private collections back in time to their Arctic encounter with man.
2 In 2009 the authors were present at Skagafjörður in northern Iceland to witness a failed attempt to capture and relocate a polar bear discovered in the area. Despite a national will to manage the situation more adeptly than had been done in the past and governmental involvement towards that end, the incident once more ended in the death of the polar bear.

References

Bennett, J. (2010) *Vibrant Matter: A Political Ecology of Things*, Durham, NC: Duke University Press.
Magnúsdóttir, L., interviewed by Snæbjörnsdóttir/Wilson, Reykjavík, 25 June 2010.
Plumwood, V. (2000) 'Being Prey', in J. O'Reilly, S. O'Reilly, and R. Sterling (eds), *The Ultimate Journey: Inspiring Stories of Living and Dying*, San Francisco, CA: Travelers' Tales Inc, pp. 128–55.
Snæbjörnsdóttir/Wilson (2006) *nanoq: flat out and bluesome*, London: Black Dog Publishing.
—— (2010) *Uncertainty in the City*, Berlin: The Green Box.

14

Becoming rhinoceros

Therio-theatricality as problem and promise in Western drama

Una Chaudhuri

> The dust raised by the animal spreads across the stage.
>
> Stage direction in Eugene Ionesco, *Rhinoceros*

The eponymous animal of Tennessee Williams's play *Night of the Iguana* is typical in one way: animals tend to be heard rather than seen on the stages of Western theater. By being located in the wings, the captivity and suffering of Williams's reptile are not only marginalized but rendered *obscene* – quite literally so, if we accept the contested etymology of the word, from the Greek *ob-skene*: off-stage (McKay 2010). This obscuring and 'obscening' of the animal is the hallmark of the dominant tradition of Western theater, a tradition that is obsessively anthropocentric, dedicated to constructing and enshrining the human as 'the paragon of animals', in Shakespeare's *Hamlet*'s famous phrase (II.ii.319), by derogating or excluding all the other animals.

However, the Western theatre's repression of animality – like the many other such repressions through which humanist cultures are produced – comes at the cost of a profound ambivalence, an anxiety that the repressed will return when least expected and will dissolve the vaulting edifice of self-importance into a quintessence of dust. To manage this anxiety, the theatre fills its spaces with animal effigies, symbols, masks, and meanings: signs of a disavowed animality that, taken together, amount to a shadow tradition, a 'therio-theatricality' through which drama peers into what John Berger so memorably calls 'the narrow abyss of noncomprehension' (1980: 5) that separates humans from the other animals. Rooted in antiquity and flourishing through early and later modernity, this shadow tradition reaches the limits of its ideological usefulness and psychological value at the dawn of the postcolonial era, when accounts of otherness proliferate to the point of shaking the rigidly binary foundation of humanist ideologies. In the 'experimental' European drama of this period (starting in the middle decades of the last century), animal visitation is often the means by which humanist certitudes are dismantled rather than confirmed.

Eugene Ionesco's mid-century play *Rhinoceros* is a paradigm of therio-theatricality, not only because of its boldly proclaimed animal subject but equally because that very subject has for so long been systematically ignored and skirted by the play's critics. The fact that the animal can *fail* to register in the critical reception and scholarly discussion even of a play *named* for an animal is in itself an interesting phenomenon, further evidence of the programmatic disavowal just mentioned. The reasons why this can happen – and the logic by which it is made to happen – have to do with the Western theatre's traditional characterology, according to which viable dramatic identity is forged in a long, lonely, and above all *verbal* journey. In this paradigm, therio-theatricality lurks in the wings, erupting periodically to activate the rich materiality that is the hallmark of theatre: its seductive traffic in bodies, objects, spaces, and sounds.

When *Rhinoceros* received a high-profile revival at the Royal Court Theatre in 2007, the critical reception closely resembled that of the equally high-profile revival of another well-known animal play, Peter Shaffer's *Equus*. The reviews were uniform in their admiration for various aspects of the *production*, especially the performances of the lead actors. On the play itself, however, there was nothing like such unanimity. Indeed, many critics explicitly framed their praise of both productions as a matter of transcending the limitations of weak and dated dramatic sources. In the case of *Equus*, this response closely resembled the one to its premiere, back in 1973, where the critical consensus could be summed up as: 'great theatre, bad play' (Chaudhuri 1984).

In 2007, both animal plays were characterized by several critics as intellectually shallow and hopelessly dated. Michael Billington of the *Guardian* said Ionesco's central metaphor was 'so vague as to be meaningless: you could apply it with equal validity to Nazism, communism or capitalist consumerism' (2007). The play was not only 'intellectually woolly and predictable', he said, it came dangerously close to implying that 'every minority is right'. Paul Taylor, writing in the *Independent*, agreed that this 'Absurdist fable about the pressure to conform (Fascist, Communist, you name it) has not survived its own occasion' (quoted by Terri Paddock 2007). It was, he said, 'dated and dead'. Similar dismissals greeted 'the hokum' of *Equus*, with its R.D. Laing-ian 'creed' – now long out of vogue – 'of communing with one's inner madman' (Brantley 2008).

Yet for all their intellectual inadequacy the plays were, the critics agreed, vastly entertaining and engaging as theatre events. There was no dearth of explanations: sometimes, the credit went to 'the wit of Martin Crimp's translation', at others to the production values, which were widely praised as exemplary, 'from the trumpeting noises of Ian Dickinson's sound design to the rhino-heads by Jonathan Beakes' (Billington 2007). Or it could have been that this 'stunning production' was directed by 'possibly the most gifted director in the country', who knew how to 'whip up the tension brilliantly', making for a staging that 'was lucid, expertly orchestrated' (Taylor 2007). In the case of *Equus*, the theatrical triumph was also explained, according to the critics, by another non-dramatic feature: the pitch-perfect casting of child- and teen-idol Daniel Radcliffe as Alan Strang, a character who is, as Ben Brantley cleverly observed, 'a tidy inversion of Harry Potter. Both come of age in a menacing, magical world where the prospect of being devoured by darkness is always imminent' (Brantley 2008).

The critical values underlying this reception tacitly privilege text and ideas over performance and production elements. In so doing, they are building on a key feature of therio-theatricality: the equation of animality with physicality and materiality – with bodies and things – and the correlated association of the human with mind, soul, thought, ideas, and meaning.

It is not surprising, therefore, that the one feature of these plays that never came up in the discussions of their theatrical strengths and dramatic weaknesses was the fact that each has, at its thematic center as well as a key element of its scenic discourse, the figure of an animal. Ionesco's play is, of course, about an 'Invasion of the Body Snatchers'–style epidemic that overtakes and transforms all but one inhabitant of a French town. *Equus* is famously based on the horrifying true story of a young man who inexplicably blinded a stableful of horses (Shaffer 1973: 9). Both play-texts incorporate extensive directions for and descriptions of what I have elsewhere called 'zooësis',[1] both mimetic (shown) and diegetic (narrated). Ionesco's pachyderms are first heard, then seen, and the climactic scene of the play features a protracted becoming-rhinoceros,[2] one of the most famous transformation scenes in world drama. Shaffer's horses are also first heard – 'the Equus Noise' is described in a prefatory note as 'a choric effect [. . .] composed of humming, thumping, and stamping' (Shaffer 1973: 16) – and then seen. Here too, the climactic scene is one in which a becoming-animal occurs, though here the terms of transformation are ritualistic and psycho-sexual rather than – as they are in *Rhinoceros* – political and emotional. Yet in spite of this extensive and explicit animal presence in these plays, the subject of animals and animality, and of the human-animal relation, has never been afforded a central critical focus. Never has a critic asked the question Shaffer's tortured protagonist, Dr Dysart, famously identifies: 'Why me? Account for me' (108).

This is, of course, 'the question of the animal',[3] a question that has haunted philosophy at least since Aristotle, the question that was bluntly shut down for several centuries by Descartes and that was taken up with increasing vigor and urgency in the course of the past century. It is a question that *Rhinoceros* engages with far more extensively than the critical tradition on the play suggests. To approach the play from the perspective of the animal and its 'question' – which is, essentially, the question of animal alterity, of the similarities and differences between animals, including human animals – is to discover in it an encapsulated history of the modern animal as well as a lively demonstration of its role in the production – and later dismantling – of Enlightenment subjectivity. The history spans the period from the Cartesian *cogito* to the post-Darwinian naked ape, though these and other of that history's formulations are presented not discursively or chronologically but rather as a stampede of more or less contradictory ideas, rushing at and around each other. This intellectual disorder is a crucial element of the play's zooësis, which amplifies these historical fragments with a powerfully chaotic scenic and performative discourse.

The animal presence that builds steadily and confidently in the course of the play culminates in an aesthetic transformation that contributes as much to the play's discourse of species as do the tortured existential heroics of its protagonist. In the last scene of the play, a stage direction tells us that

> Powerful noises of moving rhinoceroses are heard, but somehow it is a musical sound. On the up-stage wall stylized heads appear and disappear; they become more and more numerous from now on until the end of the play. Towards the end they stay fixed for longer and longer, until eventually they fill the entire back wall, remaining static. The heads, in spite of their monstrous appearance, seem to become more and more beautiful.
>
> *(Ionesco 1960: 94)*

The 'stylized heads' that gradually fill the walls of the stage recall an influential artistic movement of early twentieth-century modernism: primitivism. Cultivating an aesthetic based on the monstrous and the savage, primitivism was, as Philip Armstrong has argued, a version of the modernist turn to *wildness* as a valued ideal, which included a radical revaluation of animality:

> Rejecting the complacencies of Victorian modernity, the modernists aimed also to dispense with – or in some ways, reform – the legacy of the last great literary revolution,

> Romanticism, which they felt had been tamed by intervening generations. The wildness of Romanticism has been domesticated; its sublime potency reduced to mawkishness; it was time for art to break loose, go feral, and return to a revitalizing savagery. As this rhetoric suggests, the modernist break with the past entailed, and in many ways depended upon, a revaluation of human-animal relations.
>
> *(2008: 134)*

The result of this development is what Armstrong calls a new 'therio-primitivism' (142), a distinctively modernist mode of relating animality to an archaic, essential humanity. More or less explicit versions of therio-primitivism inform such foundational discourses of modernism as the Nietzchean, Darwinian, Marxian, and Freudian, each bequeathing a different equation between their respective normative human subjects and animality, and thus prescribing a different trajectory – towards or away from – that animality.

In both its affirmative and its skeptical versions, this therio-primitivism is one of the underlying discourses of *Rhinoceros*'s ambivalent account of animality, efficiently correlating with each of its two opposite allegories. *Rhinoceros* is the theatrical equivalent of the famous duck/rabbit optical illusion: looked at one way, it is an indictment of the human capacity to adapt to horror. This is certainly how the play's leading critics read its narrative of a mass transformation of humans into animals, in which the lure of conformity and the willingness to compromise with coercive mass movements sweeps over all but one of the inhabitants of a town, leaving him defiantly proclaiming his humanity on an empty stage. Looked at another way, it is a drama of liberation from the bonds of bourgeois conformity, a joyful return to nature and a cathartic embrace of animality. While the dominant reading is supported by and relies on the play's intellectual discourse – in particular the ideological awakening of its protagonist Berenger – the latter perspective emerges when attention is paid to the play's scenic discourse, with its proliferating parades of animal heads and its oddly exhilarating animal chorus, as well as to its most famous scene (and most challenging performance): the physical transformation of Berenger's pompous friend Jean into a rhinoceros.

The negative version of therio-primitivism, which 'conceives of animality as a state out of which the human must be forged, or from which it must travel' (Armstrong 2008: 142), underlies the allegory that has dominated the critical tradition of the play, according to which the embrace of animality is politically and culturally regressive, a betrayal of all the hard-won achievements of modern civilization. A different, 'redemptive' therio-primitivism subtly informs the characterization of the protagonist, Berenger, complicating the meaning of his later conversion to the position of defender of civilization. A second complicating factor in reading Berenger's story is the species of animal selected to represent the ideology that proves so seductive to every single one of his human companions. Had Ionesco chosen to mine traditional bestiaries and anthropomorphic fables for his satire on cowardice and conformity, the play might have been entitled 'Sheep', or 'Lemming', or even 'Chicken'. These options might have produced a more straightforward allegory, but they would also have forsaken much that is enigmatic and original – not to mention dark and disturbing – in the play.

The choice of an animal with few (if any) specifically cultural associations spins Ionesco's allegory in an unexpected way, making it, in a performatively important way, uncontrollable. Much of the ambivalence that the play registers – and elicits from its audience – is a function of its animal choice. The rhinoceros is the wild animal par excellence, rarely seen, rarely evoked in myth or fable, the subject almost exclusively of arcane natural-history taxonomies, like the one Ionesco makes brilliant theatrical hay with in Act I: 'The Asiatic rhinoceros has one horn and the African rhinoceros has two. And vice versa' (1960: 32). This importation of wildness into the space of civilization pushes the play's zooësis beyond political allegory, affording instead (or in addition) a kind of encounter with *animal ambivalence* that theater, as an art form, is particularly well suited to provide.

Ionesco's animals gain much of their meanings and effects because they – like all stage animals – are fundamentally *out of place*[4] on stage, in the theater, and in what has been, since the sixteenth century, the classic scene of comedy, the town square.[5] This archetypal setting is vividly realized in the play's first scene, its grocery and café sharply etched against a '[b]lue sky; harsh light; very white walls' (1960: 4). This quintessentially urban scene is the ideal setting for a drama of animal ambivalence, the city being, by many accounts, the epitome of de-animalized space. The ancient myth of Protagoras, for instance, posits a strictly causal relationship between wild animals and cities:

> [M]an lived at the beginning in scattered units, and there were no cities. So they began to be destroyed by wild beasts, since they were altogether weaker. Their practical art was sufficient to provide food, but insufficient for fighting against the beasts – for they did not yet possess the art of running a city, of which the art of warfare is part – and so they thought to come together and save themselves by founding cities.
>
> *(quoted in Ridout 2006: 115)*

The intrusion of animals into this exclusively human space puts them in a very special category of animals: those who encroach upon areas designated off-limits to them, spaces from which they have been deliberately and vigorously excluded. This is the category of the pest. Artists Snæbjörnsdóttir and Wilson, writing about their project on pests, entitled *Uncertainty in the City*, note:

> Long ago, settlements and therefore latterly cities were predicated on the concept of refuge, and a physical division of culture and nature. Clearly such division has proved increasingly porous as more and more animals and birds consider concentrations of human population an attraction rather than a deterrent because of the opportunities such cultures provide in terms of habitat and feeding.
>
> *(2011: 5)*

Reactions to the many animals who return to areas marked human are many and varied, ranging from fear and disgust to fascination and delight. In general, however, the pest elicits negative reactions, and the list of these provided by Snæbjörnsdóttir and Wilson is an admirable summary of many of the feelings expressed in Ionesco's play:

> Fear of disregard of individual territory
> Fear of a disregard of manner and protocols
> Fear of an implicit erosion of boundaries
> Fear of disruption of our own patterns of behaviour
> Fear of those setting up residence without permission
> Fear of aural disturbance
> Fear of damage to property. . . .
>
> *(5)*

The choice of species to play the role of pest – the rhinoceros – is as different as possible – in size, in familiarity, in provenance – from the species who usually play that role: rodents and insects. This 're-casting' of the pest is key to the play's therio-theatricality, its use of the animal as a lens to see human sociality in new – non-anthropocentric, non-theriophobic – ways.

The play begins, then, in a phobically clean and bright space. As the first act proceeds, the pristine space of civilization is engulfed with the dust of stampeding herds. (Other versions of this space of civilization in later scenes of the play – first an office, then a home – will suffer even worse damage, splintering and collapsing in a way that seems to characterize animal visitation in drama.)[6] At present, however, the only animals here, besides the human ones who will soon reveal their animality in all sorts of ways, are the unseen ones in the grocery freezer and the one who accompanies the first person to enter the stage: a housewife clutching an exemplar of that obligatory humanized animal of modern life, the pet.

The housewife's pet cat makes an ideal point of departure for the play's soundings of modern animality because it exemplifies the cultural animal practice, pet-keeping, through which modernity has increasingly compensated for the growing 'ontological and material separation of human and animal' (Armstrong 2008: 13).[7] The housewife's treatment of her cat is nothing if not obsessive, and includes an absurdist version of the pet-keeper's delusional anthropomorphizing: 'He was so gentle, just like one of us. [. . .] He was devoted to us. [. . .] He could almost talk – in fact he did' (Ionesco 1960: 31–2). Fittingly, the housewife's cat – this token of modern humanity's presumed control over the natural world – will be the first victim of the wild animality that is soon to engulf the world of the play.

The archetypal set of comedy and the humanized animal of modernity frame the ironized Nietzschean dyad of the play's two main characters. The unkempt, hungover Berenger is a debased instance of the Dionysian ideal, with the appearance and comportment of an exhausted satyr. He lives, according to his censorious friend, 'in a thick haze of alcohol' which is 'clouding [his] brain'. A rampant physicality – 'I'm conscious of my body all the time,' he says (1960: 18) – and uncontrollable appetites link him to animality and alienate him from the de-animalized world that (in the person of his punctilious friend Jean) criticizes and scolds him:

> You're in a bad way, my friend. [. . .] You're dropping with fatigue. You've gone without your sleep again, you yawn all the time, you're dead-tired [. . .] You reek of alcohol [. . .] Your clothes are all crumpled, they're a disgrace! Your shirt is downright filthy, and your shoes [. . .] What a mess you're in!
>
> *(5–6)*

Jean's contrasting rendition of the Apollonian ideal is even more pitiful and attenuated, its intellectual clarity and moral order now reduced to a smug and self-satisfied conformism: 'The superior man is the man who fulfils his duty. [. . .] His duty as an employee, for example' (7). When Berenger confesses that he sometimes wonders 'if I exist myself', Jean confidently proffers a Cartesian solution: 'You don't exist, my dear Berenger, because you don't think. Start thinking, then you will' (19). Jean's faith in the *cogito* links him to the most notorious source of the binaries – reason/passion, thought/emotion, mind/body, human/animal – that came to structure Enlightenment thought. Rene Descartes's denial of animal sentience produced the 'beast-machine' against which a new ideal of humanity would be counterposed: man as rational animal, ruling his passions and in control of his body.

Besides his Dionysian aspects, Berenger is also the play's link to another modernist discourse with a complex and ambivalent relation to animality: Freudian dream theory. Although Berenger explicitly rejects the word 'instinct' in favor of 'intuition' to describe the source of his convictions ('No, that's not what I mean, it's the rhinoceros which has instinct – I feel it intuitively, yes, that's the word, intuitively' [1960]) his association with sleeping, dreaming, and daydreaming throughout the play suggests that he, more than any other character in the play, is struggling not only with demanding social beasts but also with the monstrous animals of the unconscious. The final act begins with Berenger asleep in his room:

> *He seems to be having a bad dream, and writhes in his sleep.*
>
> BERENGER: No. [Pause] Watch out for the horns! [. . .] No! *He falls to the floor still fighting with what he has seen in his dream, and wakes up. He puts his hand to his head with an apprehensive air, then moves to the mirror.* [. . .] *He heaves a sigh of relief when he sees he has no bump.*
>
> *(71)*

Berenger's self-scrutiny in the mirror reaches an obsessional pitch in the last section of the play, after his last human companion, his girlfriend Daisy, has abandoned him, preferring 'the ardor and the tremendous energy emanating from all these creatures around us' to 'what you call love – this morbid feeling, this male weakness. And female, too' (103). Indeed, it is precisely while Berenger 'continues to examine himself in the mirror' (105) that Daisy leaves.

From that moment onwards, until the end of the play, Berenger will return to the mirror repeatedly, checking his face against the proliferating animal heads all around him: 'A man's not ugly to look at, not ugly at all! *He examines himself, passing his hand over his face.* What a funny-looking thing! What do I look like? What?' (1960: 106). This classic image of self-absorption – the individual scrutinizing his reflection – takes on a different connotation in the animalized context of the last scene, where the human face is somehow being expected to 'face down' its animal counterpart. Berenger even attempts to address the quantitative superiority of his antagonists by putting up pictures he claims to be of himself. Chillingly, the stage direction reports: 'When he hangs the pictures one sees that they are of an old man, a huge woman, and another man. The ugliness of these pictures is in contrast to the rhinoceros heads which have become very beautiful' (106). Once again, an aesthetic intervention complicates the existential allegory emphasized by most critics of the play. Berenger's trajectory, it suggests, is not so much a triumph of individualism as a fraught and unfinished passage through modernism's ambivalent animalities.

The mirror of the last scene returns us to the mirror of scene one, held up to Berenger not by himself but by the censorious Jean, who carries it in his pocket, along with a tie and a comb. One of the most striking things about the play is the way its two protagonists appear to change places as the play goes on. Jean's transformation is, of course, the most extreme and astonishing, taking him from cartoon Cartesianism ('I'm master of my own thoughts, my mind doesn't wander. I think straight, I always think straight' [1960: 61]) to Darwinian caricature:

> JEAN: Moral standards! I'm sick of moral standards! We need to go beyond moral standards!
>
> BERENGER: What would you put in their place?
>
> JEAN: Nature! [. . .] Nature has its own laws! Morality's against nature.
>
> BERENGER: Are you suggesting we replace our moral laws by the laws of the jungle?
>
> JEAN: It would suit me, suit me fine. . . . We've got to build our lives on new foundations. We must get back to primeval integrity. [. . .]
>
> BERENGER: Just think a moment. You must admit that we have a philosophy that animals don't share, and an irreplaceable set of values, which it's taken centuries of human civilization to build up . . .
>
> JEAN: When we've demolished all that, we'll be better off!
>
> *(67)*

Jean's transformation from Man of Reason to Naked Ape is not merely a change of mind; Ionesco's confident theatricality renders it as a change of just about everything, including, of course, of body. En route to that final animalization, Jean – whose first moments on stage had marked him as a fastidious dresser and stickler for social codes – tears off his clothes, shouting: 'Hot, far too hot! Demolish the lot. Clothes itch, they itch. [. . .] The swamps! The swamps!' (68).

Of course, Jean is hardly the only rationalist to succumb to the beasts. The play provides a range of positions that correlate with various versions of modernist species discourse, beginning with a Swiftian satire on the fetishization of certain modes of thought:

> LOGICIAN: Here is an example of a syllogism. The cat has four paws. Isidor and Fricot both have four paws. Therefore Isidor and Fricot are cats.
> OLD GENTLEMAN: My dog has four paws.
> LOGICIAN: Then it's a cat. [. . .]Another syllogism: All cats die. Socrates is dead. Therefore Socrates is a cat.
> OLD GENTLEMAN: And he's got four paws. That's true. I've got a cat named Socrates.
> *(1960: 18–19)*

A more familiar ideal of rationality is articulated by Dudard, who clothes his passivity and cowardice in the guise of moderation and tolerance:

> My dear Berenger, one must always make an effort to understand. And in order to understand a phenomenon and its effects you need to work back to the initial causes, by honest intellectual effort. We must try to do this because, after all, we are thinking beings. I haven't yet succeeded, as I told you, and I don't know if I shall succeed. But in any case one has to start out favorably disposed – or at least, impartial; one has to keep an open mind – that's essential to a scientific mentality. Everything is logical. To understand is to justify. [. . .] I'm trying to look the facts unemotionally in the face. I'm trying to be realistic. I also contend that there is no real evil in what occurs naturally. I don't believe in seeing evil in everything. I leave that to the inquisitors.
> *(83)*

Berenger's own transformation, from chronic outsider ('I feel out of place in life, with people . . .' [17]) to reluctant defender of human civilization follows an opposite trajectory of therio-primitivism, from sensual oblivion to a kind of instinctual self-preservation.

Berenger begins his journey in a state of extreme alienation, so bored as to seem in need of something like a herd of rhinoceroses to revitalize him. As the thunderous off-stage noise announcing their arrival grows in volume, and as all the characters on stage, one after the other, abandon their activities and conversations to look at the amazing sight unfolding just out of the audience's view, Berenger alone remains unaware and unmoved. The only animalistic human in the group seems oblivious to the animals invading his environment. Later, in sharp contrast to the repeated exclamations and ejaculations that follow the animal visitation, Berenger alone offers a few plausible explanations: perhaps the rhinoceroses have escaped from the zoo, or from a traveling circus, or perhaps they've simply been hiding 'in the surrounding swamps?' (1960: 14). Jean's increasingly impatient and bizarre retorts ('there's been no zoo in our town since all the animals were destroyed in the plague' [14]), along with his ridiculous repeated pronouncement that 'it shouldn't be allowed!' (13), quickly move the animal figure in the direction of allegory that has been highlighted in criticism of the play, making it a stand-in for all that disturbs the surface calm of a mediocre and self-satisfied society, a kind of corporate antagonist for anyone loyal to the civilization of the bright town square where the play begins. And indeed this strain of meaning runs throughout the play, signposted periodically by such observations as the following: 'You get used to it, you know. Nobody seems surprised anymore to see herds of rhinoceroses galloping through the streets. They just stand aside, and then carry on as if nothing had happened' (91).

But the Nietzschean theme introduced through the main characters cuts across the allegory that has dominated the critical tradition on the play. Berenger's animality, while being a far cry from ecstatic Dionysian embodiment, is nevertheless on a continuum with the rapturous wildness that increasingly defines the world of the play, captivating one character after another. His resistance to it requires him to draw upon reserves of restraint and self-regard that have previously been the hallmarks of the insufferable Jean. He even begins to sound like Jean, huffing that 'your duty is to oppose them, with a firm clear mind' (1960: 93). Is Berenger's final 'triumph', then, actually a regressive return to the humanism of the Enlightenment subject, compulsively 'cleansed' of animality, disembodied, and denatured? And what is it that has routed Berenger's Dionysian instincts and embodied sensualities? The answer may lie in the characterological opposition itself, the sharp contrast between the two protagonists established at the outset. Their professions of friendship, especially Berenger's, suggest a commonality and kinship that is belied by the schematic difference called for in their performances, which demand strongly contrasting appearances and behavior. When Jean holds a mirror up to Berenger in the first scene, it is pointedly to show him how different he, Berenger, is from Jean himself. Yet the evolving dynamic of this central relationship seems to contradict the very recognition – of personal identity, of essential individuality – that lies at what has been assumed to be the philosophical – humanist – heart of the play.

The epic scene of Jean's becoming-animal begins with a moment of mutual non-recognition, forecasting the massive case of mistaken identity to come:

> JEAN: I didn't recognize your voice.
> BERENGER: I didn't recognize yours either.
>
> *(1960: 59)*

As the extraordinary scene proceeds, with the Jean actor performing human animalization through every semiotic register available in theatre – voice, costume, gesture, movement, physicality, vocalization – Berenger's responses mark the transformation not as un*natural* but as un*characteristic*: 'I'm amazed to hear you say that, Jean, really! [. . .] It's not like you to say a thing like that. [. . .] You're not yourself!' (68). Berenger, who had earlier confessed to being unsure of his own existence, becomes increasingly focused on the question of identity: 'I'm frightened of becoming someone else' (73), he now confesses. Indeed, his desperate colloquy with the mirror in the play's final moments suggests that his is not so much a drama of individual self-discovery as it is one of tenuous *human construction*, made arduous not only by the presence of animality but of a particularly challenging and seductive kind of animality: the kind produced by modernism's therio-primitivism. This is, after all, a *wild* animality, bringing with it the sultry breezes of 'The swamps! The swamps!' (68), and inspiring a dreadful and a joyful awe: says Daisy, 'They're beautiful. [. . .] They're like gods!' (104).

Above all, they are *many*, and it is this feature of the animals that Ionesco's dramaturgy realizes most vividly. Long before Berenger tries to match their numbers by putting up pictures of human faces on his wall, their *quantity* has been terrifying him. As he tries to escape from Jean's apartment after Jean has become a rhinoceros, he finds all his exits blocked:

the porter has turned into a rhinoceros, and so have Jean's neighbors. He tries to escape through the window between the stage and the auditorium '*but gets back again quickly, for at the same time, crossing the orchestra pit, move a large number of rhinoceros heads in line.* "There's a whole herd of them in the street now! An army of rhinoceroses, surging up the avenue !" (1960: 70). As theatrically realized in this play, from the momentous first sighting in Act I to the 'trumpetings, hectic racing, clouds of dust' that accompany the proliferating animal heads of Act III, it is as a *multiplicity* that the animal most seriously challenges human identity. As Jacques Derrida argued in his late animal texts, the animal has the paradoxical effect, in the realm of autobiography, of correlating the individual to multiplicity:

> [M]y animal figures multiply, gain in insistence and visibility, become active, swarm, mobilize and get motivated, move and become moved all the more as my texts become more explicitly autobiographical, are more uttered in the first person.
>
> *(2008: 35)*

The singularity that might be assumed to be the quarry and ideal of the autobiographical project – the discovery of the uniqueness that distinguishes the self from all others – is contradicted by animal presence as multiplicity. When animals enter into autobiography, it would seem, they do so in numbers and with urgency, 'lunging more and more wildly in my face' (35). Their appearance seems to invite the autobiographical self *to rethink itself beyond singularity*, to seek more pluralistic modes of self-knowledge. It is an invitation that lies on the far side of the excruciating choice that appears to torture Berenger in the last moments of *Rhinoceros*. His 'either/or' concept of human animality has him lurching painfully between a feral primitivism and a 'purified' humanism, both equally impossible to inhabit permanently.

One conceptual foundation for an alternative account of the self, a 'zoo-autobiography' as Derrida calls it, is found in Alphonso Lingis's challenge to the habitual association between singular form (such as a body) and individuality:

> How myopic is the notion that a form is the principle of individuation, or a substance occupying a place to the exclusion of other substances, or that the inner organization or the self-positing identity of a subject is an entity's principle of individuation! A season, a summer, a wind, a fog, a swarm, an intensity of white at high noon have perfect individuality, though they are neither substances nor subjects.
>
> *(1999: 39)*

Lingis extends his pluralized vision to human identity, recasting individuality and agency as vast collaborative processes:

> Human animals live in symbiosis with thousands of species of anaerobic bacteria, 600 species in our mouths which neutralize the toxins all plants produce to ward off their enemies, 400 species in our intestines, without which we could not digest and absorb the food we ingest. [. . .] The number of microbes that colonize our bodies exceeds the number of cells in our bodies by up to a hundred fold. Macrophages in our bloodstream hunt and devour trillions of bacteria and viruses entering our porous bodies continually. They replicate with their own DNA and RNA and not ours; they are the agents that maintain our borders. They, and not some Aristotelian form, are true agencies of our individuation as organisms.
>
> *(1999: 38–9)*

Another generative concept in animal philosophy, Gilles Deleuze and Félix Guattari's 'becoming-animal', offers another perspective on identity beyond singularity. Allan Smith explicates the concept as involving a condition 'of acute intensity, of demonized volatility; [. . .] of contagious transport of impersonal affects and teeming multiplicities' (2007: 160). In Smith's reading, as in others, becoming-animal is closely related not only to excess and intensity but also to *multiplicity*, and to Deleuze and Guattari's interest in the aggregative or collective nature of animals, to the fact that, as Philip Armstrong puts it, 'animals are never one but always "as one"' (2008: 7).

The human response to this perceived pluralization of animal identity is deeply ambivalent. On the one hand, there is a long tradition of recognizing, even celebrating it in language, as captured in James Lipton's much-loved volume *An Exaltation of Larks*, a delightfully illustrated collection of animal group names – or 'terms of venery', so-called because of their origin in hunting culture. The impulse to capture something essential about a species in the group name assigned to it leads from the familiar 'a pride of lions' to the chilling 'a murder of crows' to a host (so to speak!) of witty coinages that leave the animal world far behind ('a flush of plumbers', 'an annoyance of cellular phones').

More recently, a new representational technology has brought the exhilaration of animal multiplicity to film animation. Boids, invented by Craig Reynolds, are distributed behavior models that simulate the behavior of the individual birds in a flock. They are hi-tech, vastly amplified versions of the kinds of synchronized performances, both human and animal, that have always and everywhere delighted audiences, from the ritual dances of early cultures to the Rockettes of Radio City Music Hall. Indeed, the practice of training humans and animals to perform in unison could itself be a consequence of ambivalent feelings towards animals: on the one hand, a desire to imitate the dance-like beauty of the natural movements of their herds and flocks; on the other, an effort to master and control the energy generated by their multiplicity and bend it to our will and pleasure. Spectacles of animal multiplicity then, be they boid animations or Lipizzaner stallions, exist on a continuum with phenomena like animal collections (actual or virtual) and even odd events like the 'cow parades' that have sprung up around the world in recent decades.[8] Part-civic celebration, part-commercial racket, these so-called 'public art' events can be read as safe and domesticated versions of the kind of animal visitation Ionesco envisions in *Rhinoceros*.

Lingis argues that animal multiplicities 'exert a primal fascination on us' because they echo and mirror 'the multiplicities in us. What is mesmerized in us are the inhuman movements and intensities in us, [. . .] the micro-organic movements and intensities in the currents of our inner rivulets and cascades' (1999: 166–7). Animal multiplicities charm and fascinate. As frequently, however, they provoke unease and anxiety, even terror. Though avian flocks may exalt the spirit, they also threaten, as Hitchcock demonstrated, to swoop and smother. As for hordes, packs, and swarms: they are often, in fiction as well as in political theory, figures for dangerous crowds: 'fringe groups, nomad armies, raiding parties, gangs, cabals, crime societies' (Smith 2007: 160). This is undoubtedly the association that is uppermost for Ionesco's Berenger, as for most critics of his play. But when *Rhinoceros* is unfolding on a stage, with all channels of theatrical semiosis fully engaged, this disturbing connotation of multiplicity is only one among several folded into the play, and not necessarily a strong one.

Ionesco repeatedly reminds us that the herds that so frighten Berenger are not the only ones in his town. The behavior of his fellow citizens at many junctures in the play can most accurately be described as herd behavior, beginning with the identical and reflexive reactions they all have to the first sighting: phrases like 'Well, of all things!' and 'Oh, the poor thing!' (this in reference to the housewife's crushed cat) are repeated endlessly, with every character taking his or her turn, and often by groups of characters in unison. That human beings are pack animals is established long before they start jumping species. Ironically, it is this very feature of the human animal that also underlies the bio-dynamics of Ionesco's plot: as character after character joins the herd, the biological foundations of social behavior come into view. The social behavior we humans prize and regard as a sign of our superiority is a product of natural selection, 'found in members of species with better survival chances in a group than in solitude. The advantages of group life can be manifold, the most important being increased chances to find food, defense against predators, and strength in numbers against competitors' (de Waal 1996: 9).

Multiplicity – both human (as sociality) and animal – is the challenging context of the drama of human recognition that lies at the heart of *Rhinoceros*. And it is closely related to the play's greatest challenge to humanism, its demonstration of the instability of identity. When Berenger comes to Jean's house at the beginning of the transformation scene he knocks on the door and calls Jean's name. A man answers, but it is not Jean. In a pitch-perfect use of illogic that was to become one of the hallmarks of absurdism, Ionesco has the man, who is Jean's neighbor, inform Berenger that 'I thought it was me you wanted. My name's Jean too' (1960: 58). The conflict between Berenger and Jean turns out to be something other than the two versions of the human – (cartoon) Dionysian and (cartoon) Apollonian – with which they began. Rather, their battle unfolds in the ideological and aesthetic arena of which a key feature was the unraveling – and eventual death, as theorized by Elinor Fuchs – of the traditional notion of dramatic character.

The reading of *Rhinoceros* that I have offered here shows the extent to which that unraveling was bound up with a resurgent therio-theatricality. Long exiled to the wings of Western theater, the animal returns to join late modernism's posthumanist project. As for the human animals who must now – at long last – share the stage of Western representation with other species, I hope I have shown here that they will find few better resources than those of theatricality – of live, embodied performance – for joining in the great drama of animal alterity.

Notes

1 I have proposed the term zooësis to refer broadly and comprehensively to the discourse of species in art, media, and culture (Chaudhuri 2007: 8–9). The term echoes both Platonic *poiesis* and Aristotelian *mimesis*, both commonly used in literary and dramatic theory to designate modes of construction and representation. More particularly, the term 'zooësis' is inspired by analogy from 'gynesis', a term proposed in the 1970s by feminist theorist Alice Jardine to refer to 'the putting into discourse of "woman" [. . .] [and] the valorization of the feminine, woman, and her obligatory, that is, historical connotations, as somehow intrinsic to new and necessary modes of thinking, writing, speaking' (1985:25). Replacing the prefix 'gyne-' (woman, female) with 'zoo-' (animal, the Greek *zoion*), zooësis refers to the ways the animal is put into discourse: constructed, represented, understood, and misunderstood. In proposing the term I also share Jardine's progressive ambition of contributing to new modes of thinking and writing that would valorize the animal and bring a heightened ethical attention to human-animal relationships.

2 Gilles Deleuze and Félix Guattari's notion of 'becoming-animal' is one of animal philosophy's most aesthetically productive concepts, as well as one of its most enigmatic. In an article reporting on a sustained theatrical exploration of the concept, my co-author and I wrote:

> the most challenging of Deleuzian definitions [. . .] was the idea that becoming is antithetical to imitation: 'We fall into a false alternative if we say that you either imitate or you are. What is real is the becoming itself, the block of becoming, not the supposedly fixed terms through which it passes.' Becoming resists metaphor and mimesis. It courts fleeting synecdoches, momentary metonymies, shifting interstices. For actors, it offers an opportunity to indulge and unleash creative impulses without pointing them towards externally (conventionally) settled images.
>
> (*Chaudhuri and Enelow 2006: 4*)

3 This phrase, 'the question of the animal', appears as the subtitle of at two influential works of animal philosophy (Calarco 2008; Wolfe 2003).
4 As Nicholas Ridout puts it, 'The impropriety of the animal on the theatre stage is experienced very precisely as a sense of the animal being in the wrong place. In the circus there are still a few tawdry reminders of nature. [. . .] The theatre, by contrast, rigorously excludes nature' (2006: 98).
5 In 1545, the influential Renaissance architect and theorist Sebastiano Serlio proposed three stock settings for drama, one each for tragedies (palaces), comedies (a town square), and pastorals (country scene).
6 Elizabeth Egloff's wonderful animal play *The Swan* concludes with the following stage direction: 'There is a huge noise: glass breaking, the world breaking, a tree cracking' (54). In Edward Albee's extraordinary exploration of the animal roots of tragedy, *The Goat*, the dialogue is punctuated by one character smashing one decorative object after another on the floor.
7 Keith Thomas notes that by '1700 all the symptoms of obsessive pet-keeping were in evidence. Pets were often fed better than servants [. . .] and they became an increasingly regular feature of painted family groups' (1984: 117).
8 'CowParade is the largest and most successful public art event in the world. CowParade events have been staged in over 75 cities worldwide since Chicago in 1999' (CowParade).

References

Armstrong, P. (2008) *What Animals Mean in the Fictions of Modernity*, London: Routledge.
Berger, J. (1980) 'Why Look at Animals?' *In About Looking*, New York: Pantheon.
Billington, M. (2007) Review of *Rhinoceros*, *Guardian*, 27 September.
Brantley, B. (2008) 'In the Darkness of the Stable', *New York Times*, 26 September.
Calarco, M. (2008) *Zoographies: The Question of the Animal from Heidegger to Derrida*, New York: Columbia University Press.
Chaudhuri, U. (1984) 'Drama in the Spectator/The Spectator in Drama', *Modern Drama*, 27.3: 281–98.
—— (2007) '(De)Facing the Animals: Zooësis and Performance', *TDR: The Journal of Performance Studies*, March: 8–20.
Chaudhuri, U. and Enelow, S. (2006) 'Animalizing Performance, Becoming-Theatre: Inside Zooësis with the Animal Project at NYU', *Theatre Topics*, 16.1: 1–17.
CowParade, available online at www.cowparade.com, accessed June 2012.
Deleuze, G. and Guattari, F. (1987) *A Thousand Plateaus: Capitalism and Schizophrenia*, B. Massumi (trans.), Minneapolis: University of Minnesota Press.
Derrida, J. (2008) *The Animal That Therefore I Am*, M. Mallet (ed.), D. Wills (trans.), New York: Fordham University Press.
Fuchs, E. (1996) *The Death of Character: Perspectives on Theater after Modernism*, Bloomington: Indiana University Press.
Ionesco, E. (1960) *Rhinoceros and Other Plays*, D. Prouse (trans.), New York: Grove Press.
Jardine, A. (1985) *Gynesis: Configurations of Woman and Modernity*, Ithaca, NY: Cornell University Press.
Lingis, A. (1999) 'Bestiality', in *Animal Others: On Ethics, Ontology, and Animal Life*, H.P. Steeves (ed.), New York: State University of New York Press, pp. 37–54.
Lipton, J. (1968) *An Exaltation of Larks; Or, The Venereal Game*, New York: Grossman.
McKay, C. (2010) 'Murder Ob/Scene: The Seen, Unseen and Ob/scene in Murder Trials', *Law Text Culture*, 14.1: 79–93.
Nagler, A.M. (1952) 'Serlio's Three Scenes', *A Source Book in Theatrical History*, New York: Dover Publications.
Reynolds, C.W. (1987) 'Flocks, Herds, and Schools: A Distributed Behavioral Model', available online at www.red3d.com/cwr/papers/1987/boids.html, accessed July 2012.
Ridout, N. (2006) *Stage Fright, Animals, and Other Theatrical Problems*, Cambridge: Cambridge University Press.
Shaffer, P. (1973) *Equus*, New York: Penguin.
Shakespeare, W. (2008) *Hamlet*, New York: Oxford University Press.
Smith, A. (2007) 'Bill Hammond's Parliament of Foules', in *Knowing Animals*, L. Simmons and P. Armstrong (eds), Leiden: Brill Press, pp. 155–81.

Snæbjörnsdóttir/Wilson (2011) *Uncertainty in the City*, Berlin: The Green Box.

Taylor, P. (2007) Review of *Rhinoceros* in the *Independent*, quoted in 'Review Round-up: How Do Critics Ride Rhinoceros?', Terri Paddock, *What's on Stage?*, available online at http://www.whatsonstage.com/west-end-theatre/news/09-2007/review-round-up-how-do-critics-ride-rhinoceros_20477.html, accessed July 2012.

Thomas, K. (1984) *Man and the Natural World: Changing Attitudes in England, 1500–1800*, Harmondsworth: Penguin.

de Waal, F. (1996) *Good Natured: The Origins of Right and Wrong in Humans and Other Animals*, Cambridge, MA: Harvard University Press.

Williams, T. (2009) *Night of the Iguana*, New York: New Directions.

Wolfe, C. (ed.) (2003) *Zoontologies: The Question of the Animal*, Minneapolis: University of Minnesota Press.

15
Bestial imaginings

Kathy Rudy

There's a 1999 short story that circulates in many women's studies classes around the country called 'Confessions of a Bioterrorist', by Berkeley women's studies professor Charis Thompson Cussins.[1] In those feminist classes that focus largely on reproductive technologies, this fictional essay is virtually canonical. It tells the story of Mary, mother of two, a successful reproductive animal physiologist who works in a 'frozen zoo', collecting and freezing sperm, eggs, and embryos from animals on the brink of extinction. A madcap, pseudo-scientific adventure about Mary and her two friends Gabriella and Eva, the story is written in a detailed, terse, and quite funny style, with lots of comic inference about ejaculation and other bodily functions. As a result of the influence of her politically correct and environmentally conscientious friends (and a few other chance encounters with Western cowboys and African bushmen), in the end, Mary decides to implant bonobo embryos into her own uterus. The story ends just as the bonobo baby is being born. It's been Xeroxed and PDF'd and passed around at feminist conferences and women's studies departments for years now. Students love it.

I come to the work of human-animal studies most recently from the position of a women's studies professor; hence the interest in this document. But to give you a map of where the essay is going, my deep background (i.e., my graduate training and most of my writing) comes from the world of theological ethics. Thus, I am interested in the question of the ethical – what is right and what is wrong, and why. But I am equally interested in questions of spirituality, mysticism, animism, and biophilia: what kinds of things connect us to the more-than-human-world? What falls outside the measurement of science? How can those things help us live fuller lives? These are the questions I bring to women's studies and human-animal studies.

In the past, when I've taught this Cussins piece or talked with other women's studies teachers who've used it in class, it was always about the limits, if there are any, to reproductive freedom and women's choice (Squier 1998). As Cussins claims in the preamble, she wrote it to shed light on 'who and what gets to reproduce where and under what conditions. [. . .] The story suggests that what is reproductively subversive or liberatory varies from one situation to another' (1999: 189); the narrative certainly makes that point beautifully. But three things have haunted me about the story for over a decade now: precisely *why* Mary implanted the embryos into herself, *what* happened after the baby is (or babies are?) born (four embryos are implanted, we don't know how many gestated), and *how* would we think about justifying such an extreme act? I understand these unanswered questions are openings and devices for readers to use their imaginations, places that can foster debate and deliberation. And that is how the story works in the feminist classroom. But when I started reading and writing in the field of human-animal studies, new ideas emerged in relation to these questions. With animals positioned as the central characters, 'why', 'what happened', and 'how can this be justified' take on very different inflections. Grounding myself initially in ideas about evolution generated by theorists Henri Bergson and Gilles Deleuze, I want to answer here *why, what, and how* in fantastical and bestial ways. My aim, ultimately, will be to uncover modes of hidden and institutionalized 'human exceptionalism', i.e., the erroneous but widespread belief that human beings are made of different 'stuff' than all other animals, and to help us imagine ourselves into a new and different future.

Why did Mary impregnate herself with bonobo embryos?

The full-scale destruction of most of the earth's wild animals is imminent. Some people actually argue that the ecological disaster has already happened, and we are now simply witnessing the after-effects (Morton 2007). Environmentalists publish ongoing evidence that we are in the middle of the sixth and greatest extinction event in the earth's history, with an unprecedented number of species lost every day. The conservation of wild animals gets funding and attention from ideologically diverse organizations, including World Wildlife Fund, International Union for Conservation of Nature, United Nations Environmental Program, World Wide Fund for Nature, Born Free Foundation, and scores of others; yet by most accounts, all these efforts are failing. By some estimates, the rate of species loss is 10,000 times greater today than previously occurring background extinctions.

And bonobos are among the most endangered, 'red listed' animals on the planet, with fewer than 25,000 individuals still living in the wild. A serious review of the history surrounding bonobos is well beyond the scope of this chapter, but a few things are worth noting. Most significantly, the bonobo has emerged as the darling of the great apes in recent decades. Touted by Frans de Waal as the 'kama sutra' ape, bonobos resolve their internal conflicts through sex, not violence (de Waal and Lanting 1997). They have sex with each other like we humans shake hands, perhaps even more frequently. They have sex face to face (dorsal), they have sex with their children, they practice homosexuality and group sex, and they seem to be some of the happiest creatures on earth. They eat very little meat, they are matriarchal in structure, and they pass the mirror test for full self-recognition. In captivity, they learn to communicate with humans through sign language and lexigrams more quickly and effortlessly than either chimps or gorillas (Savage-Rumbaugh and Lewin 1994). Until about 40 years ago, they were miscategorized as 'pygmy' chimps. They only live wildly in one part of the world, in central Democratic Republic of Congo, where they are losing their habitat to loggers, conflict, mineral mining, and human resettlement at astonishing rates. When they are squeezed out of their pocket of forest, they are usually shot. Although illegal, bonobo 'bushmeat' is a prized delicacy throughout both Africa and in some parts of Europe (Peterson 2003).

In this frame, it makes complete sense that Mary, the bioterrorist, would want to save such a marvelous and magical species from extinction. There is no question that they are very lovable creatures. The text itself explains that Mary is deeply invested in her frozen zoo as the final haven for this precious animal: 'by freezing germ plasm in perpetuity, it would buy time for human civilization to decide what they wanted/needed to save and how to save it' (Cussins

1999: 191). But the conservation motive spawns many questions. How exactly will the four embryos Mary implants stave off extinction for the entire species? Will Mary's offspring be equipped to live a natural life in the forest? Will there be any forest left for them? And why exactly did Mary have to use her own body to promulgate the embryos when she could have found another animal to be a surrogate or perhaps even developed an artificial womb?

Several years ago, my colleagues primatologist Brian Hare and science writer Vanessa Woods invited Claudine Andre, owner of the Lola Ya Bonobo sanctuary, to visit Duke. Andre is not a scientist or even an academic; during political unrest in the Congo in the mid-1990s, Andre was charged with caring for the animals in the Kinshasa zoo. There she encountered her first infant bonobos, and as the wars progressed, more and more infants poured into her care. Their mothers were victims of the bushmeat trade; they were to be sold as pets. Andre marshaled resources to buy land and hire local women to care for the babies. Almost 20 years later, Lola ya Bonobo ('paradise for bonobos' in Lingala language) is still operating, taking in orphaned bonobos and doing soft releases for older adolescents into controlled and monitored wild areas. Both Brian and Vanessa work at Lola and have written extensively about Claudine and bonobo conservation. As luck would have it, I was teaching an introduction to animals and ethics class the semester Claudine came to Duke, and Brian suggested that she should visit our class. The minute she started talking about her work at Lola and showing pictures of the infants being held and loved by the local women, I thought of the Cussins essay.

Claudine Andre was very clear right from the outset that the mission of the Lola sanctuary is not really about saving a few hundred infant bonobos from death; it is about changing the shape of humanity itself. In her thick French-accented English, she portrayed how the babies altered the reality of everyone they encountered. The love that passes between the infants and their caregivers transformed both parties permanently, she said. When you can see the world through the eyes of a bonobo, it becomes a very magical place. My notes from that class say: 'Through Claudine's eyes, I can imagine the possibility and attraction of changing places with a bonobo.'

Cussins hints at a similar reality in her short story: 'Mary had been seized by the boldness of the possibility [. . .] of moving to and with other places and points of view' (1999: 211). What if humans and other animals could share common lives in which perceptions of both worlds could be altered, on both sides? What if we could move beyond our current relationships with animals – most of which are manifested in terms of pets or meat – to achieve a deeper and truer intersubjectivity? Would having a bonobo grow inside your womb or suckle at your breast change the shape of your world? And how might we think about such shifts in terms of evolutionary theory?

***

Common understandings of Darwinian evolution understand life as progressing in a fairly straight line of progress. 'Survival of the fittest', after all, means organisms just get fitter and fitter, better and better. Thus, the popular story goes, we all started out as single-celled beings, then we became fish, then reptiles, then mammals, and so on, until we get to the pinnacle of evolution, the human, *Homo sapiens sapiens*, the wisest of the wise men. In this model, human beings surpass bonobos and all other great apes and end up the grand-prize winners of the evolutionary game. Indeed, the story seems like nothing more than common sense in a world saturated with the ideology of human exceptionalism.

A century ago, Henri Bergson challenged this model. Bergson didn't see evolution as a single, progressive trajectory, but rather as a proliferation of sensuous biology in many different directions. As Claire Colebrook suggests, 'Rather than consider living forms to be the purely statistical result of a blind variation resulting in contingent stabilities, Bergson argues that difference or variation is creative. It is not directed toward the realization of forms, and so it is not a closed finalism' (2010: 113). The way that organisms proliferate is not through mechanistic variation, for Bergson, but through the possession of different tendencies that allow and foster multiple adaptations. Bergson called this explosive creativity 'spirit', and by it he meant to highlight multitudes of biological formation and their sensual differences (2007). For Bergson, the central difference between humans and animals is not one of teleology or 'survival of the fittest', but one of different trajectories of development. For him, one organism makes gains in one direction but may also simultaneously lose talents, or capabilities, in another.

While Bergson maintained the human as a superior life form due to our unique abilities of creativity, Gilles Deleuze follows closely on Bergson's path but highlights more dramatically the multiplicity of ways that animals also display creativity or spirit. Deleuze develops these ideas into his thesis of becoming, 'a verb with a consistency all its own; it does not reduce to, or lead back to, "appearing", "being", "equaling", or "producing" (Deleuze and Guattari 1987: 239). Rather than moving in a straight line, evolution explodes in an infinite number of directions at once. Deleuze sees 'becoming' as a proliferation of changes and associations, pushing in lots of different directions at once, always moving toward openings that promulgate diversity, but without any particular notion of the good or progress. He describes a method of life that disavows human exceptionalism; he rejects the idea that humans are higher than other great apes, great apes higher than monkeys, monkeys higher than lemurs, etc. It's all mixed up and there is not a place to stand in Deleuze's framework to determine things like fitness, greatness, or superiority. Every evolutionary trajectory has something new and different to offer. Deleuze is invested in the force of metamorphosis, in the idea that life is essentially active and constantly transformative. He details many different kinds of becomings, and one consistent theme in his writing is 'becoming animal'.

Darwin himself never used the term 'survival of the fittest'; such a formulation for him would have put too much emphasis on random mutations in species and not enough on shifts and fluctuations in the environment. Darwin described the process as 'adaption'. It is always, for him, about the intermingling between organism and environment, never only about the fitness of the animal. Bergson's 'spirit' and Deleuze's 'becoming' emphasize the wonder of such interactions even more fully. It is almost as if the world itself (environment) calls new life forms into being.

In my own mind, I translate these insights into a loose history that looks something like this: proto-humans started evolving big brains and language skills over 100,000 years ago. About 10,000 years ago we developed writing and agriculture, and with those tools we began to craft the world around us to answer and meet our every need and desire. With the problem of starvation minimized for many parts of the world, we began to create science, literature, art, logic, music, poetry, medicine, culture; in the West, we discovered/named/created a god who thought we alone were special and should claim dominion over the earth. Everything else in the world simply became a backdrop for the development of our human capacities. By the time we reached the Enlightenment, the human was unparalleled in the natural world. We thought we were made of different 'stuff' than the rest of nature. The idea of human exceptionalism had taken hold. But we neglected to realize that the environment/world was calling forth lots of beings and becomings, not just us.

For us humans, there was a cost to our evolutionary trajectory. So much of the way we perceive the world is filtered through our minds, through language, writing, reason, and culture; our other senses diminished in power and utility, especially in relation to other animals. On the surface, there are many animals with more highly developed senses of sight, smell, hearing, taste, and touch than humans. Scent hounds can smell something two miles away; elephants can hear their herds from the same two-mile marker, etc. Bats and dolphins possess a sense called echolocution that many philosophers argue we humans will never fully understand (Nagel 1974). Fish breathe water, birds can fly. Other animals live in the world differently than humans because their primary way of knowing is through their bodies, not their minds.

It's possible, and this is what I think Deleuze was driving at with his 'becomings', that other animals possess many senses that we cannot really even imagine, except perhaps dimly in the realm of fiction. Animals often know when an earthquake, tornado, big storm, or tidal wave is coming long before any of our technology detects anything abnormal. While we continue to know through our brains, they know through their organs and pores and muscles. Something almost electric happens from one animal to another in a wolf pack or dog pack; the salient subject is not the individual, but the group that moves as one. Insects are even more amalgamated: in the beehive or the anthill, solitude is unimaginable; the organism itself is the swarm.

Animals, I think, have senses and powers that we have lost, that have atrophied from lack of use in most humans.[2] Animals may remember their dead more fully, perhaps they even negotiate with them. They communicate with trees and rivers and winds. They reproduce in relation to their food supply, rather than force the food supply to artificially compensate when overpopulation occurs. They sense a change in the mood of a place, they register affects and auras at lightning speed. They don't use technology to alter the world around them; they accept and respect their environments, and adapt themselves to the worlds they are given.

For the most part, most of us ignore these talents in animals. We live so far inside our own human reality that we have learned how not to see their mysterious senses; quite literally they can be sitting in front of us, lift their head to smell something, the hair rises on their back. But because we do not smell that thing, we think it's not there. We conclude that the animal is just like us, only less intelligent. We do not entertain the possibility that other animals are different, that they live in different worlds with altogether different shapes of perception. We think our scale of being is the only one that matters.

It is the things that animals know that we humans can no longer recognize or remember that I call 'spirituality'. It's akin to Bergson's notion of spirit, but my sense of meaning honors the unknown animal knowledge more robustly. It's not about god, or angels, or reciting prayers. It's about the way your skin prickles when you pass a certain someone, and it's about knowing what to do with that kind of information. The funny smell that could be something burning in the distance or could be your own imagination. The way a dream makes you feel good all day even if you can't remember the details of it. It's the glimpse of something you see out of the corner of your eye that, when you turn your gaze-filled-with-reason toward it, disappears. It's not that it's gone, really. More that something inside of you filtered it out.

Animals know things we don't know. But instead of watching them and trying to learn from them, we lock them in cages, breed them to conform to our desires, stick knives and needles in them to see what happens, eat them with no attention to the conditions in which they are raised or slaughtered, no awareness of their families or happiness. But because they do have alternate realities, they maintain their goodwill. Sometimes they smile for us, or even love us. They know through their special senses that our time is almost over, and their time is about to come. Or at least that is what I like to think.

From a perspective that sees animal talents as different from but not inferior to the attributes of the human, the reasons why Mary and Claudine share their space and their bodies with bonobos is almost the opposite of altruism. They are involved with these animals as a way to regain some of what has been lost in the hypertrophied development of human reason. They are seeking a way to have a different relationship with the world through the acquisition of a different set of senses. They want to experience things a new way, not as typical humans but as part of the process of becoming animal. They share their bodies with the other in order to remember some shadow of their own animality.

I think that this kind of connection is what many of us today are looking for. To take one simple example, the set of practices that constitute the Occupy movement are, I believe, calling us all into a new way of life; that is, Occupy is not about making demands or electing leaders but about motioning humans toward inhabiting a different reality: bringing us forth into a more connected way of living. Occupy wants us to hear each other into fuller speech, share time and space and resources with others who have previously been almost invisible to us; it wants us to focus on what we need and what we can contribute, and in doing so, become more in tune with our neighbor's needs and talents. Consider this from Open Democracy commentator Anthony Barnett:

> Occupy sees America's existing democracy as responsible for the ills of our day: financial crisis, environmental degradation, war, identitarian strife and corporatization of the State. All permitted by the 'thick-skinned' nature of U.S. democracy, 'thick' because it relies on everyone believing that 'whatever is permitted is both right and good.' Against this democracy of personal liberty that is indifferent to others, the Occupiers want a society where we are 'thin skinned' and the experience of others is our concern.
>
> *(2011)*

I believe that physical and psychical intersubjective encounters with animals can answer some of the hopes and desires produced within Occupy. And they may be what evolution itself is calling for.

Of course, we should support traditional forms of conservation. Human overconsumption and encroachment are driving the demise of many habitats, and we should try to self-correct these problems; we need to try to save the planet. I stand firm in my conviction that we can build a better world; environmentalism and good stewardship ought to be central human concerns. But I also want to note that, from the long view of evolutionary time, things can look quite different. In a world where huckleberries call bears into being, how do we know that we are the kinds of agents who could save a planet? What makes us think we are the kinds of subjects who control their environment, rather than the other way around? If, as Bergson's and Deleuze's understandings of evolution demand, new subjects emerge from collaborations and connections that seem unimaginable beforehand, isn't it possible that we residents of the earth are being called into a new form of subjectivity? If we are to welcome this transition, don't we have to let go of the old notions of human exceptionalism? Isn't conservation really a way of granting priority to that which exists now over that which is to come?

Mary and Claudine Andre may appear to be about the business of keeping bonobos on the planet, but, from a different angle, they are also inviting humans to become a different kind of creature, a being that is enmeshed and interconnected, that intuits or feels the experience of the other, and in doing so recovers or reinvents some forgotten animal talents and senses. The ways they are engaging animals – by sharing their space and bodies with them – pull us away from our minds and back into our senses. They are modeling for us a different kind of evolutionary progress, one that opens the way for humans to change and become, literally, more like animals. This kind of conservation may inadvertently be leading humans into the next phase of life on earth.

But *what happens* to Mary's baby?

Animal hybridity surrounds us. It's not just mules, red wolves, or grizzly-polar bear mixes found in nature, but ligers, tigons, zorses, wholphins, blynxes, savannahs, beefalos, camas, and many others as well. Darwin understood the importance of hybridity, especially when it resulted in animals that were not sterile. Often, the hybrid is healthier than either of its parents – the condition known as hybrid vigor – and such blendings can rapidly change the course of an organism's evolutionary path.[3]

A few scientists have toyed with questions of human/chimpanzee hybridity – either as explanations for evolution's past or as macabre experiments for the advancement of science, such as Ilya Ivanov's turn-of-the-twentieth-century attempts to create a 'humanzee' (Cohen 2010). Debates rage among scientists about whether humans and chimpanzees could successfully hybridize; some suggest that the 3 or 5 percent of difference in DNA would block successful gestation; others think such experiments are too unethical to even consider.

Many works in science fiction grapple with these topics. In many of these stories, humans and other great apes merge their bodies through, for example, technology (Dickinson 1990), use of psychotropic drugs (Self 1997), or reinterpretation of species categories themselves (Høeg 1996). But the story that most interests me here is *Lucy*, a 2010 novel by Laurence Gonzales. *Lucy* is a scientific thriller that warrants human-animal studies' attention; the minute I started reading it, I was once again reminded of Cussins' short story. Indeed, midway through my first reading of the novel, I felt as if I were imaginatively engaging with potential answers to the question that had plagued me for a decade: what happened to Mary's offspring?

The fictional settings are very different. Mary wasn't really gestating a hybrid; the embryo she implanted was fully bonobo. In this sense, even though the story insinuates she is growing a hybrid, in fact she is really acting as a surrogate mother. Gonzales, conversely, writes the story of a full human/bonobo hybrid. The narrative starts in the Congo with field primatologist Donald Stone, who falls in love with a particular female bonobo, and with the help of in vitro genetic manipulation, they conceive a child together, a girl named Lucy. For over a decade, Lucy lives a Tarzan-like life, engaging in the best behaviors of both humans and bonobos. When Lucy is about 14, war breaks out in that area of the Congo, Dr. Stone is murdered, Lucy is rescued by another primatologist, Jenny Lowe, and brought to Chicago to live. At first, Lucy has serious problems fitting into city life; Jenny Lowe does not know about her biological constitution and so chalks her behavior up to unfamiliarity. But when Lucy shows strength, skills, and sensitivities found only among animals (i.e., traits and talents humans have forgotten), Jenny begins to realize that Lucy may not be fully human. Lucy is able to intuit things from something she calls 'The Stream', a way of knowing that we humans might call 'a sixth sense'. When Jenny reads Dr. Stone's notebooks, she learns the truth behind Jenny's background. Stone writes in his journal:

> I understand that what I have done will seem beyond the pale to some. But to that charge – and to history – I have this to say: Humankind has destroyed most species with which

> it has come into contact and is rapidly destroying itself. Something must change in our human nature. And I offer Lucy as proof to the world that, even though the ethics of what I've done may be questioned, the results are unequivocal. Anyone who meets this fantastic, intelligent, and beautiful girl will have to marvel at her, no matter the means of her creation. Lucy, in short, is the best argument in my defense. The way she has blended human intelligence with a bonobo's ability to process the richness of sensory signals from The Stream, along with her gentle and loving social instincts, prove that I was right: Lucy is love made manifest. And as her offspring and their offspring continue to reproduce, a new kind of human – more human than human – will evolve.
>
> *(Gonzales 2010: 50)*

Like Mary, Stone began his efforts with a simple impulse toward conservation. The text says:

> It had all begun with his passion for the bonobos and his certainty that people were going to drive those marvelous creatures extinct. The only way to save them, he believed, or some of their best qualities, was to selectively breed them with humans. Although he saw their extinction as inevitable, at least part of their unique and brilliant character might be preserved if he could safely lock the bonobo genes inside humans.
>
> *(Gonzales 2010: 49–50)*

Indeed, Gonzales's conviction that he is writing this novel as an advocate for conservation is also addressed in an online interview about the book. Here, it is not about keeping them alive in the wild, but fundraising for better enclosures in captivity:

> I heard that the largest colony of bonobos in the world was just an hour from my home in Milwaukee. So I went there to meet them. I fell in love with them. They're sexy and clever, and they have complex language and a matriarchal social structure in which the guys do what the women tell them to do. As they got to know me better, they would come to the wire at the back of the enclosure and put their fingers through the fence, imploring me to touch them. Their hands are beautiful and so very human. Like Lucy herself, these bonobos are caught between two worlds. They can't go back to Congo, even if we allowed it. They're not fit for living in the wild and even if they survived, they'd be killed by bush meat hunters there or by the civil war. And yet it is so sad that they are kept in a cage. I am working to make it possible for people who read *Lucy* to donate money to improve their living quarters.
>
> *('Questions')*

And yet, the story of *Lucy* tells a tale that is about something much bigger than better living quarters for zoo animals and quite a bit different from what we conventionally think of as conservation.

For a couple of hundred pages, Lucy has problems fitting into high school and making friends. When her genetic identity is revealed, she is harassed and harangued by many community members; her only friend is killed simply because of her association with Lucy. Eventually her story hits the national spotlight and everyone from scientists to libertarians to the religious right gets involved. She runs but cannot hide; they capture her and perform ghoulish experiments on her. The caper is a page-turner, so much so that the reader almost forgets what's at stake. Until the end.

Lucy finds refuge on Native American sovereign land; neither the scientists nor the religious zealots can touch her. There, she finds a home with people who have always shared their lives, intimately, with animals. Her caregiver looks into Lucy's eyes with complete recognition: 'my great-grandfather was one-quarter wolf', she tells her (Gonzales 2010: 304). It takes Lucy two years to heal from the physical and emotional wounds our world inflicted on her. Eventually, she meets a man who speaks to horses and calls wild deer to his side. They conceive a child together. The novel closes just as Lucy, the hybrid, is giving birth. In a conclusion reminiscent of 'Confessions of a Bioterrorist', we are left wondering about what happens to Lucy's baby, just as the future of Mary's baby was left unknown. But we are one step further along. One step off of the trajectory of a human-centered notion of progress. One step away from human exceptionalism. One step more deeply imbricated in a Bergsonian and Deleuzian understanding of evolution. One step closer to becoming animal.

How could this be justified? One more story before we end

Mary and Dr. Stone, and the authors of these fictional people and acts, are in some sense simply going about business-as-usual for humans. While they all seem to recognize that humans have lost their way as good citizens on a limited planet, they solve that problem by appropriating bonobo embryos and wombs to restore humans to a proper relationship with nature. They have done what white Westerners have done for centuries: when we perceive a lack, we colonize someone else to fill the deficit. There's a way in which the stories themselves are embedded in the narrative of human exceptionalism; interbreeding is simply another event in a long history of white human domination of nature and animals. (To be fair, both authors recognize this and use various techniques to justify the actions of their characters. But still.) Indeed, capturing bonobos and forcing them to comingle their genes with us in whatever form would have the opposite effect of the one I am striving for in this essay, rather than saving humanity from destruction, such hubris would almost guarantee it.

To proceed with human animal hybridity like these stories suggest would be to end up with a world much like H.G. Wells's *The Island of Dr. Moreau*. Moreau's is a universe where interbreeding is controlled by humans who more or less do not want to change, adapt, or evolve, but simply want to appropriate animal talents for their own use. That approach serves as a classic model of horror. What I am looking for is a way for humans to invite change, to open ourselves to the possibilities of living very differently on this earth, to the possibility of creative evolution. The world cannot stay the same; the planet cannot bear it. We humans number seven billion and are growing; for us to have a future in this world, something must radically shift. I believe we will have to give up some aspects of what we might call 'civilization' – the thing that sets us apart from animals – because many practices associated with it are unsustainable. We are being called upon to pay the price for the damage we have already done. It is not a question of adding a few characteristics or senses to that which is currently human, but a radical transformation of the human that is called for. A transformation rooted in us becoming more like animals.

Octavia Butler's *Xenogenesis* trilogy of novels (particularly *Dawn* and *Adulthood Rites*) can help us imagine how a different approach to hybridity might help push the change in the right direction. The novels are set in a post-apocalyptic earth where the remnants of humanity struggle with war, violence, cancer, disease, and food and water shortages; nuclear poisoning has rendered all humans sterile. It is a world filled with suffering. The Oankali are polyamorous, reptilian-like aliens from outer space who land on earth and colonize humans by interbreeding with us. They are genetic engineers and gene traders by their natures: by making us into 'trading partners' with them, the Oankali satisfy their hunger for new and exotic genes; we are little more than breeding stock for them. And although they cure our cancers and diseases, they also transform us into very different kinds of creatures. They take away our propensity toward violence and self-destruction, our desires to eat meat, our practice of heterosexual monogamy as the basis of reproduction, our commitments around individuality and freedom. And because all humans are sterile, the choice for us, in these novels, is between pairing with the Oankali or human extinction.

At play in these fictions are the questions of human constitution and human exceptionalism; how much could we interbreed with another species, whether Oankali or bonobo, before we ceased being human and became something else? Lucy is presented to us as more or less an enhanced human: she hasn't given up any characteristics we would consider essential to humanity; she has just added on some good animal traits and talents. Not so in breeding with the Oankali: here, we look different, we act different, we are no longer monogamous, we no longer breed in pairs, we don't use technologies to manipulate the natural world, we don't have hierarchies or leaders or winners of any sort, there is no such thing as politics in the traditional sense and no way to gain power in any group. Instead, the heightened senses and sensibilities that the Oankali splice and breed into our genes force us to live in harmony with that planet. We have a concrete internal feeling for what plants and soil need and our new biology calls us to respond to that; in other words, we share the pain of living things around us, we no longer see ourselves as special or unique or above creation, we don't strive for heroic greatness or social power, and instead we see ourselves fully as one small part of a vast organic system. And, as a result, we sacrifice our commitments to notions of freedom, self-determination, and human exceptionalism. We become slaves not only to the Oankali, but to the natural world itself.

While some critics see the loss of human individualism and human freedom as dystopic (Peppers 1995; Zaki 1990; Butler *et al.* 2001), I suggest that these stories provide something of a blueprint for Bergson's creative evolution and Deleuze's becoming animal. Put succinctly, the new creation cannot be produced out of a colonizing act where one species pulls in the positive attributes of another without risking massive change themselves. When Butler's Oankali and humans partner, both groups are radically altered; this is what evolution and becoming animal mean. This hybridization is mutually beneficial in that it satisfies Oankali hunger for diversity and the human desire to survive, but it also creates a new kind of subject that can move forth into the future. In other words, the lesson behind Butler's narrative, I believe, is that to be truly open to allowing ourselves to adapt and be radically changed by the other lies at the core of survival and constitutes the essence of the people to come. This open orientation toward difference leads us back to the question of how human/bonobo hybridity might be justified outside the frame of colonization.

Granting that bonobos today are not in a position to demand that we partner with them, let's imagine for a moment that they are much closer to the Oankali on these several counts: they have the power to transform us into a species that can live in harmony with nature, they can give us back lost senses and connections, and mating with them in their polyamorous practices is the only way anything even partially human can stay on this planet. What kinds of characters would we humans need to *become* in order for bonobos to want us to breed with them? What kinds of creatures would we need to turn ourselves into to make other animals want to share their flesh and blood with us? Is there enough humility, hope, and wonder left in humanity that we could change ourselves into such a people? Is there a way for us to meet other animals not as dominators, but as partners invested in negotiating a new world together? Can we allow ourselves to feel bad about the many ways we have treated other animals in the past and set ourselves on the task of seeking their forgiveness? Can we imagine eliminating the parts of ourselves that bonobos would find objectionable? What would it take for us to want to live in connection and harmony with the environment like bonobos and most other animals do? How do we eliminate the parts of ourselves that ignore, transcend, and dominate nature? Why did we stop listening to other animals, to the trees, the wind, and the voices of our ancestors, and how can we start again?

Whether or not we actually reproduce with bonobos, the point here is that humanity as we know it is moving rapidly toward an end, and to continue on in any form requires an evolutionary leap. Evolution, as we saw, is rarely about adding something to what already exists, it is rarely about 'progress'. Instead, at least according to Bergson and Deleuze, it is about change. It is about difference. To adapt to the coming world, we must embrace the kind of difference that can bring us back into balance with nature. Indeed, the greatest and most fundamental distinction between humans and Oankali is precisely their perception of future difference. In general, humans fear difference and police the borders of our species to maintain human exceptionalism; the Oankali crave difference and know that their survival depends on adaptation and evolutionary change. As the main character of *Dawn* instructs her hybrid child in *Adulthood Rites*:

> Human beings fear difference, Oankali crave difference. Humans persecute their different ones, yet they need them to give themselves definition and status. Oankali seek difference and collect it. They need it to keep themselves from stagnation and overspecialization. You'll probably find both tendencies surfacing in your own behavior. When you feel a conflict, try to go the Oankali way. Embrace difference.
>
> *(Butler 1988: 80)*

Embracing difference and transforming ourselves into creatures that can listen and understand the languages of our closest relatives, that can respond to their desires, and that can be open to new possibilities of life on earth together, are, I believe, the central tasks confronting human-animal 'studies'.

Notes

1 The story was published in 1999, three years prior to the Homeland Security Act and its implications for the word 'terrorism'. While much could be said about animal activism and its complicated relationship to the concept of terrorism (most specifically, the Animal Enterprise Terrorism Act of 2006), I sidestep these important issues in this chapter to take up a different line of thinking. The short story, though, begs for an analysis in the frame of what constitutes terrorism.

2 There are, of course, many exceptions, and I do not want to speak for something called 'all humanity' here. Despite the fact that we have evolved as a species, many indigenous and aboriginal peoples work hard to maintain their historic otherworldy relationship to nature. Additionally, many philosophers and thinkers today, including David Abram (1996), Jane Bennett (2010), and others, seek ways of recovering these lost modes of being and knowledge. When I say that most of us have lost these animal senses, I mean that we cannot recover them by seeking enlightenment only within industrialized, rationality focused society. New worlds may be possible inside practices of relational ontology.

3 While popular understandings of hybridity suggest that all hybrids are sterile, recent scientific understandings counter this assumption: for example, Williamson and Vickers (2011) suggest that starfish and butterflies are the result of hybrid matings.

References

Abram, D. (1996) *The Spell of the Sensuous: Perception and Language in a More-Than-Human World*, New York: Pantheon Books.

Barnett, A. (2011) 'The Long and the Quick of Revolution', *openDemocracy*, 16 December, available online at www.opendemocracy.net/anthony-barnett/long-and-quick-of-revolution, accessed 9 February 2012.

Bennett, J. (2010) *Vibrant Matter: A Political Ecology of Things*, Durham, NC: Duke University Press.

Bergson, H. (2007) *Creative Evolution*, New York: Palgrave Macmillan.

Butler, O.E. (1988) *Adulthood Rites: Xenogenesis*, New York: Warner Books.

—— (1988) *Dawn: Xenogenesis*, New York: Warner Books.

Butler, O.E., Mehaffy, M., and Keating, A. (2001) ' "Radio Imagination": Octavia Butler on the Poetics of Narrative Embodiment', *MELUS* 26.1: 45–76.

Cohen, J. (2010) *Almost Chimpanzee: Searching for What Makes Us Human, in Rainforests, Labs, Sanctuaries, and Zoos*, New York: Times Books.

Colebrook, C. (2010) 'Creative Evolution and the Creation of Man', *Southern Journal of Philosophy* 48 (Spindel Supplement): 109–32.

Cussins, C.T. (1999) 'Confessions of a Bioterrorist: Subject Position and Reproductive Technologies', in S. Squier and A. Kaplan (eds), *Playing Dolly: Technocultural Formations, Fantasies, and Fictions of Assisted Reproduction*, New Brunswick, NJ: Rutgers University Press, pp. 189–219.

Deleuze, G. and Guattari, F. (1987) *A Thousand Plateaus: Capitalism and Schizophrenia*, B. Massumi (trans.), Minneapolis: University of Minnesota Press.

Dickinson, P. (1990) *Eva*, New York: Laurel Leaf.

Ferreira, A. (2008) 'Primate Tales: Interspecies Pregnancy and Chimerical Beings', *Science Fiction Studies* 35.2: 223–37.

Gonzales, L. (2010) *Lucy*, New York: Alfred A. Knopf.

Høeg, P. (1996) *The Woman and the Ape*, New York: Farrar, Straus and Giroux.

Morton, T. (2007) *Ecology without Nature: Rethinking Environmental Aesthetics*, Cambridge, MA: Harvard University Press.

Nagel, T. (1974) 'What Is It Like to Be a Bat?', *Philosophical Review* 83: 435–50.

Peppers, C. (1995) 'Dialogic Origins and Alien Identities in Butler's *Xenogenesis*', *Science Fiction Studies* 22.1: 47–62.

Peterson, D. (2003) *Eating Apes*, Berkeley: University of California Press.

'Questions for Laurence Gonzales: Amazon.com Review', available online at www.amazon.com/Lucy-Vintage-Laurence-Gonzales/dp/0307473902/ref=sr_1_1?ie=UTF8&qid=1328807651&sr=8-1, accessed 9 February 2012.

Savage-Rumbaugh, E.S. and Lewin, R. (1994) *Kanzi: The Ape at the Brink of the Human Mind*, New York: Wiley.

Self, W. (1997) *Great Apes*, New York: Grove Press.

Squier, S. (1998) 'Interspecies Reproduction: Xenogenic Desire and the Feminist Implications of Hybrid', *Cultural Studies* 12.1: 360–81.

de Waal, F.B.M. and Lanting, F. (1997) *Bonobo: The Forgotten Ape*, Berkeley: University of California Press.

Williamson, D. and Vickers, S. (2011) 'Interspecies Hybrids', in L. Margulis, C. Asikainen, and W. Krumbein (eds), *Chimeras and Consciousness: Evolution of the Sensory Self*, Cambridge, MA: MIT Press, pp. 183–99.

Woods, V. (2010) *Bonobo Handshake: A Memoir of Love and Adventure in the Congo*, New York: Gotham.

Zaki, H.M. (1990) 'Utopia, Dystopia, and Ideology in the Science Fiction of Octavia Butler', *Science Fiction Studies* 17.2: 239–51.

16

Embodying the feral

Indigenous traditions and the nonhuman in some recent South African novels

Wendy Woodward

South African fiction has long been crucially preoccupied with human identities, their performative qualities, the legalized contingencies of their racialized classification, as well as their potentially redeemable transformations. In the last decade or so, however, such literary considerations have extended in particular in post-apartheid fiction to the fluidity and porousness of human identities in relation to those of the nonhuman. With one exception the writers considered here imagine such ferality in relation to indigenous worldviews, some of which have been reclaimed in notions of 'heritage' since the first democratic elections in 1994 in South Africa. Indigenous magic is rendered unexceptional in the novels, even if the human body is reconfigured, sometimes quite literally, beyond rational limits. As Zakes Mda, whose writing has been erroneously labelled magical realist, asserts: 'I am a product of a magical culture. In my culture the magical is not disconcerting. It is taken for granted. No one tries to find a natural explanation for the unreal. The unreal happens as part of reality' (quoted in Fincham 2011: xxii). Similarly, representations of 'reality' in many of the narratives under discussion are not 'unreal' but an assertion of indigenous knowledges which were either disregarded or actively suppressed during the colonial and apartheid eras.[1]

When Lauren Beukes, K. Sello Duiker, Thando Mgqolozana, and Don Pinnock deploy indigenous traditions in their novels, they locate their narratives in the recognizable present rather than nostalgically questing for a pre-industrial or prelapsarian moment. Expanding human knowledges through recourse to experiences of the nonhuman, possibly of the unreal, thus suggests ways of coping with negative aspects of our unsettled, adolescent democracy. In *Zoo City* (2010), Beukes has Zinzi and her sloth seek solutions to urban crime, 'muti' (traditional medicine) murders, and xenophobia. Sello Duiker has the girl-child Nolitye, a brave hero in *The Hidden Star* (2006), who rescues herself and other children from adult abuse. Mgqolozana's Lumkile in *A Man Who Is Not a Man* (2009) suffers physically and psychically as he copes with the rigidity of Xhosa initiation rites. Pinnock's *Rainmaker* (2010) imagines a Bushman spirituality contributing to transforming urban violence and gangsterism. In *Ninevah* (2011), Rose-Innes does not deploy indigenous tradition explicitly, but she has her protagonist turn feral as city space is redefined in relation to homelessness and ecologically destructive urban development. The feral in contemporary South African fiction thus marks a transitional space for bodies in resistance to the conventional notions of 'progress' and 'civilization' as ordering urban-industrial spaces.

If embodying the feral in relation to indigeneity and tradition provides no unrealistic solutions, the potential for creative transformation flickers through all the narratives. Common to all these novels is the spatializing of such transformation within the human body itself. The protagonists of *Zoo City, Ninevah, Rainmaker,* and *A Man Who Is Not a Man* all embody the feral themselves; the protagonist of *The Hidden Star* interacts closely with shape-shifters who are both human and nonhuman. With the liminality of feral embodiment located in the quotidian, the novelists overlay and contradict the conventional dualistic splits between the unreal and the real, nature and culture. The divide between country and city which has the built environment seem an impoverished one in relation to the nonhuman is also re-imagined. As Jennifer Wolch notes in relation to the city, '[c]orporeal identity may [. . .] become increasingly destabilized as understandings of human embodiment traditionally derived through direct experience of live animal bodies/subjects evaporates or is radically transformed' (1998: 129). This 'direct experience' of nonhuman animals carries extraordinary power and effect in the actual merging of human 'corporeal identity' and those of animals.

This extending of the human body, 'the geography closest in' to borrow a term from Adrienne Rich (quoted in Nast and Pile 1998: 2), is ritualized, if not always celebrated in *Zoo City, Ninevah, The Hidden Star, A Man Who Is Not a Man,* and *Rainmaker.* Extending human body space to incorporate the nonhuman in these novels generally suggests a sincere attempt to augment the epistemology of being human. Birke's sense, in this volume, of the 'slipperiness' of the categories of wild, feral, and tame recurs here. At times, the very boundaries of human bodies waver, with the feral manifesting as outer beast or savage in order to enact an inner wildness, a reconnection with the natural. This mutability of the human/feral embodiment within the unreal/real is transformative on many levels. It dispenses with the restrictive dualisms of self and other, prehistory and modernity, spirit and body, even life and death.

Embodying the feral is usually a key component of horror narratives, where crossovers between human and nonhuman tend to be brutal and uncanny with the underlying urgent imperative to restore the human and nonhuman to their discrete categories. In the provenance of indigenous African and Bushman traditions,[2] however, such blurring of identities is shamanistic with its intertwining of human and nature, and with the nonhuman as an inherent part of human subjectivity. Even dark magic, which abuses the powers of animality, eliciting extreme fear in its potential targets, is not necessarily motivated epistemologically by divisions between human and nonhuman. The adult novels depict the-other-within either metaphorically or as a separate being: in *Zoo City*, the animal other who is always close by is spiritually attached to the human; in *Ninevah* the protagonist, Katya Grubbs, approximates an insect herself; in *A Man Who Is Not a Man*, Lumkile empathizes so acutely with the vulnerability of his dog that he enacts canine aggression. The two adolescent novels, *Rainmaker* and *The Hidden Star,* represent humans magically being-within-the-other. In the former, the protagonist shape-shifts into the body of a leopard; in the latter, good and evil spirits conceal themselves in animal bodies.

All these writers have us imagine bodies as uncertain and contingent, as porous and intersubjective. In challenging fixed notions of corporeal identities through the merged, the overlapping, the in-between, they dislodge speciesism and anthropocentrism from their humanist pedestals. Anat Pick confirms that embodiment 'provides a critical space for thinking of the human outside Cartesian abstractionism' (2011: 6), but I would add that these novels more precisely require recognition that, in southern Africa, these ways of thinking predate as well as persist beyond the histories of European colonization. Reconfiguring the human body in relation to that of the nonhuman has profound ontological implications. For Cora Diamond, '[t]he awareness we each have of being a living body, being "alive to the world", carries with it the bodily sense of vulnerability, sheer animal vulnerability, the vulnerability we share with them' (2008: 74), a condition that varies across spaces, times, and categories of human-animal relationships. Such a sense of vulnerability, in these narratives, may paradoxically generate empowerment as identities are strengthened by being at the threshold of human and animal, with the usual 'equation of animality with physicality and materiality – with bodies and things – and the correlated association of the human with mind, soul, thought, ideas, and meaning' (as noted by Chaudhuri in this volume) rendered irrelevant.

Urban dystopia, the 'animalled' and the 'unlovely'

Zoo City (set in Johannesburg) and *Ninevah* (set in Cape Town) both have feisty female protagonists who live in disintegrating urban spaces and who both overcome dangers to reach, finally, a qualified state of resolution. Here the parallels end. *Ninevah* is a fairly gentle narrative about Katya Grubbs who owns and manages the Painless Pest Relocations Company. She is assigned the job of clearing Ninevah, a new housing development, rendered uninhabitable by plagues of mysterious and aggressive insects. Here she discovers a literal underworld of smuggling, masterminded by her indigent, estranged father who has maliciously nurtured the larvae of the beetle *Promeces palustris* so that they will reappear in the rainy season. *Zoo City* is playfully described by Beukes as a 'muti noir' novel (2011: 7); undoubtedly 'noir', the muti or traditional medicine is also dark.[3] Intertextually, it includes apparently legitimate scholarly texts on aspects of the novel. The narrator Zinzi December is a 'zoo', so-called because she is rendered feral, 'animalled' as a sign of past criminal activity. She attempts to make a living, accompanied always by her animal other, Sloth,[4] by finding objects for her clients and by writing scam emails eloquently convincing naïve readers to contribute money to a 'worthy' African cause. She lives with her Congolese lover, Benoît, in Hillbrow, colloquially known as Zoo City, where she is drawn, unwittingly, into a bizarre, evil underworld of muti murders, where people are killed for their body parts.[5]

Cities, as Jennifer Wolch reminds us, banish 'wild lands and wild things – as well as people deemed wild and "savage"' (1998: 119). To counteract this, Wolch proposes a Zoöpolis which fosters 'an ethic, practice and politics of caring for animals', arguing that 'we need to renaturalize cities and invite the animals back in, and in the process re-enchant the city' (1998: 124). Zoo City, with its 'zoos', lacks any such ethic of care or any such enchantment as an urban location. The 'real' Hillbrow, a suburb of Johannesburg, was '[a]bandoned gradually from the mid-1980s by property owners fleeing its demise[;] its economic and social base is now substantially outside the law' (Bremner 2010: 217). Overcrowding is exacerbated by refugees who 'are expected to integrate in South African society' with most fetching up in 'high-density inner city residential suburbs like Hillbrow' (Bremner 2010: 171). The prevailing xenophobia directed at African people from north of South Africa has become endemic; for Bremner, '[a]partheid is not over, it has simply been deferred' (2010: 171) To such prejudices, Beukes adds the discrimination against 'zoos', which are vilified as savage and lawless, characteristics that a 'civilized' city cannot countenance.

Zinzi, a one-time drug addict, had acquired Sloth because of her unwitting responsibility for the death of her brother during a bungled hijacking she had organized. Her Sloth, Zinzi admits, is her 'own personal scarlet letter' (Beukes 2010: 50), and she notes that in China 'they execute zoos on principle [. . .] [N]othing says guilty like a spirit critter at your side' (9). No consistent symbolic links obtain between the animal's characteristics and that of the human they are attached to. The 'giant' Benoît has a diminutive Mongoose (who irritates Zinzi by sleeping on her warm laptop). The vicious Mark has a Maltese which does not seem congruent, yet Amira, who is one of the masterminds behind the muti murders, has Marabou, a stork who exists on carrion, as her 'avian Siamese twin' (68), and Odi Huron, a fellow evil genius, is 'animalled' with a massive, human-eating, albino crocodile.

Relationships between humans and their animals are close. The usually heartless Amira shows great attachment to and compassion for her stork, who lives in her rucksack because he lacks legs. Zinzi and Sloth lovingly tolerate each other's quirks, with Sloth expressing emotions of fear and anxiety which Zinzi rarely articulates. Sloth seems to understand human language, and they form a partnership in the world of crime and violence with Sloth protecting her with his claws. They are, essentially, familiars, beginning the day together:

> Sloth blinks up at me sleepily from his roost. [. . .] He's not good at mornings. There's a mossy reek that clings to his fur and his claws. [. . .] 'Come on, buddy', I say to Sloth [. . .] [who] gives a sharp sneeze of disapproval and extends his long downy arms. He clambers onto my back, fussing and shifting before he finally settles. I used to get impatient but this has become an old routine for the pair of us.
>
> *(Beukes 2010: 2)*

Zinzi is commissioned by Maltese and Marabou (Zinzi names some people after their animals) to find the missing Songweza, who with her twin brother S'busiso forms a pop duo sensation. She would prefer to leave the incriminating Sloth behind at times, but 'the feedback loop of the separation anxiety is crippling. Crack cravings have nothing on being away from your animal' (124). The attention of 'zoos' to their animals is not disinterested. As a 'spirit critter' the animal protects the 'zoo' from being dragged into the void, known as the Undertow; thus Sloth 'is the only thing between [Zinzi] and the rising dark' (299). An 'extract' from 'Aposymbiot Counselling Resources' on SAPSYCHWEB, a fictional scholarly website, rationalizes the Undertow phenomenon as 'shadow self-absorption' (158), reducing it to an analysable occurrence rather than recognizing it as a manifestation of excruciating spiritual annihilation.

Generally, the accompanying animals of 'zoos' are judged by those who are not 'animalled' as external manifestations of their savage criminality, and, while this figures for Odi Huron, Marabou, and Maltese, it is not relevant for many 'zoos' who may have acquired animals through unwitting violence (like Zinzi) or been more victim than perpetrator, for in the Democratic Republic of Congo, Benoît was forced by rebel soldiers to kill his close friend. The ontological shifts imposed on 'zoos', who are now entirely intersubjective beings, are not necessarily experienced by them as negative in relation to their animals, to whom they are, with few exceptions, deeply emotionally attached. Their ferality is thus embodied in a kind of trans-species, which, Beukes seems to suggest, is complex and dependent on the individuals themselves.

At the same time, being 'animalled' can attract perverse behaviour. Some 'zoos' are so demoralized by the lifelong anxiety about their animals' safety that they prefer to stage orgies of animal killing 'to actively invoke the terror and rapture of shadow self-absorption' (Beukes 2010: 158). Odi Huron drugs S'busiso, forcing him to lay claim to the crocodile as his animal so that Huron can rid himself of the reptile who hampers his mobility. In addition, the muti murderers tap into the powers of animal others by ritually sacrificing them after having killed their 'zoos'. In the bloody and melodramatic climax of the novel, Zinzi and Benoît attempt to rescue the twins, Songweza and S'busiso, from Huron, Marabou/Amira, and Maltese/Mark. Outfaced by their evil, Benoît is dragged below the water by the crocodile and Zinzi is shot as she attempts to rescue him. While the wounded Zinzi manages to resuscitate Benoît, she is forced to watch, from her hiding place, while the drugged S'busiso, thinking he is playing a video game, is goaded to stab his sister to death. Having released the special magical powers embodied by twins, the adolescent is then cold-bloodedly murdered.

Such a narrative denouement might seem standard 'noir' fare, but African magic is extensively involved here in ways that further complicate and localize the plot. All the main characters are possessed by 'mashavi', which is defined in an 'extract' from 'Bibliozoologika: An Entymology [*sic*] of Animalled Terms' in the following way: 'Mashavi – A Southern African word (spec. Shona) used to describe the preternatural talents conferred by an aposymbiot and the aposymbiot animal itself' (Beukes 2010: 177). It is further defined traditionally as a 'homeless spirit' of a foreigner who has not been properly buried and laid to rest with the ancestors. This spirit then searches for a willing, living host; those possessed of mashave develop particular talents, which may be healing or entirely random (177). Zinzi, whose talent is to find lost things which she can 'see' psychically, complains that 'the problem with being *mashavi* is that it's not so much a job as a vocation. You don't get to choose the ghosts that attach themselves to you' (10).

'Zoos' may be complicit in the extent to which they can be exploited for their 'preternatural talents', with the magic generated by their animals highly desirable for unscrupulous buyers. The adolescent 'zoo' whom Zinzi encounters living on the street is entirely committed to this trade:

> Nasty [. . .] pats his leg, as if calling a dog, and a Porcupine hauls itself out of the darkness, limping forward on three paws, its quills rattling. It nudges his knee with its stubby snout in wary affection. Thick ropes of drool hang from its jowls. Its eyes are dull. Its back foot is missing. The stump has healed badly, the tissue grey [. . .] It smells of damp and rot, like the broken concrete of the hole it crawled from.
>
> 'What the fuck did you do to that animal?'
>
> 'It's good money', Nasty wheedles, mocking me. 'You want some? We can get a good price for that Sloth. Rare animal hey? Start with a finger. Or a paw.'
>
> *(Beukes 2010: 182)*

As Mandlakazi, the crime reporter for the tabloid *The Daily Truth,* whom Zinzi has inveigled into investigating the muti murders, observes, 'even before you add *mashavi* into the equation [a]nimals are already some heavy magic shit' (267). Embodying the kind of ferality which crosses over into the realm of magic may connect body and spirit but, given the belief in body parts for powerful medicine, it can also mean great risk for 'zoos', especially those without family and friends who will not be missed if they are murdered.

Beukes thus has the 'zoos', with their animals, serve as metonyms for the fragility and vulnerability of life in the inner-city metropolis, as it is experienced by the marginal and the dispossessed – drug addicts, criminals, sex workers, and refugees from war-ravaged countries to the north. Feral embodiment is ideally suited to the inner-city spaces of Zoo City, the apartment blocks with concealed routes and warrens between decaying floors, the underground system of sewers which Zinzi and Sloth trawl for lost objects and use as escape routes. 'Zoos', with their animal/human boundaries remaining so porous, are not threatening per se but they can be victims of those who plot to use supernatural power over others or the perpetrators themselves. Ferality, while it might include loving relationships between 'zoos' and their animal others, can be vulnerable or empowered, ethical or evil. Most dramatically, Amira and Mark perfect the ability to transform into demons that commit murders unseen, and both finally vanish into nothingness.

That such supreme evil cannot be eradicated or even contained elides with the novel's primary representation of African magic as dark and dangerous. For 'zoos', inhabiting the real and the unreal is risky; the liminality of feral embodiment locates 'zoos' quite literally at the edge of a 'civilized' city as they become complicit or do battle, not always successfully, with malign forces and beings. Even when a *sangoma*, a traditional healer, attempts to help Zinzi solve the muti mystery, she is forced to drink his potion and the resulting trance is terrifying.[6] If the supernatural has skilfully evolved strategies to make use of modernity – the spirits contact the *sangoma* via his cell phone, the shades of those murdered for muti write to Zinzi on e-mail – such humour itself is barbed.

The ferality in *Ninevah* is similarly objectively real and subjectively connective of humanity and animality. Katya Grubbs, the protagonist and focalizer of *Ninevah*, experiences intense connections with ostracized creatures who live in the interstitial spaces of the built environment. As in *Zoo City*, ferality serves as a metaphor for alternative modes of urban living as it veers between vulnerability and empowerment. Homelessness is a recurring trope in the novel as both humans and animals lack spaces of belonging. Rose-Innes's Cape Town differs from the city of pristine beauty marketed to tourists. Instead, it is a city which lacks solidity, where walls crack and foundations shift, where urban development located in wetlands is threatened by creaturely denizens, where shackland with its ongoing, tentacled expansiveness might constitute the 'real city' (Rose-Innes 2011: 123).

The city is populated by '[t]he unlovely. The unloved' (16), the 'pests' Katya carefully and lovingly relocates – caterpillars, mongooses, beetles, rats, pigeons, squirrels, geckos, snakes. Such nonhuman beings do not respect urban space or domestic demarcations; neither wild nor tame, these feral creatures exist mostly invisibly as '[s]urvivors, squatters and invaders' (19). In Katya's analysis, they do 'no real harm. They're objectionable only because they have wandered from their proper zones, or because they trigger human shudders' (19). Initially, Katya regards herself as '[p]olicing borders' (19), relegating the wild and the tame to their proper spheres, which she does with respect and compassion. Unlike her estranged father, who treated these 'pests' as vermin and exterminated them, she has a no-kill policy, imagining a swarm of caterpillars she is called in to remove as putative subjects:

> Once you get some movement going [by manipulating the swarm into a capture-box], it's easier: caterpillars, like migrating wildebeest, have a strong herding impulse. They sense a stirring, they start to push. Perhaps they feel some dim invertebrate anxiety that the swarm has not yet been consummated, that this is not the right tree, that a better tree awaits, that they will be left behind. This is as far as her study of caterpillar psychology goes.
>
> *(Rose-Innes 2011:12)*

If Katya conceives of some animality in these caterpillars, she is also hyper-aware of their agency and their collective will. As Jennifer Wolch observes, so-called pests in the city are defined inimically in relation to pets; at the same time their agency is acknowledged in their 'affecting the urban environment' because of the 'social or economic costs they impose' (1998: 128). For Katya, on the other hand, these creatures deserve to lead unthreatened lives. She mourns the death of fish that someone had relocated into tanks on the mountain and feels remorse at her hastiness to dump cold-blooded creatures into a canal. She celebrates the 'in-between' in a parking garage inhabited by city pigeons and possibly rats. For her this space is not 'wild [. . .] but not civilised, either' (Rose-Innes 2011: 34) in its confirmation that 'resilient' and 'dark-adapted' creatures are present.

Katya's feral embodiment also has elements of the hidden and subterfuge and derives from her strong contact with the 'unloved' and the 'unlovely'. When she imagines escaping from an intruder in her flat in Ninevah, she could 'slither out' insect-like from under the bed covers, 'crawl through rooms' and 'scuttle downstairs' (112) but she 'revert[s] to small-animal mode' (112) and stays put. She not only creeps like an adapted animal through subterranean tunnels in the climax of the narrative, but she has a strong sense of 'downness – of space under the surface' imagining 'raw earth, elemental' which is 'alive with a million worms, with buried things'(29) beneath the built environment. The boundaries of Katya's body are not extended into intersubjective realms with an animal other as in *Zoo City*. Instead, she herself approximates a member of the insect or amphibian world in her sensibilities and responses. Her father had called her Katyapillar; the uniform she wears is a 'poison-toad green, boomslang [tree-snake] green' (10), as though she herself is an embodiment of feral nature. While Rose-Innes does not specify any traditional indigenous origin for Katya's feral embodiment, the character's openness to the nonhuman as well as her interconnectedness with what she calls 'the beasts' (74) echo traditional African beliefs in the sacredness of all nonhuman creatures. Credo Mutwa, for example, describes how in precolonial Africa '[w]e regarded ourselves as part of all [. . .] living things [. . .] [I]nstilled in our souls [was] a deep dependence upon the living nature around us' (1996: 13).

At various stages of the novel, Katya metaphorically embodies the feral in differing forms which enlarges her subjective, emotional dimensions. In their discussion of theorizing affect, Seigworth and Gregg mention that one approach is located in 'the sometimes archaic and often occulted practices of human/nonhuman nature as intimately interlaced, including [. . .] investigations into a body's incorporative capacities for scaffolding and extension' (2010: 6). This strand of investigation replicates Rose-Innes's fictional representation of her character's embodiment which both scaffolds and extends the human. In the tunnel under Ninevah the 'proper approach' is to 'get down on all fours, emulate the beasts. Pest's eye view' (Rose-Innes 2011: 109). Before the rains Katya is irresistibly drawn to the frogsong. 'She feels half-frog, half-girl, lapping at the moisture in the air. [. . .] Her frog skin is wet and alive. She bounds over to the giant gates on frog legs, clutches the bars with frog fingers, throat pulsing with excitement' (150). Then she becomes so highly attuned to the coming metamorphosis with the beetles about to swarm that initially she becomes beetle: 'The beetles smoothly lap her wrists, covering her like iridescent scales' (174). Soon, however, their stings are too painful and she flicks them off, aware that attempting to relocate their huge numbers is not a job 'for someone with only two arms and two legs' (174). Her father's strategy (he has recently re-emerged) of killing the beetles is ineffectual and heartless; examining the beetles' corpses, Katya 'feels the violence in her body' (179):

> She has seen many small deaths in her life, and perhaps there is no great tragedy in the destruction of an insect. Or hundreds of them. But it is still a death, and she mourns the undoing of a creature as finely made, as beautifully wrought as this one. It seems hammered out of some rare metal, chased and moulded. Knight at arms, tiny samurai: its suit of armour crushed and pulled apart.'
>
> *(Rose-Innes 2011: 179)*

Insect death, tragic for its own sake and including an unacknowledged grandeur, is echoed in Wisława Szymborska's poem, 'Seen from Above'. Here 'the dead beetle in the road / [which] gleams unlamented in the sun' is metonymic of human responses to animal deaths: 'For our peace of mind, animals do not pass away, / but die a seemingly shallower death / losing – we'd like to believe – fewer feelings and less world' (2002: 66).

In Katya's embodied empathy with the dead beetles, however, she not only laments their deaths but experiences Diamond's sense of the 'sheer animal vulnerability we share' with the nonhuman. Raimond Gaita implicitly amplifies this sense of vulnerability as he writes about feeling compassion for insects, suggesting that even if we cannot conceptualize how an insect might feel, to be unmoved by insects 'betrays a failure of the imagination' (2004: 121–2). In his analysis, attributing agency to insects is significant in our acceptance of them as objects of pity. Still, he concludes, 'speculation about the inner lives of insects plays no part, and should play no part, in the pity we sometimes feel for them' (Gaita 2004: 122).

The pity that Katya feels for the beetles killed by her father takes place at Ninevah, a housing estate built by the unscrupulous Mr Brand, into the Noordhoek wetlands in greater Cape Town. Called Ninevah because of its faux Babylonian décor, its sterility contrasts initially with the vibrancy of the indigenous *fynbos* (fine bush) growing rampantly outside. Katya's assignment, to rid the place of the feral beetles and other creatures and to make the place safe for human habitation, is utterly subverted – partly through her father's plans to make the place uninhabitable in revenge for Mr Brand underpaying him to eradicate the beetles a season before (he had set up 'trade routes' [Rose-Innes 2011: 110] of smuggled fittings and décor, and nurtured the larvae of the beetle so that the infestation would recur like a biblical plague) and partly through nature itself. With the annual rains, the wetlands inundate the estate; thus the wilderness triumphs, with Ninevah, a quasi-ancient colonial city, breached and ruined beneath the mud and the bodies of the live and dead beetles.

Ferality and the wilderness, Rose-Innes suggests, are programmed, in the long term, to triumph. Ninevah, subsequently, regresses even more dramatically into an in-between place and an extension of the surrounding bush. When Katya returns some months later with more caterpillars to be relocated, Ninevah, partially subsiding into the wetlands, has been inhabited by nearby shack dwellers who have transformed the sterility of the complex into a vibrant living and working space. Only Pascal, the security guard from the Democratic Republic of Congo, perhaps sinisterly, is missing and untraceable, an unwanted foreigner in the new economy.

In the denouement of the narrative, Katya has installed her father into a new rented apartment assigned to her after her rented house was destabilized by the building of this same apartment block. Katya now lives in her van, accompanied by Soldier, the black and tan guard dog from Ninevah, whom she had feared. Reduced to an emaciated and traumatized animal, he claimed her on her return visit to the housing estate. Like a feral creature herself, she and her canine other live in the margins of the city, the van painted with cockroaches, a metal carapace protecting her and Soldier. In Katya's homelessness, she joins vagrants and shack dwellers as the fate of Ninevah becomes proleptic of the 'overlap' of different '[z]ones where the world is taking form: where things get mixed up and wander from their positions' (Rose-Innes 2011: 207). Similarly, in her work, she has rescinded her belief in keeping the wild and the tame in their proper places. In her own embodiment, she has accepted the lack of hierarchies between subject and object: 'Even human skin, Katya has read, is porous and infested, every second letting microscopic creatures in and out. Our own bodies are menageries. Short of total sterility, there is no controlling it' (207). For Rose-Innes, 'the geography closest in' serves as the ultimate in-between space, a location without the dualism of self and other, without barriers between human and insect. The human, implicitly following after the ways of African tradition, needs nature to complete her as she lives adaptively in the city.

Becoming dog, becoming leopard

A Man Who Is Not a Man by Thando Mgqolozana, *The Hidden Star* by K. Sello Duiker, and *Rainmaker* by Don Pinnock, like the novels discussed above, engage with notions of conventional human subjectivity in relation to animality, but they include indigenous tradition far more directly in their narratives. All deal with problematized identities supported by feral embodiment, either through Xhosa or / Xam tradition.[7] In its witnessing of personal trauma, Mgqolozana's novel is a testament, in a realist mode. In the adolescent novels by Duiker and Pinnock 'the unreal happens as part of reality', as Mda would put it, and both begin in edgy urban locations, the former in an African township where children are disappearing, the latter in coloured suburbs riven by gang warfare.

In *A Man Who Is Not a Man* animality is represented dualistically, as lesser than human, yet, ironically, it is through asserting courage in relation to dogness that Lumkile, the narrator, restores his sense of self in opposition to Xhosa traditions of masculinity. Lumkile, after eschewing a life of crime in Cape Town, returns to his mother in the rural Eastern Cape. Having written his final school exams, he pledges to go to the mountain, to become initiated into manhood. In Xhosa tradition, the young man is circumcized without anaesthetic by a village elder. Painted with clay to look like a goat, he then lives solitarily and remotely for eight days without basic sustenance or water until he is welcomed back to the village as a man. The ceremony, the building of the ritual hut, the general monitoring, and the specific care of his 'limb', which is bandaged in a leather thong, all depend on older male relatives who are his attendants throughout the process, but because Lumkile's grandfather is an irresponsible alcoholic and his uncle an absent sheep-shearer, he is without such support. On the third day, a severe infection in his penis, his 'defeated limb', which reeks like 'the decaying body of a stray dog' (Mgqolozana 2009: 13), sends him to the local hospital.

Mocked by his grandfather and even a nurse, he is blamed for his affliction although the medical staff diagnose the serious possibility of gangrene and amputation. After a partial recovery, he returns to his hut, but when the ritual time is up the local men are merciless. Rather than escorting him triumphantly back to the village, they arrive en masse with their dogs who are implicitly encouraged to attack Lumkile's dog.

> My dog gave me *that look* again; this time there was *fear* in his glassy eyes. He was scraping the dust, causing turmoil with his hind legs while he made *heartbreaking moans*. Urine dripped from his member. *He was calling me to his defence.* [. . .] He was not aware that the chest people were wishing with all their might that these dogs would dare to attack him, so that I could teach them a lesson they would not forget. [. . .] All they saw was him, their *victim*.
>
> *(Mgqolozana 2009: 147, emphasis added)*

Lumkile prepares to come to the defence of 'man's best friend' (147), but then the largest, most beautiful dog goes for him instead. He sets upon the animal so violently with a knobkerrie, a large stick symbolic of manhood, that he paralyses the dog (whom he subsequently sees hanged by his handler in the village). The men are furious but his uncle dissuades them from taking revenge:

> 'Madoda [men], since when do you mind a fight between two dogs? A dog is a dog is a dog. Dogs fight each other every day. Dogs have ripped each other's testicles since I was a young boy. I say again, what is new in a fight between two dogs?'
>
> That's what he said, my uncle. My own flesh and blood was calling me a dog.
>
> *(Mgqolozana 2009: 149)*

When Lumkile washes himself ritually, his uncle mocks him further in front of the men, shouting that the brand of soap he had for his nephew was 'good for animals' (153).

All initiates are dehumanized as goats with their white, painted bodies but then reintegrated into humanity through having completed the circumcision trials. Denied this ritual socially, Lumkile remains animal in the eyes of the village men, but this final public humiliation effectively galvanizes him to claim a masculine identity outside of Xhosa culture so that he can conceive 'what manhood is really about' (Mgqolozana 2009: 183). Rather than being abjected by his body, he feels burgeoning pride in 'true manhood' (183), yet he remains too constrained by his culture to acknowledge that identifying with his dog when he is in jeopardy, in a way which recalls Diamond's observations about common human and animal vulnerabilities, is transformative. Ironically, he is proud that his circumcision experience 'made me more of a human not less' (183). Because becoming animal is debased and seen as anti-human, animality has to be contained, not celebrated. Any embodied feral contiguity between human and nonhuman can only result in victimhood or extreme violence.

K. Sello Duiker, on the other hand, represents this overlap between animality and humanity as multi-faceted and indicative of either good or evil. In *The Hidden Star*, set in Phola township, realism obtains in the narrative with Nolitye as a young schoolgirl connecting with her friends, standing up to the school bully, struggling with poverty and an unsympathetic mother who barely feeds her. In terms of the 'unreal', the pack of street dogs and Mandla the donkey speak human language, and Nolitye is visited by the shape-shifter Nomakhosi who is the Spirit of the Stone – a sacred object which needs to be put together again so that Nolitye can rescue her father, her real mother (as opposed to the impostor she lives with), and several missing children who have been enslaved in an underworld inhabited by monstrous and evil creatures as well as helpful animal guides. This quest is hampered by MaMtonga, a dark witch who covets the stone for sinister purposes, by the false mother who is actually a Night Rider, and by Vera, an evil Renegade who 'work[s] with the dark forces', although good Renegades are 'healers in the service of light' (Duiker 2006: 100).

A dog called 'Mister', a large and opinionated German shepherd, guards the final piece of the sacred stone. Formerly an ancestral spirit, but a meddling one, he was demoted to being a good Renegade. Now Mister has a 'private-school accent' (Duiker 2006: 170) and puts on airs. The disparity between Mister's situation at the end of a chain in a butcher's dusty backyard and his assumption of upper-class arrogance and expectations of obeisance from the children is represented as humorous rather than tragic. Once Nolitye has removed a thorn from his paw, he gives her the final piece of the stone. Although Mister has shape-shifted into a breed of dog deployed by the police to hunt out and restrain criminals (167), he behaves like a gentleman with the children, and he serves as a beacon of reliability and dependability in his ferality, in contrast to trickster humans who may be darkly magic with shape-shifting identities. He is both animal and human, holding both identities in tension. With a human consciousness and an animal body, he refers to himself as a 'so-called dog' (171), for embodying the feral has been his punishment.

If 'species indeterminacy', to borrow a phrase from Anat Pick (2011: 85), aptly but negatively describes the being of Mister, it is a state almost neurotically rejected by Lumkile, which implies that in African culture such 'indeterminancy' may be positive (to some extent) in fable but not in a realistic mode. Mister's ferality, his cross-over between humanity and animality, suggests that shape-shifting or the blurring of human and animal identities tends to be more of a multi-faceted process than Pick's notion of 'the binary rationale of metamorphic transformation, by which one thing becomes another, in an act of conversion' (2011: 85). Certainly in *Rainmaker*, the metamorphosis of Ky into a leopard dramatizes the ongoing transformation of his identity from one steeped in urban violence into an acceptance of his Bushman heritage and his calling as a shaman.

As the novel opens, Ky Rahl, the narrator, is a member of the Blood Brothers Gang in greater Cape Town. In a revenge killing spree against the Tough Livings Gang, Ky shoots Akker, a gang kingpin, in the leg after the latter insulted him for being 'the son of a Boesman [Bushman] mother' (Pinnock 2010: 3). With a price on his head, Ky flees to Auri, an acquaintance from school. In hiding, he dreams of hunting eland and being taken by a lion who licks his tears, which signifies to Auri's father that Ky might have an aptitude for shamanism. At night he is surreptitiously taken from Cape Town to the Cederberg, a mountainous wilderness area far from the city, to be the putative apprentice of Zimry, a /Xam shaman who works as a ranger. The urban Ky slowly learns to embrace his Bushman heritage, accepting that the term 'Bushman' does not have to be one of opprobrium. He becomes skilled at calling animals, at identifying the local flora for its edible and medicinal properties, with his education culminating in shape-shifting and rain-making as a *!gixa* (a /Xam shaman).[8]

Ky connects with his ancestors, the First People, by providing a body for them in this world. Calling animals, a strategy used in hunting, means 'you can enter their world with your spirit', as Zimry explains (Pinnock 2010: 41). To his delight, Ky learns to call frogs, dassies, and even a leopard. In an extension of becoming with animals (Haraway 2008), shamans customarily inhabit the bodies of animals to usher them into the spirit world. When Ky goes into a trance he becomes a jackal, a fish, and a leopard. To develop these details, Pinnock has relied on authentic traditional Bushman beliefs. As David Lewis-Williams explains, Bushman shamans, through a trance dance, 'activate a supernatural potency so that they can enter an altered state of consciousness [in order to] cure people of known and unknown illnesses, go on out-of-body travel, and visit the abode of god' (2003: 27). The shamanic spiritual experience depends on animal embodiment as s/he 'dies' and crosses over into the spirit world, with the behaviour of the person in a trance emulating that of a dying antelope (2003: 33–8).

Feral embodiment, then, is fundamental to Bushman spirituality, with its species indeterminacy and the permeability between human and nonhuman animal, as well as between the real and the unreal. In Bushman creation stories, humans and animals interconnect; thus animals were once First People who only subsequently became animals. Rock paintings, which Zimry uses as a pedagogic tool, depict humans with animal heads, denoting the shamans' spirit animals:

> Some *!giten* [shamans] work only with animals. They see that the herds are healthy. Help them to migrate to where rain has made the grass sweet. [. . .] Or they work with plants or birds or fish. These are *!giten* who can become birds or pawed animals. They use the animals' skill to hunt or the birds' wings to travel.
>
> *(Pinnock 2010: 95)*

Zimry is reticent about teaching Ky shape-shifting because of the risks involved, but after a leopard, contrary to natural lore, visits their fire a number of times, he agrees to escort his apprentice to the Other Side.

Ky negotiates with a leopard to interchange bodies but only until dawn, at the leopard's request. Ky's sensations are transformative as he embodies leopard, and he is incredulous at having whiskers, paws, and claws as well as an acute sense of smell and sight. When he meets and plays with a female leopard, the experience is therapeutic; he sloughs off the human fear he had experienced constantly in the gangs to celebrate a profound feeling of safety and a mystical, ecological belonging. Not only is feral embodiment empowering and psychologically transformative, it also serves to guard the animal herself. When Ky shape-shifts again he does so in order to save both this leopard and Zimry, who have been shot by poachers. He heals the leopard's wound, fatally attacks one of the men, but is unable to save Zimry, who has already gone to the Other Side. As in the archetypal hero's journey, Ky, left without his mentor, has to be heroically assertive, but he is helped by Zimry's spirit and those of the First Races in his mission 'to restore the balance of the world' (Pinnock 2010: 129). Rather than leaving Ky, now united with Auri who is training as a shaman, in an idealized wilderness, Pinnock has Ky return to the city to work with gangs. Fittingly, the ancient spirituality of the /Xam will have to engage with and be tested by a violent modernity. Ky has been prepared for his mission in the ganglands, partly through having accepted the risks and the vulnerability attendant on inhabiting the body of an animal.

The place of feral embodiment in relation to traditional spirituality in these novels, as Pinnock demonstrates here, is not one divorced from contemporary urban existence with its pressing social issues, nor is it an individualized or autistic pursuit. To expand Jennifer Wolch's observation and suggest that (human) corporeal embodiment may become increasingly stabilized through a reconnection with the nonhuman via the boundary crossing of feral embodiment would however be naïvely optimistic. Only in *The Hidden Star*, which is an adolescent fable, is such a possibility brought neatly to a closure. The endings of the other novels are more reticent, with the narratives more contingent and less firmly concluded. Diamond's notion that we share an embodied vulnerability with animals recurs, exemplified in Lumkile's physical affliction and his identification with his dog's fear, in Katya Grubbs' interstitial living with her dog, in Zinzi December's failure to rescue the twins and the animals of those murdered for muti, in the deaths of Zimry and the poachers.

Yet some potential for personal or social transformation in connection with feral embodiment also manifests in the endings of the novels: Lumkile takes pride in an alternative masculine identity, Katya experiences the joy of living close to nature as she re-visions the city, Zinzi and Sloth undertake a selfless journey to the DRC to attempt to locate Benoît's family, Ky plans to return to the city to work with the gangs. Certainly, Pick's sense of embodiment as a 'critical space for thinking of the human outside Cartesian abstractionism' (2011: 6) is borne out by the narratives' engagement with feral embodiment as a potential metonym for the edgy vibrancy of post-apartheid South Africa.[9] Moreover, these contemporary novelists' deliberate invocations of feral embodiment often respectfully invoke pre-colonial traditions and beliefs on their own terms, suggesting that the so-called magic of shared human-animal embodiment is a critical means of mapping pathways from fractious pasts to more productively shared futures.

Notes

1 Indigenous tradition should not, however, be romanticized. It can be patriarchal (as I write, President Zuma is about to marry his fourth wife) and instrumentalizing of animals (in ritual sacrifice, for example).

2 The term 'Bushman' has been regarded as pejorative in both scholarly texts and colloquially, but is now preferred by many Bushman people themselves (Lewis-Williams 2003: 123).

3 In a first for Africa, *Zoo City* was awarded the Arthur C. Clarke Prize for Science Fiction in 2011.

4 That Beukes' representation of human embodiment with constant animal accompaniment has echoes of Philip Pullman's humans with their daemons is implicitly acknowledged by the 'Online Documentary Database' in her novel which promotes a documentary of an Afghan warlord animalled by an Antarctic penguin in body armour and then inserts Pullman's *The Golden Compass* under 'Recommendations' (65).

5 The belief in the desirability of body parts as muti is not merely a figment of Beukes' imagination. The article by Candice Bailey, 'Muti Killings Is a Way of Life in Rural Areas' (2010), is based on

research undertaken by the Human Rights League in Mozambique and supported by Childline in South Africa.

6 In the interest of valuing local traditions, payments for *sangomas* are now covered by Medical Aids in South Africa.

7 /Xam is the name for Bushmen in the Cape. In a brief 'guide to /Xam click symbols' Pinnock explains that '/ is the sound made when the tip of the tongue is pressed against the front teeth and quickly withdrawn' (2010: frontispiece).

8 '! [is] the sound when the tongue is pressed against the roof of your mouth and released sharply downwards, like when a cork is pulled from a bottle' (Pinnock 2010: frontispiece).

9 In this chapter, feral embodiment has been discussed very much from human points of view. Not only have animal perspectives not been developed in the narratives, but the human and nonhuman may be inextricably interconnected. Mister in *The Hidden Star* is both human and dog. In Ninevah creatures labelled as 'pests' have collective rather than individual lives, and in *A Man Who Is Not a Man* Lumkile's dog is represented as a generic rather than a specific, named dog. Nonhuman subjectivities are more discernible in *Zoo City* but even here if Sloth seems assertive his very existence, like that of the leopard in *Rainmaker*, is dependent on human fates.

References

Ansell, G. (2011) 'Behind All the Monkey Business', *Mail and Guardian*, 6 May, available online at http://mg.co.za/article/2011-05-06-behind-all-the-monkey-business.

Bailey, C. (2010) 'Muti Killings Is a Way of Life in Rural Areas', available online at http://www.iol.co.za/news/south-africa/muti-killings-is-a-way-of-life-in-rural-areas-1.470603#.UngTt0L7X3A, accessed 22 March 2012.

Beukes, L. (2010) *Zoo City*, Auckland Park: Jacana.

Bremner, L. (2010) *Writing the City into Being: Essays on Johannesburg 1998–2008*, Johannesburg: Fourthwall Books.

Diamond, C. (2008) 'The Difficulty of Reality and the Difficulty of Philosophy', in S. Cavell *et al.* (eds), *Philosophy and Animal Life*, New York: Columbia University Press.

Duiker, K.S. (2006) *The Hidden Star*, Cape Town: Umuzi.

Fincham, G. (2011) *Dance of Life: The Novels of Zakes Mda in Post-apartheid South Africa*, Cape Town: University of Cape Town Press.

Gaita, R. ([2002] 2004) *The Philosopher's Dog*, London: Routledge.

Haraway, D. (2008) *When Species Meet*, Minneapolis: University of Minnesota Press.

Lewis-Williams, D. (2003) *Images of Mystery: Rock Art of the Drakensberg*, Cape Town: Double Storey.

Mgqolozana, T. (2009) *A Man Who Is Not a Man*, Scottsville: University of Kwazulu-Natal Press.

Mutwa, C. (1996) *Isilwane The Animal: Tales and Fables of Africa*, Cape Town: Struik.

Nast, H.J. and Pile, S. (1998) 'Introduction: Making Places Bodies', in H.J. Nast and S. Pile (eds), *Places through the Body*, London: Routledge, pp. 1–19.

Pick, A. (2011) *Creaturely Poetics: Animality and Vulnerability in Literature and Film*, New York: Columbia University Press.

Pinnock, D. (2010) *Rainmaker*, Auckland Park: Jacana.

Rose-Innes, H. (2011) *Ninevah*, Cape Town: Umuzi.

Seigworth, G.J. and Gregg, M. (2010) 'An Inventory of Shimmers', in M. Gregg and G.J. Seigworth (eds), *The Affect Theory Reader*, Durham, NC: Duke University Press.

Szymborska, W. (2002) 'Seen from Above', in *Miracle Fair: Selected Poems*, J. Trzeciak (trans.), New York: Norton, p. 66.

Wolch, J. (1998) 'Zoopolis', in J. Wolch and J. Emel (eds), *Animal Geographies: Place, Politics, and Identity in the Nature-Culture Borderlands*, London: Verso.

17

Reconfiguring wild spaces

The porous boundaries of wild animal geographies

Henry Buller

Like all zoos, by definition, Paignton Zoo in Devon, South West England, is a physically bounded space. Not only do the different animal exhibits live in variously designed enclosures within the general zoo space – including a Hagenback-style moat-and-concrete baboon display, a glass-walled and video-surveilled lion park, as well as a total-immersion hanger-like super-heated reptile environment – but the zoo itself, which lies in a wooded valley, is also enclosed by a large, secure fence separating it, to the north, from a large supermarket and, to the south, from a housing estate replete with sea views. This external fence defines the zoo as a distinctive space and the animals within it as fundamentally exotic to this temperate, generally rainy, coastal, and densely humanized region.

Yet every three or four metres along the perimeter fence is a small ground-level box, open at both ends, permitting passage from the outside to the inside, and vice versa. These boxes are for the badgers to pass through. Living in the surrounding woods, and in the wooded periphery of the zoo itself, the badgers move easily between the two worlds. Perhaps less taxonomically 'wild' than the lions and elephants whose enclosures they burrow into, the indigenous badgers are nevertheless relatively free to come and go as they please and to feed when and where they want to. They move to and from spaces beyond the human gaze and escape most direct human interference.

Staying with Paignton Zoo, there is a moment every afternoon when the pelicans are fed. On a purpose-built feeding platform, adjacent to a spectator viewing area, a zoo 'keeper' throws fish, imported in wooden crates from South America, at the gathering pelicans, who, with clipped wings, float almost stationary on the water's surface. No sooner are the fish released from the keeper's hands, however, than gulls swoop down from the Paignton skies to seize them before the pelicans can scoop them into their distinctive beaks. Frenzied scuffles break out between the foreign water-bound pelicans and the wholly familiar airborne gulls.

In 2011, an Egyptian cobra 'escaped' from New York's Bronx Zoo. Newspaper vendors in the Manhattan subway stations soon launched into an original refrain, 'Cobra, cobra, one bite, it's over', while anxious passengers stared up the northbound 2 Train tunnel as if expecting the reptile to emerge from the dark at any moment. The cobra was later found; it had never left its enclosure in the zoo park. This incident recalls the famous 1874 Central Park Zoo escape story, when the *New York Herald* published an entirely fictitious article claiming that many wild animals had broken out of the zoo and were savaging the New York population. 'The scene at the Fifth Avenue Hotel', claimed the account, 'where the Malayan tapir that killed the two policemen burst in among the mob of gentlemen standing in the portico, can never be forgotten' (*New York Herald* 1874). Like the fictitious tiger in Jonathan Lethem's novel *Chronic City* (2010), which stalked the municipal population of lower Manhattan, these animal stories challenge the reassuring boundaries between the city's otherwise largely fictionalized fauna and, in doing so, reanimate the threat of the 'true' wild.

The 'placing' of animals within human-ordered and human-defined spaces has been a foundational step in all human-animal relations, yet one that has long since moved away from a singular reliance on straightforward ecological or natural locational predispositions. The distribution of many categories of key species of the world's fauna has far less to do with Sclater-Wallace's bio-regional classification than with the spread of various human desires, whether material or metaphysical. Under Leroi-Gourhan's mytho-grammatical analysis of the Upper Palaeolithic cave art, the physical spacing and patterning of animal species depictions constituted an early form of language and thus spatialized world-ordering (Lewis-Williams 2002), echoing Berger's (1980) contention that animals first entered the human imagination as metaphors rather than as flesh and bone. Similarly, Morris's (1998) account of the power of animals in Malawian culture draws particular attention to the symbolic distinction between the chaos of the exterior, in this case, the 'woodland', where the animals live, and the ordered human space of the village, a distinction that has come to structure their thought, their cosmology, and their social practices.

The terms 'wild', 'domestic', 'companion', 'feral', 'pet', 'invasive', 'alien' all contain implicit – and sometimes explicit – spatial categorizations that ultimately say less about the animal than about us. 'That is why', write Deleuze and Guatarri, 'the distinction we must make is less between kinds of animals than between the different states according to which they are integrated into family institutions, State apparatuses, war machines, etc.' (1987: 268). Zoos, homes, parks, cities, the countryside, parade grounds, the 'wilderness' are all spatial categories that encompass normative orderings in which animals are both materially and semiotically 'placed'.

As a geographer, I am fascinated by these in-placings and out-of-placings and more particularly by their active subversion and contagious transgression. The city and its defining segregation from the 'wild' stands as perhaps the most robust of these exclusionary spatial forms, and the countryside, to a lesser extent. Both, however, are arguably becoming more and more the sites of animalian transgression. Whether this is because the transgressions are themselves multiplying or because we are becoming more simply aware, and more critical, of the explanatory limitations (and inherent porosity) of pre-fixed categories and structures and more attentive to things placed within them as 'knot[s] in a field of relatedness' (Haraway 2000: 290) is moot.

In this chapter, which draws upon contemporary geographical scholarship, I look at these various incursions and boundary transgressions that challenge our conventional understanding of the category of the 'wild' animal and its distanced spatialized expression. Today's truly 'wild' creatures are closer than we might think.

The wild is not us

Conventional wisdom, and popular representation, has cities as first and foremost actively de-wilded and de-natured spaces in which humans are, partly as a consequence of that de-wilding, 'condemned to live' (Park 1984: 3). Tuan explores the idea that cities are 'artifacts and worlds of artifice placed at varying distance from human conditions close to nature' (1978: 1); while Anderson (2003) links the realization of 'full humanity' to such distanciation. There are few if any trees or animals in Fritz Lang's *Metropolis*. Deliberate manoeuvres of careful eradication and exclusion, coupled with a wholly anthropocentric, two-leg apartheid, have sought to draw an unequivocal and 'foundational' (Hinchliffe 1999) boundary between cities and towns on the one hand and nature and the 'wild' on the other.

Escaping from Central Park Zoo, the animals in the 2005 film *Madagascar* know their place:

> 'Do you ever see any penguins running free around New York City?'
> 'Of course not. We don't belong here. It's just not natural.'

The city, by definition, has no place for wild creatures, unless they are carefully domesticated or enclosed and displayed as purposeful others, messengers from a spatial and temporal elsewhere, whether they be the 'live' animals of a central city zoo or the taxidermized fauna of early twentieth-century natural history museums. Of the former, Braverman remarks that zoos materialize the traditional separation between humans and animals and between nature and culture: 'Without the city', she quotes from a zoo manager, 'there would also not be a zoo' (2013: 29). Of the latter, Haraway (1989) makes the point that the purified nature, truth-telling, and faux realism of Carl Akeley's famous African Hall in New York's Natural History Museum came at a critical time when civilized, masculine urbanity felt itself increasingly threatened by new forms of human wildness. Only after animals have been shot, skinned, and then recreated as taxidermized re-presentations of the wild is the 'hygiene of nature' able to 'cure the sick vision of civilised man' (Haraway 1989: 30).

For Jane Jacobs, urban environments are 'as natural as colonies of prairie dogs and the beds of oysters' (1992: 443), while the geographer David Harvey famously proclaims that 'there is nothing unnatural about New York' (Harvey 1996: 186). Though cities and nonhumans might ultimately be 'inseparable in thought and practice' (Lynn and Sheppard 2004: 54), it is nonetheless the city that seeks to define the parameters of its naturality: no penguins running free, for a start. A city is above all an act of placing, a spatial ordering and selecting where 'civilization' seeks both to remove 'nature' from itself and itself from 'nature'. The gradual removal of livestock from the streets of the modern Western city (Philo 1995) has purified and foreshortened what was once more of a continuum of human-animal relationality into a starker and more contrasting taxonomy. In contradistinction, we might think of the non-city, the countryside, as its faunistic foil, where nature roams free. But even though 'animals (wild, companion, domestic) are players in the construction of the material ecological rural, and also the imaginative and economic rural' (Jones 2013: 1), the countryside too is an ordered space (Buller 2004, 2008) and many of the animals that inhabit it are largely preselected through centuries of human activity; few penguins run free there, too.

Over the last fifteen or so years, animal studies – and animal geography in particular – has questioned many of the assumptions that underlie such narratives of exclusion and anthropocentric ordering. Writing in 2002, Wolch called for scholars of human-animal interaction to 're-imagine the *anima urbis* – the breath, life, soul and spirit of the city – as being embodied in its animal life' (721). Her project has been to rethink urban theory and try to unsettle its anthropocentric heritage: 'to create a new political ecology of people and animals in the city' (734–5). In an earlier paper (1998), she draws attention to the persistent duality between 'pets' and 'pests'; city animals, it seems, are either one or the other.

But such binary distinctions have become too simple. Many non-companion animals live and have always lived in cities. What is needed is a way of accounting for their newly acknowledged co-presence. As I have argued elsewhere (Buller 2013) – paraphrasing Richard Leakey's famous response to Jane Goodall's observations of chimpanzee tool use – animal presences in the city call for us either to redefine 'city', redefine 'wild', or accept such animals as citizens. Of the latter, de Planhol (2004: 399) recounts the example of the Paris-based group 'L'Ecole du Chat', which has successfully lobbied the city's administration for official recognition of the 300 or so feral cats of Père Lachaise cemetery, as well as cats in other parks and cemeteries, as 'animal citizens' of the capital. Of course, companion animals may have already reached the lofty heights of citizenship (Donaldson and Kymlicka 2011). The explosion of dog parks in US cities, including one in New York's Washington Square – a focal point for many historical movements for the formal recognition of the rights of various 'others' – might be seen as testament to the growing role of canine New Yorkers in determining the spatial form of the city.

Taking up Wolch's (1998) call for a recasting of the metropolis as a 'Zoopolis', Braun argues that it is time to 'de-fetishize the city', redefining it as a 'more than human' place (Braun 2005: 646). This is exactly what Hovorka's (2008) 'transspecies urban theory' achieves. Looking at urban livestock in African cities and, in particular, chickens in Botswana, she develops an urban theory that 'incorporates animal actors and recognizes that interspecies mingling is fundamental to city life' (2008: 96). In a recent paper, Srinivasan (2013) takes this further by considering the complex status and lives of Indian 'street dogs'. Accepted, and resolutely not 'strays', and therefore indispensable, these dogs exhibit a form of 'interspecies cohabitation' that is neither that of companion nor domesticated animals. 'In this kind of "living with"', she argues, 'animal autonomy does not necessarily imply spatial separation or wildness' (2012: 116). Yet these are still our categorizations. Arguing forcibly that different 'stories' contribute to the making and meaning of places, be they cities or elsewhere, Van Dooren and Bird Rose ask:

> Whose stories come to matter in the emergence of a place? [. . .] What might it mean to take storied-places seriously as multispecies achievements? More concretely, what would it mean to take seriously the way in which some specific animals story their specific places?
>
> *(2012: 3)*

In the city, human/nonhuman encounters are ordered by what Luther (2013: 36) calls 'narratives of socio-spatial belonging'. Dogs, chickens, cats, these are all, to some extent, familiar urban dwellers (Kean 2011). Urban wild animals too are actively and materially reconfiguring our notions of urban space and, in doing so, challenge the very category of 'wild'. In this, argue Donaldson and Kymlicka (2011), they become 'denizens' (rather than 'citizens') of the multi-species metropolis, an interesting compromise which, while it confirms 'otherness', nonetheless opens up a framing for potentially transformative ethical engagement (Luther 2013).

The rhetoric of the urban *animalia* is frequently one of spatial and temporal movement: of invasion, or re-colonization, of decline or explosion, of coming or going. Yet some of the animals concerned were here first and have simply remained, often unseen. Others have been displaced by urban expansion and are adapting to their new circumstances, re-colonizing old spaces if not former ecologies. Others still are genuine newcomers, migrants, like the human inhabitants of so many cities.

Leaving aside zoo animals for the moment – one is far more likely to see a tiger in an urban zoo than anywhere in the wild – there are a number of particular dimensions to the contested notion of 'wild', both within and beyond the city, that I want to explore in this chapter: first, a new and reinvigorated observational and empirical recognition of the city as redolent with wild spaces and species; second, a transgressive neo-wildness that derives from adaptive behavioural and physical animal responses to living in the presence of humans; and third, a less assured sense of the 'wild' animal that emerges from the processes and practices of wild animal reintroduction.

Urban wildlife

In her recently published book *Field Notes from a Hidden City*, Esther Woolfson (2013) describes her daily encounters with her home city of Aberdeen's wildlife, from pigeons and cormorants to spiders and worms. Rescuing a young pigeon from the winter snow and contemplating its return to the 'wild', she writes:

> I began to think about wildness in relation to creatures who live in cities, about whether or not we consider them less wild than creatures living elsewhere, or think of them as somehow lesser parts of nature itself. [. . .] Their presence may be the only contact many urban people have with the natural world but our relationship with them seems changed by proximity, diminished by the fact of their being here among us.
>
> *(2013: e-book location 99/4891)*

'Urban wildlife' is defined as 'non-domestic vertebrates and invertebrates of urban and urbanizing areas' (Adams 2005: 139). There exists a long tradition of both scientific and popular urban wildlife conservation, through the designation of reserves and wilderness areas by both statutory and voluntary agencies (Adams and Leedy 1991; Lorimer 2008). Much of this has been driven by a sense of 'human social need' (Adams 2005) and arguably by an enduring paradigm of boundary making (Harrison and Burgess 1994).

More recently, however, the status of urban wildlife is receiving renewed attention; 'things are brewing', observe Hinchliffe and Whatmore (2006: 123), as the 'urban green' becomes revalued politically, aesthetically, conceptually, even ethically (Luther 2013), into what Lorimer refers to as a more 'fluid biogeography' where fixed territories are replaced by 'open geographies of interpenetrating and overlapping networks (2008: 2056). As Thomson points out in her penetrating study of Melbourne's bats, for these animals 'the city is just another habitat opportunity' (2007: 89). Suddenly, urban wildlife, from voles and bats to peregrines and redstarts, is everywhere, no longer confined to labelled 'nature' spaces but recognized as an active co-presence on tower blocks, sewage plants, brownfield land, old cars, and abandoned sites. In Hinchliffe and Whatmore's words, cities have become co-inhabited 'with and against the grain of urban design' (2006: 128).

Alongside the renewed interest in urban wildlife amongst bio-geographers and others from animal studies, there has been a notable explosion in recent years of what we might term urban nature writing. A significant number of 'nature writers' have rediscovered the city as a haven of hidden wildness, with both popular and 'natural history' accounts of the often unacknowledged wild spaces and species of the urban environment flourishing over the last decade (Bennett and Teague 1999; Dixon 2002). Richard Mabey's groundbreaking book *The Unofficial Countryside* (1973) stands as a precursor to all of these. Described as a 'Doomsday

Book of a topography too fascinating to be left alone' (Sinclair 2010), it charts the urban wilderness of abandoned construction sites, backyards, and central reservations.

In many ways, the new urban nature writing is a challenge to the nostalgic, romantic, yet previously dominant countryside nature writing so prevalent in the UK, where the wild is essentialized as remote, apart, and, in some obscure way, 'natural'; where animals burrow below ground, out of sight, not into subway tracks and along service conduits but into the earth and rock – genuinely natural environments; where birds nest in trees and hedgerows, not on apartment house ledges and chimney stacks; where dead animals fall and are consumed, rather than being bagged and taken out with the rubbish. Macfarlane observes that 'if the wild were to come close to extinction, its final fastness would be the mountain tops, and the valleys they protected' (2007: 58). In his wild, yet hopeful, geography of the British Isles, he charts such extreme places and their 'fierce elementality'. Even Macfarlane, however, accepts that our contemporary understanding of the wild needs to be more nuanced. Wild nature is not just a product of civilization's self-distanciation, but a ubiquitous and contemporary expression of relational vitality that is more than a mere vestige of a non-anthropocentric past, a wild of miniature as well as spectacular scale.

Despite this acknowledgement of the human construction of the wild and the spectral presences of past nonhuman passage through landscapes, there is still an inexorable sense of fixity in wild nature, whether it be in the repetitive vitality of natural growth and seasonal change, the indigenousness of wildlife, the seeming immutability of landforms, or the 'gravitational pull' (Macfarlane 2007: 176) of more-than-human time. Human longing for Nature, according to Jane Bennett, is 'the longing for something solid, fixed and final, something authoritative, divinely domestic. [. . .] But this very longing', she goes on, 'is also an ache for something that is larger than life, extraordinary, unencompassable – in short, Wild' (1994: 72).

Although urban wild animals, for all their recent championing, may well remain, for many, 'unofficial' (Pyle 2002) and 'uncanny' (Kaika 2004), somehow 'lesser parts of nature' (Woolfson 2013), especially when held against the charisma and magnificence of more distant fauna, it is perhaps in their very spatial proximity and unexceptional daily encounter that a new sense of interspecies sharing may flourish. Recent urban nature writing acknowledges numerous urban species, particularly birds, as representing a critical element not only of that natural fixity that Bennett identifies, but also of a vital connectivity:

> Living in a city, we are all elements of a biological and ecological chain described by words that express the complex web of connection between us and hint of dependency and need – commensal, mutual, symbiotic, predatory, synanthropic. [. . .] In different degrees, we share our vulnerability.
>
> *(Woolfson 2013: 132/4891).*

Becoming urban wild

Urban wilding and re-wilding, be it literary, material, or conceptual, has been largely a re-wilding of the urban space that *Homo urbanis* has made his/her own through adapting and evolving social practice (and maybe even biological processes) to become urban. However, many animal species have also similarly adapted, not only through colonizing cities with their physical presence but also displaying ecological and behavioural changes in response to their adopted environment (and, in some occasions, to direct human interference). A new and distinctive urban 'wild' is arguably coming into being: a wilding that is more than merely a spatial and ecological presence; the hybrid wild of the new urban fauna (Herda-Rapp and Marotz 2005).

Pigeons and foxes, whose numbers in cities have often become higher than in their neighbouring traditional 'natural' habitats, tend to exemplify this trans-boundary status. Of the former, Jerolmack writes:

> This animal is what I would call a double hybrid. It was created by humans for domestic use but then escaped to become feral. Its physical and biological structure, as well as its reproductive abilities and habits such as dwelling on window ledges, are the product of millennia of human intervention in nature. This particular type of pigeon *never* existed 'in the wild;' its 'natural habitat' is among humans.
>
> *(2007: 90)*

That being said, as Jerolmack (2009) points out, pigeons have more recently undergone a further taxonomic reshuffle to become a despised and nuisance species, generative of social disorder in urban settings. They have crossed (and arguably recrossed) the human socio/spatial, material/semantic boundaries of 'nature' and 'culture' to be somehow 'out of place' in all. The pigeons' error was to get too close. Contrast their treatment in the representational politics of urban wildlife with that of the exotic New York parakeets (native to South America), described as the 'blameless victims' of accidental release. Though they number many hundreds in the metropolitan area, they are praised for their 'good ecological behaviour' as metaphors of successful integration (Seymour 2013).

Much has been said about urban foxes, from the poems of Ruth Padel (2004) to the scientific investigation of their responsive sociobiology in urban settings (MacDonald and Newdick 1982; Soulsbury *et al.* 2010, 2011). Here, too, we find a species of vigorous boundary-crossers (Knight 2000). Researchers have shown, for example, how urban red foxes (*Vulpes vulpes*) alter ranging patterns to smaller areas in cities to avoid traversing roads, thereby reducing pack mortality (Baker *et al.* 2007); how they live in smaller packs than they would in the countryside; and how they learn and adapt to changing food sources and the rhythms of their availability. Such adaptive social organization, while not (yet) supported by evidence of specific genetic differentiation from non-urban foxes (Wandeler *et al.* 2003), nonetheless suggests that urban foxes might be said to constitute a distinctive urban vulpine 'society'.

> One cub has died on the road. Magpies
> have eaten her. The last two play-learn, eat solid food
> and follow their parents through dusk. Twins
> of the Greek night sky, Castor and Pollux, shine
> through damp London nights as earthworms
> leave burrows. Parents spoon crane-flies off lawns
> with their tongues, teach young to deadhead the bins
>
> on Bemerton and Havelock, lift black plates
> for frankincense, rot-lustre gems
> of sunk baconfat. To strip flaking bark
> for silverheave woodlice, listen
> for worm-bristles rasping through grass.
> If worm-tails are gripping the burrow –
> even a worm can be frantic – the grey-black lips

> pull gently taught – and pause – and pull again.
> A technique used by bait collecting fishermen.
>
> Ruth Padel, 'The Worms' (2004)

Of course, for foxes, with their new-found suburbanity has come human reprobation, just like pigeons with their new-found feral status. On occasion, this rises to the level of 'moral panic' (Cassidy and Mills 2012), comparable in many ways to the increasingly frequent reports of 'alien big cats' roaming the British countryside (Buller 2004). Estimates place the population of 'urban' foxes in London at around 10,000, provoking periodic calls for a programme of systematic culling. Yet the urban fox also has many defenders not only for its representational and symbolic potency as a messenger of rustic rural naturality in an otherwise metropolitan setting, but also, and perhaps paradoxically, for its shape-shifting urban sophistication. Commenting on the growing public awareness of urban foxes in British cities, Mugford suggests that 'they are coming closer, collecting food, rolling over, even allowing petting – in some instances the behaviour is more dog-like than fox-like' (quoted in Anon. 2013).

The growing presence in cities of species and individuals generally considered intolerant of coexistence with human animals, or unfamiliar with the ecologies and materialities of urban infrastructure, implies considerable adaptation (arguably on the part of both nonhuman and human). Long-term and permanent behavioural change, both in response to novel stresses, induced by proximity to humans and their domesticated animals and environmental affordances have been widely observed while some evidence suggests that 'micro-evolutionary' shifts are also beginning to be discerned in species of lizard, moth, and finch (DeStefano and DeGraaf 2003; Ditchkoff *et al.* 2006).

Blue tits (*Parus caeruleus*) pecking off the tops of traditional doorstep milk bottles to get at the cream have become part of British urban folklore. Other species, including the blackbird (*Terdus merula*), the magpie (*Pica pica*), and the red squirrel (*Sciurus vulgaris*), have been highly successful in their adaptation to the new ecological niches provided by towns and cities (Jerzac 2001; Rutz 2008). Scientists have given the name 'synurbanization' to these biological and behavioural adjustments of wild animal populations to urban environments (Luniak 2004). Although it remains to be confirmed whether, genetically, these urban populations differ or will differ from their non-urban confreres (Jerzac 2001) or whether such adaptations remain within the broader plasticity of the species, synurbic populations, many of which were once drawn from largely 'shy' species, now occupy spaces in which a coexisting human presence is ubiquitous and multiple. Thus, while the general impact of urbanization on wildlife has been fairly catastrophic, with declining numbers of species and ecologies, 'the growing tendency towards synurbanisation observed recently in birds and mammals is an optimistic chance for enriching the biodiversity of urban wildlife' (Luniak 2004: 53).

Introducing wildlife

If, through an engagement with the notion of multi-species encounter, our project is to challenge those 'ossified' (Jerolmack 2007) categories of 'culture' and 'nature', 'human' and 'nonhuman', then we must look at the 'city' and other wild spaces differently too. These become 'not so much an objective fact as [. . .] a specific material mode of storying [. . .] a story told and enacted by many creatures' (Van Dooren and Bird Rose 2012: 18). Hence the spatial fetishism, the taxonomic absolutism, and the nonhuman exclusivity of the 'wild' needs to be overcome. Wildlife, argue Whatmore and Thorne, should be reconceived as a 'relational achievement' that is 'spun between people and animals, plants and soils, documents and devices, in heterogeneous social networks that are performed in and through multiple places and fluid ecologies' (1998: 437). Whatmore and Thorne's own work traces the mobilization of animals (in this case *Caiman latirostris*) in global networks of science and natural history that intertwine Latin American jungles and the international boardrooms of global conservation institutions. Quoting Elspeth Probyn's assertion that 'a thing's place [is] no longer anything but a point in its movement [. . .] a space that takes for us the form of relations among sites' (1996: 11), they show how wild-ness is a constantly renegotiated performance that can rarely be encircled in thick boundary lines (Hinchliffe *et al.* 2005).

In 2009, the Scottish Wildlife Trust initiated a five-year trial reintroduction of beavers (*Castor fiber*) into an area known as Knapdale in the far west of Scotland. Having been extinct in the UK for over 400 years, the reintroduction of this initial batch of 16 genetically pure, heavily quarantined, and disease-tested Norwegian animals into the remote and sparsely populated forest area, some distance from any concentrated human population or activity, was seen as a way of restoring a lost ecosystem, encouraging biodiversity, and closely studying and monitoring the process of wild species reintroduction. Since the start of the programme, some of the beavers have died or disappeared, and a large number of scientific monitoring reports and studies have been produced following relatively intense observation of the animals and the site, leading one commentator to describe the site as an 'outside zoo' (Fisher 2011). Estimates of a re-established population of 30 to 40 beavers by the end of the trial have had to be significantly revised downwards.

During the same period, in the east of Scotland, in the region to the north of the cities of Dundee and Perth, an entirely different form of beaver reintroduction has been taking place. Unofficial, unmonitored, and unplanned, the apparent recolonization of the streams and woodlands of Tayside by beavers of unknown origin (but probably escapees from captivity) was not only occurring without human intervention but the 'free-living' beaver population in Tayside was proving to be a lot more successful than the official trial, with an early estimated number of some 100 or so individuals, later raised to 140 (Scottish Natural Heritage 2012). Various arguments were put forward by the official conservation and wildlife management organization Scottish Natural Heritage (SNH) that the unplanned recolonization needed to be halted and prevented (due to issues of animal welfare, predation, declining fish stocks, inappropriate beaver breeds, damage to water courses and private woodlands) and a decision to launch a recapturing programme was subsequently announced (SNH 2010a), though later rescinded following strong popular opposition. Significantly, SNH claimed at the time that the absence of sufficient monitoring of this recolonization, as well as the lack of information on the origins of the animals, effectively rendered it unacceptable. Nevertheless, the irony of the situation did not escape the SNH:

> Capturing beavers is not something we ever thought we would have to be involved in, nor is it something we would want to be spending resources on, particularly in such times as these.
>
> *(SNH 2010b)*

The 'bounded' and (bio)secure beavers of Knapdale, in their designated and policed reintroduction space, stand thereby in contrast to the 'free' and feral beavers of Tayside. While the former, for all their monitoring, predetermined biological suitability, and unequivocally wild location are not reproducing as well as hoped, the latter, less scientifically proper and actively sharing co-constituted space with the human occupants of the Tayside environment, are thriving. In Tayside, SNH now recommends 'trialling mitigation methods to allow people and beavers to coexist' (2012: i). Both wild beaver presences, Knapdale and Tayside, might be described as 'relational achievements', though the more successful of the two has been very much more on the beavers' own terms.

Endpoint

There is a closely defended ordering of the human world in the notion of the 'wild', as Jamie (2011) reveals in a remarkable reappraisal of one of the most widely read 'nature' books of the last half-century, Maxwell's *Ring of Bright Water* (1960). This, she argues, was not about an otter; it was about the countryside (and about class, and about sexuality). It was a nostalgic yearning not only for a fast disappearing aristocratic rurality of elite landownership and social order, but also an innocent sense of nature and of the rural *animalia* in which otters could play in baths, where nature was benign, where wild animals could be tamed and owned as trusted pets. Here, we might tentatively claim, lies Freud's notion of the 'nature reserve' as a 'place where imagination can run wild even in the context of civilized society' (Oliver 2009: 254).

Today's reconquering otters wouldn't be happy in a bath or riding in a rowing boat. They embody an altogether different nature and an altogether different set of relations between countryside and fauna, between human and nonhuman. Maxwell's otter apparently meant more to him than most human beings. Yet wild otters today, like urban foxes, pigeons, racoons, bats, and rats, stand both for a different 'wild nature', one whose 'difference', on the one hand, commands a certain respect and attention, yet on the other hand is open to, and invites, a combination of both greater multi-species conviviality – inclusivity, making room and being together, 'temporary identification with others in a shared place' (Van Dooren and Bird Rose 2012: 17) – and what Candea (2010) calls, drawing from his fascinating study of meerkat/volunteer interaction, 'inter-patience' or the allowing of things to happen. In this way, the city, for example, might become a 'space for flourishing of as many different forms of life as possible' (Van Dooren and Bird Rose 2012: 17).

The 'natural' remains, it would seem, ordered by the social, and yet it often appears to erase 'all but humans in the making of these wild places' (Whatmore and Thorne 1998: 437). Like all orders, this should invite our scepticism and critical investigation. Following Bruno Latour (2007), we might recognize that any such simplistic classification of the world into 'wild' and 'not wild' has its own effects: first – after Foucault – in creating and reinforcing hegemonic orderings; and second, in actually transgressing those orderings through everyday relational practice. And it is through such cosmopolitical practice that we ultimately generate 'response-ability' (Stengers 2005). Urbanization, with its coming together of the social and the material, with its multiplicity of publics, networks, actors, and meanings, its design and its affects, offers, according to Barnett (2012), opportunities for a new responsibility. Paradoxically, perhaps, it is within cities that we will best learn to live with the wild:

> If you get rid of all the cats in Manhattan, then the coyotes will go home to Montana or wherever it is they call home.
>
> *(Thomas Payne, forum participant, New York Times, 25 March 2010)*

References

Adams, L. (2005) 'Urban Wildlife Ecology and Conservation: A Brief History of the Discipline', *Urban Ecosystems* 8: 139–56.

Adams, L. and Leedy, E.L. (eds) (1991) *Wildlife Conservation in Metropolitan Environments*, Columbia, MD: National Institute for Urban Wildlife.

Anderson, K. (2003) 'White Natures: Sydney's Royal Agricultural Show in Post-humanist Perspective', *Transactions of the Institute of British Geographers* 28: 422–41.
Anon. (2013) 'Who, What and, Why: Are Urban Fox Numbers Rising?', available online at www.bbc.co.uk/news/magazine-21409631, accessed 11 February 2013.
Baker, P.J, Dowding, C.V., Molony, S.E., White, P.C.L., and Harris, S. (2007) 'Activity Patterns of Urban Red Foxes (*Vulpes vulpes*) Reduce the Risk of Traffic-Induced Mortality', *Behavioral Ecology* 18.4: 716–24.
Barnett, C. (2012) 'Changing Cities', in R. Tyszczuk, J. Smith, N. Clark, and M. Butcher (eds), *ATLAS: Geography, Architecture and Change in an Interdependent World,* London: Black Dog Publishing, pp. 72–78.
Bennett, J. (1994) *Thoreau's Nature: Ethics, Politics, and the Wild*, New York: Sage.
Bennett, M. and Teague D., (eds) (1999) *The Nature of Cities*, Tuscon: University of Arizona Press.
Berger, J. (1980) *Why Look at Animals?*, Harmondsworth: Penguin.
Braun, B. (2005) 'Environmental Issues: Writing a More-than-Human Urban Geography', *Progress in Human Geography* 29.5: 635–50.
Braverman, I. (2013) *Zooland*, Stanford, CA: Stanford Law Books.
Buller, H. (2004) 'Where the Wild Things Are: The Evolving Iconography of Rural Fauna', *Journal of Rural Studies* 20: 131–41.
——(2008) 'Safe from the Wolf: Biosecurity, Biodiversity, and Competing Philosophies of Nature', *Environment and Planning A* 40: 1583–97.
——(2013) '*Animal Geographies*', *Progress in Human Geography*, available online at http://phg.sagepub.com/content/early/2013/03/19/0309132513479295.abstract, accessed 27 January 2013.
Candea, M. (2010) 'I Fell in Love with Carlos the Meerkat: Engagement and Detachment in Human-Animal Relations', *American Ethnologist* 37.2: 241–58.
Cassidy, A. and Mills, B. (2012) '"Fox Tots Attack Shock": Urban Foxes, Mass Media and Boundary-breaching', *Environmental Communication* 6.4: 494–511.
Deleuze, G. and Guatarri, F. (1987) *A Thousand Plateaus,* B. Massumi (trans.), London: Continuum Press.
de Planhol, X. (2004) *Le Paysage Animal: une zoogéographie historique*, Paris: Fayard.
DeStefano, S. and DeGraaf, R.M. (2003) 'Exploring the Ecology of Suburban Wildlife', *Frontiers in Ecology and Environment* 1.2: 95–101.
Ditchkoff, S., Saalfield, S., and Gibson, C.J. (2006) 'Animal Behaviour in Urban Ecosystems: Modifications Due to Human Stress', *Urban Ecosystems* 9: 5–12.
Dixon, T. (1999) 'Inculcating Wildness: Ecocomposition, Nature Writing, and the Regreening of the American Suburb', in M. Bennett and D. Teague (eds), *The Nature of Cities*, Tuscon: University of Arizona Press, pp. 77–92.
Dixon, T.F. (ed.) (2002) *City Wilds: Essays and Stories about Urban Nature*, Athens: University of Georgia Press.
Donaldson, S. and Kymlicka, W. (2011) *Zoopolis: A Political Theory of Animal Rights*, Oxford: Oxford University Press.
Fisher, M. (2011) Blog Entry, 10 January 2011 and 17 January 2011, available online at www.self-willed-land.org.uk/articles/tayside_beaver.htm, accessed 18 May 2013.
Gandy, M. (2002) *Concrete and Clay: Reworking Nature in New York*, Boston, MA: MIT Press.
—— (1989) 'Teddy Bear Patriarchy: Taxidermy in the Garden of Eden, New York City, 1908–1936', *Social Text* 11: 19–64.
Haraway, D. (2000) *How Like a Leaf*, London: Routledge.
Harrison, C. and Burgess, J. (1994) 'Social Constructions of Nature: A Case Study of Conflicts over the Development of Rainham Marshes', *Transactions of the Institute of British Geographers* 19.3: 291–310.
Harrison, C. and Davies, G. (2002) 'Conserving Biodiversity that Matters: Practitioners' Perspectives on Brownfield Development and Urban Nature Conservation in London', *Journal of Environmental Management* 65.1: 95–108.
Harvey, D. (1996) *Justice, Nature and the Geography of Difference*, Oxford: Blackwell.
Herda-Rapp, A. and Marotz, K. (2005) 'Contested Meanings: The Social Construction of the Mourning Dove in Wisconsin', in A. Herda-Rapp and T.L. Goedeke (eds), *Mad about Wildlife*, Boston, MA: Brill, pp. 73–122.
Hinchliffe, S. (1999) 'Cities and Natures: Intimate Strangers', in J. Allen, D. Massey, and M. Pryke (eds), *Unsettling Cities*, London: Routledge, pp. 138–76.

Hinchliffe, S. and Whatmore, S. (2006) 'Living Cities: Towards a Politics of Conviviality', *Science as Culture* 15.2: 123–38.
Hinchliffe, S., Kearnes, M., Degen, M., and Whatmore, S. (2005) 'Urban Wild Things: A Cosmopolitical Experiment', *Environment and Planning D: Society and Space* 23: 643–58.
Hovorka, A. (2008) 'Transspecies Urban Theory: Chickens in an African City', *Cultural Geographies* 15: 119–41.
Jacobs, J. (1992) *The Death and Life of Great American Cities*, New York: Vantage.
Jamie, K. (2011) 'Diary', *The London Review of Books* 33.14: 38–9.
Jerolmack, C. (2007) 'Animal Archaeology: Domestic Pigeons and Nature-Culture Dialectic', *Qualitative Sociology Review* 3.1: 74–94.
—— (2009) 'How Pigeons Became Rats: The Cultural-Spatial Logic of Problem Animals', *Social Problems* 55.1: 72–94.
Jerzac, L. (2001) 'Synurbanisation of the Magpie in the Palearctic', in J.M. Marzluff, R. Bowman, and R. Donnelly (eds), *Avian Ecology and Conservation in an Urbanising World,* Dortrecht: Kluwer, pp. 403–25.
Jones, O. (2006) 'Non Human Rural Studies', in P. Cloke, T. Marsden, and P. Mooney (eds), *Handbook of Rural Studies*, London: Sage, pp. 185–200.
——(2013) 'The Animality of Rural Landscapes in Affective Registers UK', *Landscape Research.*
Kaika, M. (2004) *City of Flows: Modernity, Nature and the City*, London: Routledge.
Kean, H. (2011) 'Traces and Representations: Animal Pasts in London's Present', *London Journal* 36.1: 54–71.
Knight, J. (2000) 'Introduction', in J. Knight (ed.), *Natural Enemies: People-Wildlife Conflicts in Anthropological Perspective*, London: Routledge, pp. 1–35.
Latour, B. (2007) *Reassembling the Social*, Oxford: Blackwell.
Lethem, J. (2010) *Chronic City*, London: Faber and Faber.
Lewis-Williams, D. (2002) *The Mind in the Cave*, London: Thames and Hudson.
Lorimer, J. (2008) 'Living Roofs and Brownfield Wildlife: Towards a Fluid Biogeography of UK Nature Conservation', *Environment and Planning A* 40: 2042–60.
Luniak, M. (2004) 'Synurbanisation – Adaptation of Animal Wildlife to Urban Development', in J. Shaw *et al.* (eds), *Proceedings of the 4th International Urban Wildlife Symposium*, pp. 50–55.
Luther, E. (2013) 'Tales of Cruelty and Belonging: In Search of an Ethic for Urban Human-Wildlife Relations', *Animal Studies Journal* 2.1: 35–54.
Lynn, W.S. and Sheppard, E. (2004) 'Cities', in S. Harrison, S. Pile, and N. Thrift (eds), *Patterned Ground: Entanglements of Nature and Culture*, London: Reaktion, pp. 53–5.
Mabey, R. (1973) *The Unofficial Countryside*, London: Little Toller.
MacDonald, D.W. and Newdick, M.T. (1982) 'The Distribution and Ecology of Foxes in Urban Areas', in R. Bornkamm, J.A. Lee, and M.R.D. Seaward (eds), *Urban Ecology*, Oxford: Oxford University Press, pp. 123–38.
Macfarlane, R. (2007) *The Wild Places*, Cambridge: Granta.
Morris, B. (1998) *The Power of Animals*, London: Berg.
New York Herald (1874) originally published 9 November 1874, unpaginated full text available online at http://www.museumofhoaxes.com/hoax/text/display/the_central_parl_zoo_escape_text, accessed 3 September 2013.
Oliver, K. (2009) *Animal Lessons*, New York: Columbia University Press.
Padel, R. (2004) *The Soho Leopard*, London: Chatto and Windus.
Park, R. (1984) *The City*, Chicago, IL: University of Chicago Press.
Payne, T. (2010) forum participant, 'TriBeCa Coyote Captured!', *New York Times*, 25 March, http://cityroom.blogs.nytimes.com/2010/03/25/tribeca-coyote-captured-2/?apage=1#comments.
Philo, C. (1995) 'Animals, Geography, and the City: Notes on Inclusions and Exclusions', *Environment and Planning D: Society and Space* 13.6: 655–81.
Probyn, E. (1996) *Outside Belongings,* London: Routledge.
Pyle, R. (2002) 'The Extinction of Experience', in T.F. Dixon (ed.), *City Wilds: Essays and Stories about Urban Nature*, Athens: University of Georgia Press, 257–267.
Rutz, C. (2008) 'The Establishment of an Urban Bird Population', *Journal of Animal Ecology* 77: 1008–19.
Scottish Natural Heritage (2010a) *Feral Beavers to Be Recaptured*, press release, 26 November, Edinburgh: Scottish Natural Heritage.

——(2010b) *Tayside Beaver Recapture*, press release, 20 December, Edinburgh: Scottish Natural Heritage.
——(2012) *Distribution, Population and Activities of Beavers in Tayside*, Commissioned Report 540, Edinburgh: Scottish Natural Heritage.
Seymour, M. (2013) '"Support Your Local Invasive Species": Animal Protection Rhetoric and Non-Native Species', *Society and Animals* 21.1: 54–74.
Sinclair, I. (2010) 'Life on the Margins', *Guardian*, 28 May, available online at www.guardian.co.uk/books/2010/may/29/iain-sinclair-richard-mabey-rereading, accessed 16 March 2013.
Soulsbury, C.D., Baker, P.J., Iossa, G., and Harris, S. (2010) 'Red Foxes (*Vulpes vulpes*)', in S.D. Gehrt, S. Riley, and B. Cypher (eds), *Urban Carnivores: Ecology, Conflict, and Conservation*, Baltimore, MD: Johns Hopkins University Press, pp. 63–75.
Soulsbury, C.D., Iossa, G., Baker, P.J., White, P.C.L., and Harris, S. (2011) 'Behavioral and Spatial Analysis of Extraterritorial Movements in Red Foxes (*Vulpes vulpes*)', *Journal of Mammalogy* 92.1: 190–9.
Srinivasan, K. (2013) 'The Biopolitics of Animal Being and Welfare: Dog Control and Care in the UK and India', *Transactions of the Institute of British Geographers* 38: 106–19.
Stengers, I. (2005) 'The Cosmopolitical Proposal', in B. Latour and P. Wiebel (eds), *Making Things Public*, Cambridge, MA: MIT Press, pp. 904–1003.
Thomson, M. (2007) 'Placing the Wild in the City: Thinking with Melbourne's Bats', *Society and Animals* 15.1: 79–95.
Tuan, Y.F (1978) 'The City: Its Distance from Nature', *The Geographical Review* 68.1: 1–12.
Van Dooren, T. and Bird Rose, D. (2012) 'Storied Places in a Multispecies City', *Humanimalia* 3.2: 1–27.
Wandeler, P., Funk, S.M., Largiader, C.R., Gloors, S., and Breitenmoser, U. (2003) 'The City-Fox Phenomenon: Genetic Consequences of a Recent Colonization of Urban Habitat', *Molecular Ecology* 12: 647–56.
Whatmore, S. and Thorne, L. (1998) 'Wild(er)ness: Reconfiguring the Geographies of Wildlife', *Transactions of the Institute of British Geographers* 23.4: 435–54.
Wolch, J. (1998) 'Zoopolis', in J. Wolch and J. Emel (eds), *Animal Geographies: Place, Politics and Identity in the Nature-Culture Borderlands*, London: Verso, pp. 119–38.
——(2002) 'Anima urbis', *Progress in Human Geography* 26.2: 721–42.
Woolfson, E. (2013) *Field Notes from a Hidden City*, Cambridge: Granta.

18

Relationships between Sámi reindeer herders, lands, and reindeer

Elina Helander-Renvall

There has been much interest in the reindeer-herder relationship but oftentimes it has involved ideas about how to manage reindeer herding and how to control a herd. Many researchers, such as Robert Paine (1994), Jan Åge Riseth (2009), and Nils Mikkel Sara (1993), argue that Sámi reindeer herders establish management units which mediate the relation between herds and pasture. In this chapter I would like to elaborate the relationship which exists between Sámi reindeer herders, the land, and reindeer. My interest is in how this relationship is intersubjectively experienced and organized. The aforementioned are involved in an interdependent relationship with each other. Sámi herders are closely tied in with land and their relationship to land is in many ways reflected within herding activities through which these persons relate to reindeer inhabiting their environment.

As confirmed by researchers, Sámi reindeer herders have the ability and knowledge to adapt to the uncertain conditions of the Arctic. Throughout the last decade reindeer research has focused on discussions relating to herding management, adaptation to climate and global change, and various herding conditions in relation to diverse threats and vulnerabilities (Gaup Eira 2012). As I see it, in order to grasp this adaptive ability of Sámi herders in a more comprehensive way, one must also pay greater attention to their understanding of their realities in the sense of relationships.

Knowledge of these relationships is shaped by stories, songs, place names, myths, experiences, animal population fluctuations, and landscape changeability throughout different seasons. Despite this general lack of awareness among researchers, some native scholars, such as Nils Oskal (1995), Gregory Cajete (2000), Shawn Wilson (2001), and Margaret Kovach (2009), have paid special attention to relationships and their significance to an indigenous paradigm and conceptual outlook. Cree scholar Shawn Wilson suggests:

> An Indigenous paradigm comes from the fundamental belief that knowledge is relational. Knowledge is shared with all creation. It is not just interpersonal relationships, or just with the research subjects I may be working with, but it is a relationship with all creation. It is with the cosmos; it is with the animals, with the plants, with the earth that we share this knowledge. [. . .] It goes beyond the idea of individual knowledge to the

> concept of relational knowledge [. . .] you are answerable to all your relations when you are doing research.
>
> *(2001: 177)*

According to Gregory Cajete, a Tewa Indian scholar, through the accomplishment of various natural interconnections, indigenous people live 'in the sea of relationships' (2000: 178). Cajete also stresses that '[n]ative science continually relates to and speaks of the world as full of *active* entities with which people engage' (27). As a matter of fact, and in contradistinction to most Western perspectives, all entities in nature embody relationships.

Personhood and relatedness have their roots in the shared spiritual and physical reality that exists between humans and nonhumans. Sámi herder and writer Johan Turi[1] has given many examples of such relational aspects of the surrounding world in his *Book of Lappland.* For instance, Turi gives expression to this awareness through descriptions of lands that are so lovely that they laugh. With the same sensitivity for the relations I am concerned with here, Turi explains:

> And when you are good, and all things go well, then you think that the whole living-place rejoices, and when things are sad and you sorrow, then you think that the whole living-place weeps, and all the stones and trees and everything in the whole world, and it does not gladden you any longer. [. . .] And when it is a bad place for reindeer, then it is a bad living-place.
>
> *(Turi 1966: 106)*

The knowledge described here by Turi is an example of an experience which is shared by both the herders and the entire world around them. The same kind of ontology was articulated by a young reindeer herder in Norway, who was interviewed by American scholar Marilyn Fowler:

> 'When you are walking up in the nature now with your reindeer, it is a special feeling. Nothing else matters except you are alone up with your reindeer and it's perfect. I have a dog with me and my binoculars to look up at my reindeer and then I walk around there. I almost feel like I *am* the nature. I do not want to think about what is happening down here. You are just walking and you never have a rush. You just walk slowly and relaxed and [. . .] it's fun.'
>
> *(2012: 346)*

The herder continues:

> 'When you go up there, you are thinking like a reindeer. When the reindeer are sleeping, you are sleeping. They are thinking about food, how to survive, I am thinking how to survive. They sleep with one eye open, so if something happen, they are ready to run. It is the same with me when I am up there.'
>
> *(347)*

True to his own assessments, Turi also believes that herders and reindeer have similarly constituted personalities (1966: 65–6), and what we encounter at work here is the specific notion of relationships in which subjectivity (nature of the self), experience, and social actualization of self and other beings are manifested through their relationality (Helander-Renvall

2010). This perspective is supported by Tim Ingold (2000: 58), who has claimed that hunters and gatherers constitute the world through their engagements with it in the context of everyday activities. Also Rane Willerslev argues, in his discussion of the Yukaghir people of Siberia, that 'meaning is inherent in the relational contexts of peoples' direct perceptual engagement with the world' (2007: 20).

In traditional Sámi herding society, social interaction was kin-based and linked to the Sámi organizational units, meaning that herders lacked experiences from outside of this system. Much knowing in herding contexts still comes from what relations are and how they are performed and dealt with. An outline of a number of the core values in relationships that are important to Sámi cultural thinking and practices includes: close kinship ties, appreciation of peace, a meaningful life, a value-driven perception of life, respect of sacred places and spirits, subsistence as a matter of worldview, and love of nature and animals.

It is important too to acknowledge that there are some critical reviews and individual statements regarding the current research and testimonies concerning the Sámi relationship to nature as currently perceived and experienced. I am aware of the existence of misinterpreted statements linked to the past of the Sámi people, describing them in terms of idealized others, and that, as a result of this and in the context of the political and environmental discourse (Heikkilä 2004), the Sámi are expected to remain authentic and culturally pure (Mathisen 2004). It is also stated that indigenous or Sámi researchers and writers politicize various subject matters in relation to nature and ecology (Conrad 2004). Various conceptual power struggles may emerge from the consciousness of the failings of Western culture, which can lead to unrealistic expectations of other cultures. Yet alternative ways of perceiving nature may contribute to transformations of sensibilities to all life through 'the greening of the self' (Macy 2009: 240), for instance, by combining the mystical with the pragmatic and by transcending separateness and coextending one's sense of oneself to other beings.

I acknowledge that this type of discussion has a value of its own, but here my main intent is to analyse the relatedness of Sámi reindeer herders to their environment with a special focus on the land. To help support this, I will discuss how traditional views within the reindeer herding society are still in force, despite the fact that in some cases traditional positions are there only as some types of background factors, albeit influencing the course of opinions and events. According to traditional Sámi ethical principles, all creatures are equal and they have the right to existence and life. However, in contemporary practical herding contexts, there are many decisions to make in relation to how nature works and to contradictions linked to the social, environmental, and political discourses of modern society. The practice of herding itself as an activity and enterprise creates further problems.

Animistic relations

People everywhere are now increasingly aware of nature's processes and the role of humans regarding changes in climate and biodiversity. People make sense of living their lives in the world by focused discussion about particular ways of being related to the world (Harvey 2005: 83) and their potential for a greater understanding of values regarding nature. This may undoubtedly lead in many cases to a situation in which people are disposed to perceive and experience nature in personal terms (Milton 2002: 51). Anthropologist Nurit Bird-David writes about animism as a relational epistemology, in which 'epistemology is about knowing the world by focusing primarily on relatedness, from a related point of view, within the shifting horizons of the related viewer' (2002: 77). Other anthropologists have shown an interest in rethinking animism.

More recently, animism has actually become a term to describe 'a style of worldview that recognizes the personhood of many beings with whom humans share this world' (Harvey 2006: 205; see also Helander-Renvall 2010). Thereupon, it makes sense to claim that

> [i]n dialogue with particular indigenous ontologies, epistemologies and cosmovisions, the new animism contests modernist preconceptions and invites the widening of relational engagements generated and enhanced by gift exchanges and other forms of mutuality. In both indigenous and Western forms, animism encourages humans to see the world as a diverse community of living persons worthy of particular kinds of respect'.
>
> *(Harvey 2005: 83)*

In *Sacred Knowledge* (2008), ecologist Fikret Berkes focuses on indigenous ecological views, practices, and indigenous peoples' relationships with lands and animals. He has also brought to the fore aspects of sacredness as influencing factors of ecological thinking by describing how indigenous and local groups use rituals to create success in hunting and to show respect for nature. For example, among the Cree of eastern James Bay, animals control the hunt and people show respect for animals in a number of different ways. An increase in the hunter's hunting success goes hand in hand with his respect for animals (Berkes 2008: 81). Berkes has also expanded the notions of ecology to include more holistic views regarding nature-human relationships and how sustainability works. Furthermore, he claims that the community-of-beings worldview is particularly important to apply now because this bearing makes humans part of an ecological system (Berkes 2008: 93).

When we approach Sámi religion, too, making offerings with regard to reindeer herding is premised on the idea that the landscape is an essential entity that is alive and animated; we see how shamanism is and has always been one of the main ways to establish contact with the supernatural world. Sámi shamanism has its roots in animistic beliefs, as documented in the extensive source text called *noaiddástallan* in Northern Sámi (see Bäckman and Hultkrantz 1978; Pentikäinen 1995). *Historiae Norwegiae*, which was written in the second half of the twelfth century, offers another early account of Sámi shamanism, and shows how a shaman's drum depicts animals, thus indicating the relations between the human realm and animals (the significance of the drum is discussed below).

Sources are limited, however. Because of the culture's animistic religion and its associated beliefs and practices, Sámi shamanism and the persons who practised it were subjected to persecution through state and church courts in Finland, Sweden, and Norway, especially during the seventeenth and eighteenth centuries. Some shamans were killed and their drums taken away, and hundreds of drums were burned. As a response to the Church's policies regarding the Sámi and their traditional culture, most shamanic elements and practices were lost or went underground. Yet many features remain intact, and there is a renewed interest in shamanism in the Sámi society today. Many still adhere to the belief in close ties with nature, the concept of spirits and ancestors, animals, and interaction with the land. In fact, Sámi combine elements from different religious sources in revising and perceiving their spirituality.

Within these beliefs and practices, we encounter the sacred manifested in many ways (Pentikäinen 1995). *Sieidi* are sacred stones that are located, for instance, on islands and mountain slopes or in places where people hunt and fish and where reindeer migrate (Helander-Renvall 2005: 19; see also Äikäs 2011; Äikäs *et al.* 2009). According to Finnish archaeologist Antti Lahelma (2008: 127), in Northern Finland there are approximately 100 *sieidi* sacred sites, which have traditionally been places for worship or sacrificial ritual. To some extent even today, Sámi continue to have contact with these revelatory places (Uddenberg 2000), as they actualize and re-actualize the sacredness of each location (Bornstein 2002; Joy forthcoming).

Certain aspects of shamanism can be located that are particularly important when talking about relationships. One such aspect is to be found among the symbols drawn on traditional Sámi drums. The Sámi shaman drum (*goavddis*) is an expression of the Sámi cosmological, cultural, and spiritual world picture. According to Francis Joy (2011), Sámi drums have historical value in the Sámi culture. The pictures on the drum were positioned according to how the cosmic order, in relation to different realms, was understood. The division between different areas on the surface of the drum relates to diverse physical and psychic realities. According to Sámi traditional beliefs, there are three principal layers which make up the world: 1) the realm of the heavenly deities; 2) the middle realm of everyday activities and spirits; and 3) the realm of the dead. Joy offers an understanding of these realms in the context of the network of relations:

> It was understood that what took place on the earth was reflected in the skies and in the world of the ancestors, thus highlighting the relationship between the supernatural world, where certain deities or totemic ancestral spirits resided, and the physical world and how, for example, the animals in the physical reality were related to those in the spiritual realities.
>
> *(2011: 240)*

I mention this element because, at least traditionally, Sámi used to outline their innate experiences of nature in a more holistic and wider way. As I have already stressed, Joy's research on Sámi shamanism (forthcoming) indicates that there still exist views and practices among the Sámi people which include a series of interconnecting features that show an unbroken series of links with the past, portrayed through landscape, animals, and art.

On reindeer herding in Finland

Currently, there are approximately 9,000 Sámi in Finland. The majority live within the Sámi Home Land (defined by law in 1995), comprising the four northernmost municipalities of Finland, namely Utsjoki, Inari, Sodankylä (Vuotso), and Enontekiö. There are eight reindeer herding districts located within these areas. In total, however, there are 56 reindeer herding districts in Finland; each one has designated pasture lands and numbers of reindeer. The reindeer co-operatives or districts constitute members of the Reindeer Herders' Association, called *Paliskuntain Yhdistys* in Finnish.

In Finland, reindeer herding is practised by Finns and Sámi in the counties of Lapland and Oulu. The reindeer herding area comprises 114,000 square kilometres or 36 per cent of Finland's land area. In 2011–12, there were 196,084 reindeer within the Finnish reindeer herding area, and of this number 80,755 reindeer belong to the reindeer districts of the Sámi Home Land (Jänkälä 2012). Currently, the total number of herders in Finland is 4,530 and includes 1,260 herders who live and work within the Sámi reindeer herding districts inside the Sámi Home Land.

Since the early 1960s, the Sámi reindeer herding society has been subjected to many changes. One of the more influential changes has been the introduction of modern technology, in the form of snowmobiles, helicopters, modern phones, and computers (Helander-Renvall 2008). Modern technology has contributed to the permanent dwelling of families in villages and to increased specialization of men within various herding activities. Moreover, for the Sámi people, reindeer herding is now both a modern occupation and a way of life. Reindeer herding has been developed into a commercial meat production industry.

The formal reindeer herding administration is closely linked to the non-Sámi ways of governance while locally and informally still maintaining traditional organizational entities, called in Sámi *siida* (a kin-based group) and *báiki* (a family or household). Also reindeer herding and its social, economic, and natural environments have changed dramatically. There are reports showing that the impacts of climate change in the Arctic are greater than expected. For instance, in the Sámi herding areas, climate change is causing a greater risk of ice formation during early winter when rainfalls create an ice-layer on lichen and ground, hindering the reindeers' access to forage (Riseth *et al.* 2011; Gaup Eira 2012). It has also become difficult to predict the coming weather, meaning, for instance, that herders are more careful when moving across lakes and rivers. Animals likewise can easily fall through the thin ice covering lakes and rivers. There are also other factors that have an impact on the herders' way of life, such as competing economic and land use activities, including mining, forestry, and tourism (Riseth and Oksanen 2007). Furthermore, Sámi herders live far from areas that provide modern educational and economic opportunities, which causes many Sámi families and young people to move away from the land to cities in other geographic areas (Renko and Sutinen 2006).

There are some principles that are important to understand when talking about modern reindeer herding, including: the autonomy of reindeer husbandry (herders are their own masters); the social bonds of the extensive kinship system proceeding to the network of mutual obligations; *siida* solidarity; dialogue and consensus; and responsibility toward the land and the spirits (Riseth 2009: 128; Dana and Riseth 2011: 103). It is interesting to note that the fundamentals of this list include perceptual sensitivity to cultural, social, and environmental elements.

Predators

When the Sámi meet each other, they often request the latest news by asking: '*mii gullo*?' (what do you hear?). Usually people answer: '*mis lea ráfi*' (we have peace) or they can say, '*mii leat eallán dearvan*' (we have been in good health). *Ráfi* (peace) is a concept that has many applications within Sámi circles. When people say that they have peace, they mean absence of sicknesses, absence of predators, life without troublesome events, existence of favourable herding and weather conditions, healthy lands, good social relations, and so on. Lydia Heikkilä (2004: 147) mentions the Sámi concept *og guohtunráfi* (grazing peace for reindeer) and its importance for herders' evaluation of the grazing conditions. Quite often, for instance, wolverines or eagles cause uneasiness in the reindeer herd. Also the activities and movements of humans can disturb the peace of the grazing lands and reindeer. However, it is worth adding that unstable conditions are inherent in the North, threatening the balance and stability of the husbandry and keeping the adaptive capacities of herders alert.

During recent years, there has been much discussion in Finland about problems and challenges caused by predators. In the Sámi areas, and also inside wider geographical areas in Finland, there are many predatory animals and birds, including bears, wolves, wolverines, lynxes, and eagles, the numbers of which seem to be increasing (Norberg 2010; Ollila and Jänkälä 2012: 2). It has been estimated that in the whole reindeer herding area of Finland, predators take over 20,000 reindeer each year, and that only one-fifth of the number of those that have been killed or damaged are found (Ollila and Jänkälä 2012: 3). Some speculate that, on average, one wolverine kills 22 reindeer per year, and one wolf 100 reindeer per year (see Danell and Norberg 2010: 15–18).

Whether predators kill reindeer or not, Sámi herders tend to show attentive respect for them and acknowledge their right to take reindeer now and then (Uddenberg 2000; Sikku and Torp 2004: 37). Not all predators are regarded as wrongdoing animals (Magga *et al.* 2001: 18), but intolerance is shown for those specific individuals that maltreat and ravage reindeer, for instance, a wolf that scatters and kills several animals without eating the carcasses, or a wolverine that attacks and damages reindeer without killing them. Herders are willing to hunt such predators in order to safeguard the reindeer herds (Sikku and Torp 2004). These measures are taken because reindeer herders have some tacitly understood agreements with reindeer: herders get meat and other goods from reindeer, and reindeer ask in exchange for protection and a good life.

Within Sámi society it tends to be older people who show more understanding and reverence for the animals of the forests, whether they are predators or not. This may be because Sámi tradition-based knowledge is decreasing in modern Sámi society, and many young people's contact with nature is changing in motive from subsistence to leisure. It is also striking that young people lose faith in the possibilities to start or continue with subsistence activities (Pohjola and Valkonen 2012: 99), which can be linked to the current problems experienced by both Finnish and Sámi herders.

The number of predators seems to be greater than the reindeer herding industry can tolerate. Research conducted among Finnish reindeer herders in Southern Lapland shows that the existence of many predators within the pasture areas creates a variety of economic and social problems for those concerned (Pohjola and Valkonen 2012). The mainstream authorities, through their laws, officials, and research establishments, administer the herding regulations. It is obvious that the situation is critical from the herders' vantage point, especially as they simultaneously meet with other threats to their industry and way of life.

When herders sustain economic loss due to the attacks on their herds by animals of the forests, they have to take a stand on what to think and do about the situation. Under certain circumstances, conflict has emerged between herders and predators, as well as clashes between herders and mainstream authorities (Sikku and Torp 2004: 127–30; Pohjola and Valkonen 2012). The point here is that the struggles that may exist between Sámi reindeer herders and predators do not necessarily point to the end of relatedness. They continue to relate to each other simultaneously both in a conflicting and a 'sociable' way, because the herders' relationship to nature is ambiguous and entangled in many ways, including in mainstream thinking and discourses. From the Sámi point of view, predators have been part of their cultural and spiritual landscape, sharing lands and animals; and these relations have been portrayed through painted symbols on traditional Sámi shaman drums, being sacred (bear) or accompanying shamans (wolf); and they have contributed to the ways in which reindeer behave and how the Sámi herd and treat reindeer.

There are many stories and accounts regarding human beings' cohabitating or collaborating with wild animals. During a series of interviews I conducted in North Norway during 1998–2000 inquiring about Sámi customs, a former Sámi reindeer herder from Sirma told me that once, when he stayed overnight in a Sámi tent, a wolf suddenly showed up and remained with him in the tent throughout that night. This took place before the 1950s when there were still many wolves in the Sámi north. The interviewed herder believed that the wolf needed protection because s/he was excluded from a wolf pack. This kind of coexistence gives possibilities for humans and nonhumans to learn to know each other. In addition, trust and cooperation between the two will be enhanced. It is interesting to note that, in this context, the specific animal in question relied more on a human person than on animal persons. The particular context has much to say about how an animal can behave or how s/he can be regarded by humans. According to Sámi ethical rules, wild animals also have the right to sustenance and physical safety (Magga *et al.* 2001: 18). It seems that moral rules significantly govern the relations between herders and animals. But now, beyond the range of the Sámi traditional ways, new national and global regulations emerge that are premised on environmental thinking and discourses that place wild animals and nature outside the Sámi ways of perceiving things.

Siida – a way of life

When Sámi people meet, another one of the ways they initiate communication with each other is by asking questions about families, relatives, and relationships. These questions will elicit answers regarding a person's social ties and his or her ties to pasture lands in their environment. Traditionally there are two organizing units in the structure of a herding society, as mentioned above: the smaller unit is the household or family, called *báiki*, and the larger unit is called *siida*. Each person belongs first to their *báiki*, which in turn belongs to this larger organization called *siida*. Traditionally, the *báiki* or family consists of a group of people with close kinship ties. They have the skills necessary to manage a herd, as knowledge-related advantages are needed in everyday working contexts, such as gathering, migration, and herding. This is especially so because a central precondition for a herder's operative potential within a herding community is his or her skills of communication, cooperation, and overall sociability (Dana and Riseth 2011: 112).

Reindeer-related skills are learned first and foremost inside a family (*báiki*) and within the circle of a larger kin-based group (*siida*). The *báiki* functions as a unit with all the necessary knowledge and skills to manage herds, which is an advantage, since individual herders do not then need to have all the skills, knowledge, and understanding to successfully conduct their work. To work with a herd, the herder needs to understand how reindeer relate to landscape under various circumstances. The tasks of herders are related to the seasonal cycle, and they need to have knowledge of, for instance, the behaviour and biology of reindeer, grazing lands as terrain and sources of food and rest, and appropriate adaptive and strategic choices to deal with changes and vulnerabilities, including other *siida* groups' herding patterns. Sámi pay attention to subtle changes while working with and observing nature, and, because the grazing lands are under snow cover most of the year, an intimate understanding of snow and ice conditions is essential to reindeer husbandry (Helander-Renvall 2007; Riseth *et al.* 2011). Still, some people may know a lot about various herding conditions (ice, rough terrain, predators), but they may not understand well how to operate strategically under difficult circumstances.

Tasks at the *siida* level, such as bringing reindeer together into large herds, call for particularly good communication and other social skills among herders. Sámi children have to learn to memorize different aspects of lands, such as rivers, hills, and mountain slopes, as well as place names, customary law, and other cultural aspects of importance. Without a detailed knowledge of place names, for example, herders cannot communicate efficiently. Competence in social relationships is learned and strengthened through training received in the *báiki* and *siida*.

Another type of social ability needed is networking competence. This relationship skill is needed, for example, when moving herds consisting of several thousand animals to faraway locations, when negotiating with members of other *siida* groups for the use of pasture lands, or when making strategic choices during seasonal migrations or in bad grazing conditions. There are also various types of cultural knowledge for herders to grasp. One important type is customary law connected to herding, which regulates the use of grazing lands. Traditionally, different Sámi families or family groups use certain areas, and they consider that these areas belong to them because of the long-term use. In some Sámi areas, however, it has become hard to clarify land-use rights due to conflicting claims.

Tasks at the *báiki* level are concerned with individual animals as private property and they include slaughtering reindeer, taking care of food supplies, preparation and production of clothing, childrearing, and taking care of the family economy. *Báiki* is also a territorial concept based on kin relations. In this respect, as place, *báiki* is a well-known locality, and it is recognized as home, as an inhabited place, farm, or camp (Helander 1999: 12; Schanche 2002: 166).

Siida is a unit that is organized through close kinship relations based on *báiki* (family): a *siida* may consist of two to ten families. The *siida* system is flexible and the size of a *siida* varies over the year depending on various circumstances, including grazing conditions and various herding groups' operations. A *siida* has traditional links to a specific area, that is, a geographic basis for its operations, and *siida* also consist of a certain number of reindeer that are privately owned by *siida* members. Tasks at the *siida* level are, for instance, herding, gathering of reindeer and leading them to corrals, arranging round-ups, migrating with reindeer to seasonal lands, and building fences.

Siida refers not just to collective management of herds and lands or to a group of households; it also refers to a way of life, as herders belonging to various *siida* constellations must know the lands, reindeer, reindeer earmarks, weather types, snow terminology, other herders and their kin ties, the overall flexibility of the pastoral system, Sámi concepts of time, traditional knowledge, spiritual matters, and proper timing as a basis for different social, cultural, economic, and operational activities.

Lands, herders, reindeer

It is essential to understand that reindeer husbandry is a traditional subsistence activity of the Sámi people, and approximately 10 to 15 per cent of the Sámi are tied to this kind of life. The basic ecological elements of herding as a pastoral system of subsistence are the ever-shifting seasons and the movement and migration of reindeer within a reindeer herding district. While following the reindeer, a herder learns to know nature in many different situations and weather types. Lands are a collection and an assortment of familiar places, each with its own subsistence quality and spiritual meaning (Helander 1999; Schanche 2002).

Reindeer herders' activities have strong links to their ancestral lands, which invite a particular mode of living and behaviour based on their local characteristics and circumstances. Lands are perceived as living entities, and they are active in relation to humans and animals. The activities of the lands can be symbolically understood through the meddling of the earth spirits, called *gufihttar* in the Northern Sámi. Lands nurture and sustain human beings; they are subjectified in animistic ways; and humans, animals, and their shared environment are all actors in an ongoing social relationship (Helander-Renvall 2010).

The presence of various spirits of nature contributes to an understanding of the lands as living organisms. The spirits of the earth are believed to watch herders and therefore especially some of the older herders are concerned regarding what they themselves do and say (Oskal 1995; Buljo Eira 2002; Helander-Renvall 2010). By using rough language, a herder condemns her/himself, and as a result of this, calamities start taking place on grazing lands, to animals and herders. It is also important for the Sámi reindeer herders to negotiate with spirits, such as the earth spirits, in order to get permission to use a particular location for herding or to stay in a place (Oskal 1995; Uddenberg 2000; Helander-Renvall 2010).

Keith Basso (1996) has argued for the importance of analysing the local understanding of environment as he describes how Apache people perceive their lands. Similarly, this is important to describing the Sámi herders' spatial organization and overall activity, which is very much dependent on how reindeer move across lands and how reindeer behave in different contexts in time and space (Turi 1966). According to Ingvar Åhren (1988), a place achieves its meaning for herders through the relationship between them and reindeer. The important point here is to pay attention to how reindeer perceive a place, how the animals value a place, which places they like, what paths they follow, and which kinds of lands they travel across when they migrate. The idea here is that herders perceive places in the same way as reindeer. For instance, reindeer seem to search for places that provide food and shelter, and where there is enough coldness. Åhren (1988: 118–19) calls the lands that reindeer like *trivselland*, a Swedish word meaning 'the lands of well-being'. He also thinks that there must be some other factors, some signals that contribute to reindeers' behaviour in relation to their movement to a specific locality. He mentions peace and openness as factors that give positive signals to reindeer and also to herders (Åhren 1988: 121; see also Heikkilä 2004: 147).

Some of the places in which the Sámi herders and their reindeer stay or dwell are called *livvasadji* (resting place) in the Kautokeino municipality of Norway. These lands are today still regarded as sacred by some people, and one is not allowed to swear at reindeer, lands, or other human persons there (Buljo Eira 2002: 141). Here a proper and appropriate kind of behaviour is sought.

Through talking and listening to the lands and the spirits who protect them, the lands are acknowledged as socially oriented beings. Landscape is experienced during herding and migration, and these activities gather memories and experiences. The shamanic artiefacts, stories, place names, myths, and beliefs play a role in understanding the herding landscape. Many herders believe that if one lives and relates with caution to the land, it will continue to provide sources for survival. Reindeer function as a stabilizing force for herders in relation to lands because they like to use the same familiar paths and areas year after year.

Reindeer luck

Sámi philosopher Nils Oskal describes the Sámi concept of *boazo-lihkku* as reindeer luck (1995: 132–49, 2000). Turi claims that knowledge is not enough when ensuring success in herding because one also needs reindeer luck. With good reindeer luck a herder has a large and beautiful herd. Normally reindeer luck lasts a long time and it can be inherited from parents. If you have reindeer luck, your herd will survive over time and it does not decrease in number even if you slaughter animals regularly.

Reindeer luck improves and can be maintained if you live a value-driven life. This involves being honest, fair, and honourable as well as treating reindeer in a kind and good way. Reindeer should not be treated simply as animals that provide meat; instead, one should think of them as having their own intrinsic value. One should not think or talk negatively about reindeer, and one is not permitted to count them or talk about them with exact numbers. In addition, one has to get along with other herders, and also with lands. Oskal (2000: 178) argues that if a Sámi reindeer herder wants to be successful in his herding activities, he must honour the powers of nature, especially the grazing lands, reindeer, and the spirits dwelling there. Actually, there exists a special dialogue with herding lands consisting of prayers, ceremonials, dreaming, talking, and listening.

Paul-Anders Simma, a member of a Sámi reindeer herding family in Sweden, says that offerings are also an important element in the acquisition of reindeer luck, as they are linked to a deep respect for, and humble attitude toward, all of nature (Helander-Renvall 2010). Before using an area, one has to ask spirits (masters of lands and reindeer) for permission to do so. In some Sámi areas, when herders leave their summer camps they still give thanks to the lands for their willingness to take care of the reindeer. In reindeer herding society, in order to obtain or maintain reindeer luck, one is not supposed to control the world; instead, one tries to come to an understanding with it and move along with it.

Conclusion

I have described relationships and the concept of relational personhood and expressed the idea that Sámi reindeer herders, understanding that everything is connected, place value on respect and intersubjective ways in their relationship to all beings of their natural environment. Sámi people are prepared to alter their behaviour and adaptive activities to fit into nature's workings and to sustain their culture. However, currently, there is some kind of mixture in the ways herders perceive nature, which is partly dependent on the fact that Sámi people have been subjected to enormous pressure on their means of subsistence and worldview. At the same time, traditions persist; Sámi reindeer herders think and act in accord with a set of principles that make their culture animistic and longstanding.

However, one cannot ignore the impact of modern governance, including the discourse regarding nature and economics that exists generally in mainstream society, which undermines herders' options to make the daily choices that maintain connections and sustain underlying relationships. This means that we need multiple perspectives to examine our understanding of the world, using as many lenses as possible. Reindeer herders' knowledge is based on experiences that take place on their land in the context of their herding practices. Their relationship to animals is a product and an expression of these experiences and practices. When describing sustainability and adaptability of the reindeer herding culture, I find it necessary to include in such studies also epistemological, ontological, and spiritual sentiments to ensure a greater understanding of the challenges of maintaining reindeer husbandry as subsistence and as a way of life. Understanding animism as relatedness is one of the perspectives needed when describing Sámi herding culture.

Note

1 Johan Turi was born in 1854 and died in 1936. He lived in Kiruna, Sweden; in 1910 the book *Muittalus sámiid birra* was published, which was later translated into several languages.

References

Åhren, I. (1988) 'Det samiska rummet', in S. Fjellheim (ed.), *Åarjel-saemieh. Samer i sør,* Bodø: Offset Nord A/S, pp. 117–22.

Äikäs, T. (2011) *Rantakiviltä tuntureille. Pyhät paikat saamelaisten rituaalisessa maisemassa*, Pohjois-Suomen historiallinen yhdistys, Tornio: Tornion Kirjapaino.

Äikäs, T., Puputti, A., Núñez, M., Aspi, J., and Okkonen, J. (2009) 'Sacred and Profane Livelihood: Animal Bones from Sieidi Sites in Northern Finland', *Norwegian Archaeological Review* 42.2: 109–22.

Basso, K.H. (1996) *Wisdom Sits in Places. Landscape and Language among the Western Apache*, Albuquerque: University of New Mexico Press.

Berkes, F. (2008) *Sacred Knowledge: Traditional Ecological Knowledge and Resource Management*, New York: Taylor and Francis.

Bird-David, N. (2002) '"Animism" Revisited: Personhood, Environment, and Relational Epistemology', in G. Harvey (ed.), *Readings in Indigenous Religions*, London: Continuum, pp. 72–105.

Bornstein, A. (2002) *Den samiska vandringsrösten: jag är kunskapen*, Stockholm: Svenska Förlaget.

Buljo Eira, M.K. (2002) 'Mu kultureanadagat', in S. Andersen (ed.), *Sámisk landskap og Agenda 21. Kultur, naering, miljøvern og demokrati*, Dieđut 1, Guovdageaidnu: Sami Instituhtta, pp. 136–49.

Bäckman, L. and Hultkrantz, A. (eds) (1978) *Studies in Lapp Shamanism*, Stockholm: Almqvist and Wiksell.

Cajete, G. (2000) *Native Science – Natural Laws of Interdependence*, Santa Fe, NM: Clear Light.

Conrad, J. (2004) 'Mapping Space, Claiming Place: The (Ethno-) Politics of Everyday Geography in Northern Norway', in A. Siikala *et al.* (eds), *Creating Diversities. Folklore, Religion and the Politics of Heritage*, Helsinki: Finnish Literature Society, pp. 165–89.

Dana, L.P. and Riseth, J.A. (2011) 'Reindeer Herders in Finland: Pulled to Community-Based Entrepreneurship and Pushed to Individualistic Firms', *The Polar Journal* 1.1: 108–23.

Danell, Ö. and Norberg, H. (2010) 'Petoeläintilanteen ja liikennevahinkojen vaikutukset Suomen porotalouden teurasmääriin 2005/06–2008/09', *Poromies* 6: 15–21.

Fowler, M.E. (2012) *Wisdom from the North: Sámi Perspectives on Sustainable Leadership*, PhD thesis, San Francisco: California Institute of Integral Studies.

Gaup Eira, I.M. (2012) *The Silent Language of Snow. Sámi Traditional Knowledge of Snow in Times of Climate Change*, PhD thesis, Tromsø: University of Tromsø.

Harvey, G. (2005) 'Animism – A Contemporary Perspective', in B. Taylor (ed.), *Encyclopedia of Religion and Nature*, London: Continuum, pp. 81–3.

——(2006) *Animism: Respecting the Living World*, New York: Columbia University Press.

Heikkilä, L. (2004) 'Sami Reindeer Herding Confronted with Modern Environmental Management', in L.M. Andreassen (ed.), *Sámiske landskapsstudier: Rapport fra et arbeidsseminar*, Dieđut 5, Guovdageaidnu: Sami Instituhtta, pp. 138–52.

Helander, E. (1999) 'Sámi Subsistence Activities. Spatial Aspects and Structuration', *Acta Borealia, A Nordic Journal of Circumpolar Societies* 16.2: 7–25.

Helander-Renvall, E. (2005) *Sami Mythic Texts and Stories*, Silde: Kalevaprint Oy.

——(2007) 'Traditional Ecological Knowledge, Snow and Sámi Reindeer Herding', in P. Kankaanpää *et al.* (eds), *Knowledge and Power in the Arctic: Conference Proceedings*, Rovaniemi: University of Lappland, pp. 87–99.

——(2008) 'Logical Adaptation to Modern Technology – Snowmobile Revolution in Sápmi', in L. Heininen and K. Laine (eds), *The Borderless North. Proceedings of the Fourth Open Assembly of the Northern Research Forum*, Oulu: Ouluprint Oy, pp. 27–34.

——(2010) 'Animism, Personhood and the Nature of Reality: Sámi Perspectives', *Polar Record* 46.1: 44–56.

Ingold, T. (2000) *The Perception of the Environment. Essays in Livelihood, Dwelling and Skill*, London and New York: Routledge.

Jernsletten, J. (2002) 'Sørsamisk landskapsforståelse i et religionsvitenskapelig perspektiv', in S. Andersen (ed.), *Samisk landskap og Agenda 21. Kultur, naering, miljøvern og demokrati*, Dieđut 1, Guovdageaidnu: Sámi Instituhtta, pp. 107–14.

Joy, F. (2011) 'The History of Lapland and the Case of the Sámi Noaidi Drum Figures Reversed', *Estonial Journal of Folklore* 47: 113–44.

——(forthcoming) 'Sámi Shamanism, Fishing Magic, and Drum Symbolism', *International Journal for Shamanistic Research*.

Jänkälä, A. (2012) *Personal Information*, Rovaniemi: Paliskuntain yhdistys.

Kovach, M. (2009) *Indigenous Methodologies: Characteristics, Conversations and Contexts*, Toronto: University of Toronto Press.

Lahelma, A. (2005) 'Between the Worlds. Rock Art, Landscape and Shamanism in Subneolothic Finland', *Norwegian Archaeological Review* 38.1: 29–47.

——(2008) *A Touch of Red: Archaeological and Ethnographic Approaches to Interpreting Finnish Rock Paintings*, Vaasa: Waasa Graphics Oy.

Macy, J. (2009) '*The Greening of the Self*', in L. Buzzell and C. Chalquist (eds), *Ecotherapy – Healing with Nature in Mind*, San Francisco, CA: Sierra Club Books, pp. 238–45.

Magga, O.H. *et al.* (2001) *Dyrevelferd i samisk kultur*, Rapport: Lantbruks-og matdepartementet.

Mathisen S.R. (2004) 'Hegemonic Representations of Sámi Culture: From Narratives of Noble Savages to Discourses on Ecological Sámi', in A. Siikala, B. Klein, and S.R. Mathisen (eds), *Creating Diversities: Folklore, Religion and the Politics of Heritage*, Helsinki: Finnish Literature Society, pp. 17–30.

Milton, K. (2002) *Loving Nature: Towards an Ecology of Emotion*, London: Routledge.

Norberg, H. (2010) 'Maasuurpedot ja porotalouden kannattavuus', *Poromies* 4: 17–21.

Ollila, A. and Jänkälä, A. (2012) *Tilannekatsaus maasuurpetojen aiheuttamien porovahinkojen korvaamisesta*, Rovaniemi: Paliskuntain yhdistys.

Oskal, N. (1995) *Det rette, det gode og reinlykken*, PhD thesis, Tromsø: University of Tromsø.

——(2000) 'On Nature and Reindeer Luck', *Rangifer* 20.2–3: 175–80.

Paine, R. (1994) *Herds of the Tundra: A Portrait of Saami Pastoralism*, Washington, DC: Smithsonian Institution Press.

Pentikäinen, J. (1995) *Saamelaiset-pohjoisen kansan mytologia*, Helsinki: SKS.

Pohjola, A. and Valkonen, J. (eds) (2012) *Poronhoitajien hyvinvoinnin uhat ja avun tarpeet*, Rovaniemi: Lapland University Press.

Renko, M. and Sutinen, T. (2006) *Poroja kahta puolta puuta. Tutkimus porotilojen kehityksestä ja toimeentulosta*, Rovaniemi: Lapin yliopisto.

Riseth, J.Å. (2009) *Modernization and Pasture Degradation: A Comparative Study of Two Sámi Reindeer Pasture Regions in Norway 1969–1990*, Saarbrucken: VDM Verlag.

Riseth, J.Å. and Oksanen, L. (2007) 'Ressursøkonomiske og økologiske perspektiver på grenseoverskridende reindrift', in E.G. Broderstad, E. Niemi, and I. Sommerseth (eds), *Grenseoverskridende reindrift før og etter 1905*, Tromsø: Universitetet i Tromsø, pp. 93–113.

Riseth, J.Å., Tømmervik, H., Helander-Renvall, E., and Labba, N. (2011) 'Sami Traditional Ecological Knowledge as a Guide to Science: Snow, Ice and Reindeer Pasture Facing Climate Change', *Polar Record* 47: 202–17.

Sara, M.N. (1993) *Reindriftsnaeringens tilpasning og reindriftspolitikk*, Dieđut 2, Guovdageaidnu: Sami Instituhtta.

Schanche, A. (2002) 'Meahcci, den Samiske utmarka', in S. Andersen (ed.), *Samiske landskap og Agenda 21. Kultur, naering, miljøvern og demokrati*, Dieđut 1, Guovdageaidnu: Sámi Instituhtta, pp. 156–70.

Sikku, O.J. and Torp, E. (2004) *Vargen är värst. Traditionell shamisk kunskap om rovdjur*, Uddevalla: Mediaprint i Uddevalla.

Turi, J. (1966) *Book of Lappland*, Oosterhout: Anthropological Publications.

Uddenberg, N. (2000) *Renarna, markerna och människorna: Om svenska samers syn på natur, djur och miljö*, Nora: Nya Doxa.

Viveiros de Castro, E. (1998) 'Cosmological Deixis and Amerindian Perspectivism', *Journal of the Royal Anthropological Institute* 4: 469–88.

Willerslev, R. (2007) *Soul Hunters: Hunting, Animism, and Personhood among the Siberian Yukaghirs*, Berkeley: University of California Press.

Wilson, S. (2001) 'What Is Indigenous Research Methodology?', *Canadian Journal of Native Education* 25.2: 175–9.

19

Kinship imaginaries

Children's stories of wild friendships, fear, and freedom

Leesa Fawcett

> We have to meet the universe halfway, to move toward what may come to be, in ways that are accountable for our part in the world's differential becoming. All real living is meeting. And each meeting matters.
>
> Karen Barad, *Meeting the Universe Halfway*

Through a materialist, ecological feminist and phenomenological lens, this chapter explores Canadian children's narrative ideas about common and familiar wild animals, as a way to gain a wider angle view on imagining and teaching interspecies ethics. In keeping with Karen Barad's quote above, I am searching for ways to meet living beings and enhance our mutual becoming and flourishing in this world and others. Children have unique perspectives on other animal lives; they share being cared for at home, often with companion species, and yet when children cross the constructed polite limits of Western adult culture they are said to be wild. Feral children – those abandoned by humans and raised by animals – are rarely, if ever, asked what they learned about animal sociality and culture (Noske 1989). The porosity and possibilities of boundary crossings between wild, feral, and domestic spheres are limitless, not just for children and animals. One of the intricate assumptions of this chapter is that clusters of attitudes towards animals are culturally produced and circulated, and have serious consequences for ethics, citizenship, and education. I begin with the following theoretical and praxis-oriented assumptions: 1) this chapter is part of a larger anti-hegemonic tool kit and postcolonial project on animal ethics; 2) this is speculative work, not definitive or prescriptive; 3) the overall intent is to disrupt notions of autonomy and individualism and focus on interdependence and interrelations.

To begin, it is important to recall that the experience and observation of animals has always played a primary role in what it means to be human, including our philosophy, language, art, science, and consciousness (Midgley 1978; Shepard 1997). The first human metaphors and symbols were comparisons with other animals; the first paint was animal blood (Berger 1980). We are inextricably linked through time, co-evolution, and culture with other animals. In recent decades, human-animal studies has emerged as an interdisciplinary field with critical explorations across the humanities, law, social sciences, and arts. Even with concentrated efforts in animal ethics and scholarly attention to the drastic need for law reform to catch up to current social values about animals (Bisgould 2011), public education has lagged behind, with notable exceptions (Kahn and Humes 2009; Pedersen 2010). In all these admirable endeavours, rarely do we turn to children to learn, even though their ideas and actions mirror, refract, and reject our societal perspectives about animals in ways that point to more life-affirming possibilities that could inform studies about, and teaching of, human-animal relationships.

From a quantum physics and feminist stance, Karen Barad helps illustrate the relational aspects of human-animal studies, as she writes:

> But phenomena – whether lizards, electrons or humans – exist only as a result of, and part of, the world's on-going intra-activity, its dynamic and contingent differentiation into specific relationalities. 'We humans' don't make it so, not by dint of our own will, and not on our own. But through our advances, we participate in bringing forth the world in its specificity, including ourselves.
>
> *(2007: 353)*

In this chapter, I will follow the disappearance of wild animals in lived reality alongside their increasing presence in children's culture, drawing on examples and data from a larger research project I did with five- and ten-year-old Canadian children on common and familiar animals (Fawcett 2002).[1] The data includes: children's ideas, stories, and drawings about wildness, interspecies fear, friendship, and freedom. Then, from this work, I trace some possibilities and meanings for children's narrative ethics in the face of so-called nature-deficit disorder (Louv 2008), kincentric ideals, kinship imaginaries, and the potential of feral education.

Disappearing wildness

There is questionable value and much shaming in repeating the doom and gloom data about animals disappearing from our lives faster than they have at any other time in the known history of the world. Instead, I turn to some problems with problems. Phenomenologically, a problem can be defined as a collision between an inclusive understanding and a particular fact that does not fit into it (Jonas 1982). For example, if humans need animals for human well-being, indeed, in order to be human, then extinction is the jarring fact. By exterminating animal others we deplete our humanity, we cut away pieces of our human-ness.

Endangered species are not accidents; they are the outcomes of the ways in which we decide to know and treat other animals, as Charles Bergman (1990) learned when he studied seven of the most endangered species in North America. The damage to those animals' lives was almost always due to habitat loss because places for innumerable animals to live and eat were sold to land developers. Land 'development' is such an oxymoronic word. Why does land need development? When did development come to mean reduced diversity and more monocultures? And the overarching question is, who benefits? Not wild animals. Wild animals are increasingly endangered in our minds and in our direct experiences, long before they actually become physically, ecologically endangered (Bergman 1990), as the different children's stories will demonstrate.

Naturalist extraordinaire, the late John Livingston (2007), persuasively argued that many people struggle with 'experiential under-nutrition' whereby their sensory engagement with nature is decreased. Subsequently, as nature disappears, so do the possibilities of diverse human-animal relationships. Acknowledging that 'nature' is one of the most complex words in the English language, Raymond Williams also argues that abstract nature is the result of a 'reduction of a multiplicity to a singularity' (1980: 221). I surmise that Livingston was on to something and that the richness of being a human animal among many animals is also being diminished. Consequently, biotic reduction diminishes our political ecologies from multiplicities to singularities. Decisive political aspects of democracy have to do with beliefs and actions about the commonalities amongst us – the land, water, air – the global commons.

One of the trademarks of human exceptionalism in the Western world is the belief in democracy. Democratic participation is fundamentally important for debates and decisions about conserving habitats and inhabitants. Challenges to democracy and conservation of the commons include huge forces such as economic globalization that also sweep away local biodiversity, local memory, and autonomy (Ehrenfeld 2000). I strongly believe that the depletion of biodiversity runs counter to democratic values, and I am in good company, because in the Faculty of Environmental Studies where I work my voice is one among many arguing that biodiversity is an intrinsic public good and democratic policy must be reformed to consider long-term public preferences.

In terms of the public good, there has been a large following within educational circles and beyond of Richard Louv's ideas about the vital importance of nature and outdoor play for children's healthy mental and physical development. Louv (2008) outlines how children's awareness of global environmental threats has increased while their actual lived experiences with nature have drastically decreased, replaced by fears (of dirt, animals, etc.), risk aversion, phobias, and constant parental vigilance. Relying on the University of Illinois' Human-Environment Laboratory and its findings that children with Attention Deficit Hyperactivity Disorder (ADHD) were calmer and had more positive emotions after spending time in natural green settings, Louv proposes that many children have 'nature-deficit disorder'. Louv writes that the 'real disorder is less in the child [. . .] [T]he society that has disengaged the child from nature is most certainly disordered, if well-meaning' (2008: 109). The enthusiastic following that Louv has engendered possibly speaks to the romanticization of childhoods past, but it has also sparked the redesign of urban spaces to include more green, treed areas and an invitation to be reenchanted by nature. As strategically powerful as Louv's ideas have been, they come peppered with serious problems, such as the pathologizing of an already contested disorder (ADHD); and his problematic attention to childhood obesity, despite increasing beliefs that 'weight obsession is seen to have its roots in inequalities based on gender, sexuality, race, class, and ability; for this reason, obesity has been described as a social justice issue' (C. Russell *et al.* forthcoming 2014).

I am haunted by the eerie realization that synchronously as children's contact with nature decreases, and animals and their ecosystems are endangered, North Americans bring more and more companion animals and plants into their domestic lives. What was once wild is increasingly domesticated. Do we imagine secretly that if we care for them at home, we are enacting a twisted type of ecological protection, and that someday, if need be, they can be released and become feral? Feral animals escape the boundaries of domesticated, companion, and wild animals. I recently started feeding a feral cat who showed up hungry and crying. I live in a rural area where it is hard for anyone to just show up on someone's doorstep. This scrawny young cat is an unidentified male, un-neutered, and clearly lost if not homeless. I think he may have been swept into the river during the spring floods and somehow scrambled ashore at our home. He is definitely feral, quite capable of catching birds on the fly and small rodents; at the same time he is highly socialized with humans, choosing to drape himself around my neck, purring, whenever possible. I fed him because he was desperate and starving, yet I am sure he is decreasing the local bird biodiversity as I write.

Alongside the increase in companion animals and their feral counterparts, another paradox has emerged. As more wild animals are endangered and disappear we continue to inundate children with animal representations and symbolism. From the time they are infants we give them: stuffed animals to sleep and play with; ubiquitous zoo visits; and endless allegories, stories, movies, and cartoons with animals to teach them implicit morals and values while entertaining and anthropocentrically edu-taining them (Timmerman and Ostertag 2011). Later, as these children 'grow up', Western culture encourages them to separate from animals, to shrink the amount of animalness in their lives and to relegate their desires for physical touch and friendship into private companion species realms. It has even been claimed that it is acceptable for children to dream about animals, but mature adults should no longer have animals in their dreams (Van de Castle 1983).

In stark contrast to the dominant Western trajectories, Indigenous peoples offer a life-affirming counterpoint. For example, Anishinaabe and Haudenausnee peoples in Canada believe animals are our kin and must be treated with respect as our elders. 'All our Relations' is a kincentric ontology of Being, which details human obligations to other animals (Arquette 1999). Likewise, 'Kincentric ecology' (Salmon 2013) means that humans are kin to all of Life. Humans were the last and youngest creation and so the least experienced and most dependent on all other beings. Compare this to a commonly held belief in Western culture that human maturity involves a critical separation from the animal part of ourselves. What if the Western version of the 'death' of the child-animal relationship is one of the reasons there is a lack of humane remedies for the disappearance of other life forms?

I have found that children's stories about animals narrate particular ethical paths through interspecies neighborhoods and, as such, are particularly generative for my thinking about human-animal relationships more broadly. Drawing on some of the over 200 stories and drawings about common and familiar animals by children in kindergarten and grade five, I explore how these children narrate ideas about friendship across species, concerns about animal freedom, and fear between species. The children's storied experiences transgress in authentic and irreverent ways the boundaries between humans and other animals, and productively play with Western ideas about friendship, kinship, and anthropomorphism. What do these stories tell us about broader cultural narratives? And when we listen to these stories, how do they alter, if at all, our collective responsibility for knowledge-making and reciprocity between humans and other animals?

Storied animal encounters

The late Val Plumwood wrote that:

> the real moral task of developing an adequate ethical response to the non-human world [. . .] includes developing narrative ethics to the other [. . .] and developing the stances of openness and attention that are preliminary to dialogical and communicative relationships of sensitivity, negotiation and mutual adaptation.
>
> *(2002: 169–70)*

Animal ethics and narrative ethics are inextricably linked; ethical responses to animals, when not tightly bound by anthropocentrism, can go beyond prevailing ideologies to broaden our imaginative scope. Often, in relationships between humans and companion animals, intimate friendships are forged. Indeed, Susan McHugh's call for a narrative ethology 'affirms the ways in which ethology and fiction alike proceed from complicated operations of affect, and leads to an ethics premised on feelings honoured as concrete, intense, and shared' (2011: 218). Narrative studies of human knowledge–making about animal lives, particularly posthumanist ones, unshackle all sorts of natures and animals. Thomas King advises us, 'Want a different ethic? Tell a different story' (2003: 164).

Children's literature teems with animal stories; talking animals, wild beasts to conquer, animals to save, tame, and be kind to, and, more often than not, animals as empty bodies to fill with human societal dos and don'ts. Erica Fudge uses the fantastical world of *The Wind in the Willows* to show how characters like Toad, Badger, and Mole are humans in disguise leading to 'anthropomorphism at its most extreme, and, paradoxically, at its most invisible. We forget that the animals are animals' (2002: 72). This compulsory forgetfulness is a teaching in itself. In our utilitarian use of animals as moral teachers we erase animals as subjects of unique lives. Or, when animals have lives in literature and can talk, as in the popular *Charlotte's Web*, only a little girl named Fern can hear them. What have all the adults lost, Fudge wonders? Subsequently, she queries: 'If we could hear animals speak to each other, could we still do what we do to them?' (Fudge 2002: 74).

I am interested in why adults write children's literature, but I am less interested in dressed-up animals who speak to children to teach them the 'right' way to be a good human being. Rather, I am deeply intrigued by some people's capacity to belong to their animal-ness, not in a colonizing, imperial way, but in embodied, sensory, and imaginative ways. The size and dimensions of human empathy and imagination are what limit accounts of animal subjectivity. Children's own stories offer a rich, more diverse diet than much of the children's literature does.

In creating a space for children to tell and draw their stories, I purposely chose to avoid focusing on animals that E.O. Wilson (1992) has called 'charismatic mega-fauna', such as the large, media-popular animals like whales, bears, and lions. As a marine biologist working with whales I had become well versed in Wilson's observations about charismatic megafauna – the public adores whales, but not zooplankton. Subsequently, taking a different tack, I chose to explore human relationships with common and familiar animals long before they reach endangered status, as well as the political entanglements that shadow such a designation.

Instead, I asked Canadian children to tell me stories about three common and familiar animals – bats, frogs, and raccoons. These three wild animals are easily found in most Canadian urban, suburban, and rural settings; all are boundary creatures moving between air, water, and land. The bat is the only flying nocturnal mammal; the frog is an amphibian; and the raccoon is a masked mammal, an omnivore, an opportunist, and seen by some as an urban trickster. Each child told stories and made drawings about each of the three animals and was interviewed about their attitudes to nature, animal experiences, natural history knowledge, and problem-solving abilities when encountering other animals. For brevity, here I will focus mainly on the data from the bat interactions. Like snakes, bats are one of the 'disliked/hated' species in Canada. Bat biologists have observed for some time that the protection and conservation of bats often depends on addressing public attitudes and folklore (Fenton 2001).

One of the most disturbing, statistically significant trends in the results was that while the younger children had less overall knowledge they had more positive attitudes towards all the animals. Younger children, who did not necessarily have high knowledge scores for the animals, described very moral actions in their narrative relationships with animals. The ten-year-old, grade-five children had more knowledge about the animals (as one would expect). While older children knew more facts they also had more misconceptions. For example, most of the older children believed that bats get in your hair, are blind, and are universally dangerous. Misconceptions, of course, can lead to all sorts of fears, some quite exaggerated. The older children's fear was exhibited by the size of the bats in their drawings. More often than not the bats were disproportionately large compared to the humans. Fear can make things seem bigger than they are.

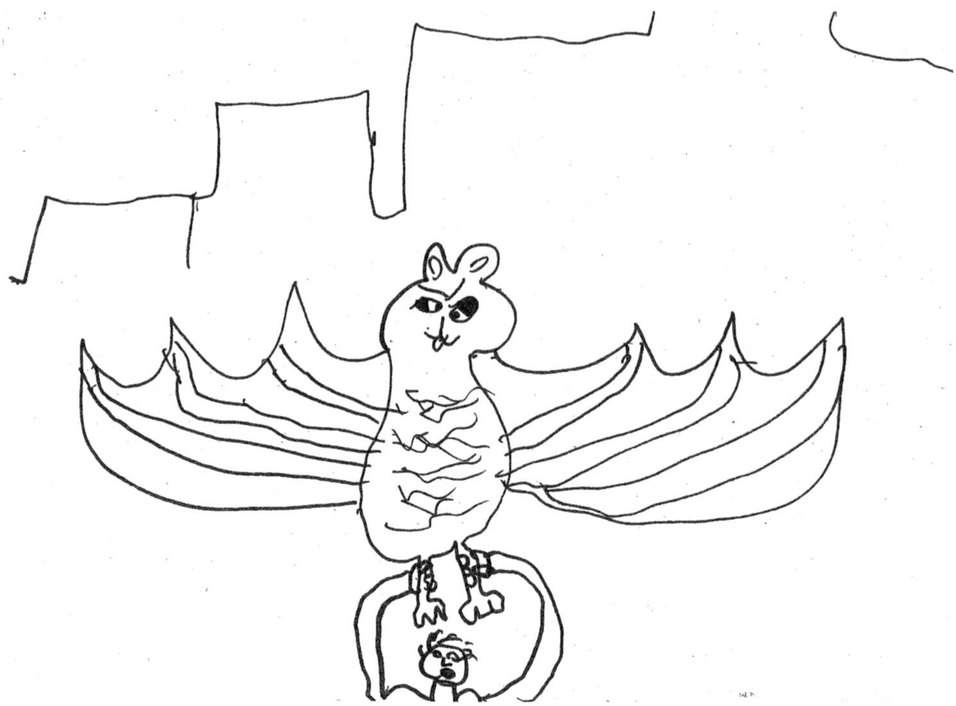

Figure 19.1 Bat carrying a young boy away. Author's collection.

This came as a surprise, given that Stephen Kellert's (1997) studies generally state that the older you are, the more knowledgeable you become, and the more moralistic attitudes you have. Instead, in my research, approaching a being ethically appears to precede knowledge. My research resonates more with that of Jim Cheney and Anthony Weston (1999), who disrupt and resist the reigning epistemology-based ethics, in which ethical action is usually in strict response to garnering knowledge first. Whereas in opposition, they posit approaching another being with a certain etiquette of humility and vulnerability from the first encounter, in order to allow knowledge to emerge from the encounters themselves. They favour developing an environmental ethics-based epistemology, where ethical action is first and predominately a way to enrich the world and create more possibilities for deeper knowing. For them, an ethics-based epistemology is preferable on the grounds that: a) the world is neither easily nor simply knowable; b) ethics is not extensionist and incremental, but pluralistic and dissonant; and c) because hidden possibilities surround us, the task of ethics is to call them out, illuminate, and improve the world. Similarly, feminist moral philosopher Lorraine Code (1987) brilliantly critiques objectifying practices that are 'epistemically irresponsible' and bring harm to others in their ethical barrenness.

In contrast to the kindergarten children, the grade-five children had feelings of fear, harm, and anxiety towards the bats. The following story is a classic example, as it reproduces the myth that bats are blind, and that they fly into people's hair. A 10-year-old boy related this story:

> The bat saw the person and the person got scared because bats like to go into people's hair and get caught. I think. And they're blind and they can't really see that good. The bat got

> caught in the person's hair and he couldn't get it out so he had to go to this other person's house to try and get it out. They had to cut the person's hair in places to pull it out. The person was kind of sad because he had to get his hair cut and it didn't look too good and it hurt because the bat was pulling – the bat was trying to get out too. The bat was pretty scared because he didn't know where he was because they're blind, almost perfectly blind.

Fear has more deadly repercussions: the fear of bats often led to the bat's death, as this 10-year-old recounts:

> Sometimes when a person meets a bat they get really scared. A bat swooped into our house and swooped around all over and I hid underneath a table. My Mom was huddled in a corner. And my Dad used his tennis racquet to swat at the bat. And he swatted it. I think my Dad killed it and put it in a bag and chucked it in the garbage. [After reflecting on her drawing, she added] I think the bat's more afraid of us than we are of him.

One of the more startling results was that the younger children thought of the animals as their friends and were rarely afraid of them. Human friendship is defined as a common social relationship between one subject and another, founded upon reciprocal feelings of affection, support, and companionship (Damon 1983). The kindergarten children's stories contained the acknowledgement of difference, alongside elements of reciprocity, playfulness, empathy, and imagination between human beings and other animals. In their stories, the younger children described the animals as other subjects, both like and unlike themselves. Although younger and smaller, the kindergarten children had much more accurate ideas about the size of the animals compared to their own size. Generally, they were not afraid and were looking for friends of any size.

Figure 19.2 Kaitlyn's bat friendship. Author's collection.

The younger children start with the relationships, the relational space between themselves and the animals, and they relate in non-violent ways. Put another way, younger children demonstrated more intense positive affect towards these animals. A six-year-old female told this bat story, emphasizing the bat's subjectivity, agency, and difference:

> I think bats are awake at night so the person would have to meeted [*sic*] it at night-time. Maybe the bat wanted to know if the person flew like bats do. Maybe the person wanted to know what bats eat. The bat told the person what he eats. The bat is happy because he has a new friend now. The person also feels happy because she has a friend. They were playing in the park.

This young girl recognized difference; there were acts of reciprocity between her and the bat before they became friends. She met the bat at night and they wondered about each other's different habits before they became friends.

A five-year-old boy told the following story, again illustrating the subjectivity of the animal:

> The person's going to be an angel, so it can fly just like the bat. The angel and the bat meet when they are both flying in the sky. They both lost their balance and then they both get up and went for a walk. And then they both went down the path to the little boy's house and they both could have peanut butter and jelly sandwiches and the bat could have bugs in his. The boy is happy because he's got a pet. The bat is happy because he got a friend too.

Why is there such a difference between five-year-olds and 10-year-olds? Increasing age appears to give children greater access to dominant Western cultural currency about animals. Yet this cultural currency seems to reproduce negative, anxious, fearful notions about wild common and familiar animals. The emotional distancing from animals that seems to happen between five- and ten-year-olds could reinforce the animal industrial complex and make it more irreproachable to exploit animals. How conscious are we of how we teach children about specific animals?

Comparing younger and older children's stories, I was struck by the disciplinary role of schooling and education. I am deeply concerned about what is lost when we school feelings for other animals out of children. Let's take the case of Barbara McClintock, who, after being scorned by her peers for decades, won a Nobel Prize in science for her research on genetic transposition in corn. When Evelyn Fox Keller (1983) asked McClintock how she accomplished such great science, McClintock discussed the importance of having a close feeling for the organism. As McClintock demonstrated, it is more than the precious otherness of the world: it is the precious relationality of the world. The younger children felt a playful and friendly relationship between themselves and the animals; in contrast, the older children did not. How does imaginative play between species intersect with questions of curiosity, friendship, and ethics between species? If 'education is the point at which we decide whether we love the world enough to assume responsibility for it' (Greene 1995: 196), then something is amiss in the way Western systems educate children about animals and responsibilities.

Closely tied to friendship is the theme of wildness and concern for an animal's freedom throughout the children's stories. Gary Nabham and Stephen Trimble (1994) argue that children need to know wild animals to understand wild places, and further that the experience of wildness is a basic human need, not a luxury. In my research, one young child said, 'We're like animals because we're wild too. Sometimes I am wild.' When specifically asked what a wild animal was, a 10-year-old student replied: 'An animal that has no owners or masters.' Other students said, 'Something that is not domesticated, doesn't have to listen to people', and, finally, 'An animal that could live on its own if we hadn't interfered.' Often, curiosity is seen as a childish trait. Curiosity is infantilized and information is prioritized. Given the vast amount of information available, deadening curiosity only serves to deaden the world. We need to be curious, willing to be surprised, able to cope with unknowns to keep wildness alive.

I took wild animals into the classrooms and the children's awe and curiosity was abundant. When I gave half the children the opportunity to experience these three wild animals alive, they were much more likely to attribute subjectivity and agency to the animals in their stories. In *The Social Creation of Nature*, Neil Evernden writes that a child's miraculous encounter with another species:

> might be marked by delight, fear or amazement. But more important by far is the impact of the realization that there is an other, something in experience, which cannot be contained in the self and is, therefore, uncanny – and *wild*. To encounter the wild other, to greet another 'I am,' is to accept the other's existence in one's life world.
>
> *(1992: 112)*

Experience is a messy word, a multi-layered thing that can range from direct to vicarious to mediated to simulated experiences of anything. Cultural stories, language, sensory perceptions all have various responsibilities in experiences. Figure 19.3 is an example of a drawing done by one of the older children before he had ever experienced the presence of a live bat. His cartoon contains a larger-than-life, ferocious, fanged bat and a screaming, frightened youth.

Figure 19.3 Before experiencing a wild bat. Author's collection.

Figure 19.4 After experiencing a wild bat. Author's collection.

Eventually, I showed the children live bats; they were able to witness the actual size of bats in comparison to their own bodies, and were able to be in the presence of bats.

Overwhelmingly, the older children, in particular, were surprised by the small size and vulnerability of live bats. I believe one of the transformative moments for the children was the acknowledgement of differences between humans and bat others, and the possibility and recognition of reciprocal fear. The experience of seeing a real live bat played a significant role in shifting the relationships. The older children shifted from fear and mastery towards wonder and curiosity in the presence of a bat. The same artist of Figure 9.3 drew another picture (see Figure 9.4) after he had experienced a live bat. Note how the bat has changed from a threatening animal to a cautious one around humans.

Kinship imaginaries

In the interdisciplinarity of human-animal studies, there are varied and rich ideas about kinship. I appreciate the cross-species conviviality that Ralph Acampora (1999: 118) discusses, and Kenneth Shapiro's (1990) notion of kinaesthetic empathy with his canine companion. Here, I am using the word kinship, as I understand it, in Maurice Merleau-Ponty's (1968) sense of a 'strange kinship' where we are strangers (étrangers) and also connected as kin through the flesh of the world. Ted Toadvine elucidates this idea further when he calls for a 'phenomenology of the impossible' requiring an 'attentivess to the resistance of what cannot be thought or perceived, to the opacity of a wild being that circumscribes our concepts and precepts' (2003: 150). Later in his work, Merleau-Ponty, who was deeply attentive to the roles of diverse sensory perceptions, also used the term 'interanimality' in reference to relations between different species and their emergent ongoing becoming with each other and their environs (Toadvine 2003: 227). To teach and learn about interanimality has profound implications for human-animal relationships and harkens to Indigenous philosophies, such as 'All my Relations' mentioned earlier. Thinking about interanimality and kinship reminds me again of feral children, and the missed opportunities when feral children were never asked what they learned from the animals who took them into their social life, as feminist anthropologist Barbara Noske (1989) questioned so well. I would argue that making embodied knowledge about animals, as subjects of their own lives – lives and deaths that intermingle with ours in diverse ways – is an ethical prerequisite to compassionate conservation action.

A kinship imaginary is a tribute to the children's stories, a place to hold and nurture the children's imaginative relationships with other wild animals in a dominant culture that largely denies this possibility. While compiling the children's data, a kinship imaginary emerged that comprised the following components: 1) a recognition of differences, similarities, and unknowns between humans and other animals, combined with a lively curiosity about it all; and 2) a developing sense of interspecies empathy. One of my guides as I think about kinship imaginaries is Donna Haraway's work on significant otherness, including her commitment to:

> emergent practices: i.e., in vulnerable, on-the-ground work that cobbles together non-harmonious agencies and ways of living that are accountable both to their disparate inherited histories and to their barely possible but absolutely necessary joint futures.
>
> *(2003: 7)*

Emergent narrative practices from the children's voices helped me to envision concepts of radical otherness that include an appreciation of what it means to be alive, which is closely related to the children's respect for animals' actions and inner lives. For example, in terms of aliveness, the children said: 'Animals are alive and so are we'; 'Like us they live'; 'We live in the same world'; 'I know I'm an animal because I'm alive'. The children's voices complement Hans Jonas' (1982) philosophy of aliveness, in which he proclaims that only Life can know Life.

As outlined above, the first component of a kinship imaginary contains the recognition of difference and similarity in beings and bodies relating to one another. A young five-year-old boy said we are different from animals 'because we have different faces'. Perhaps implicit in this young child's acknowledgement of difference is a budding sense of what it means to encounter another animal, face to face. Levinas (1999) has persuasively argued that recognizing the face of the other signifies the separate, infinite life and expressions of another being, and it is this face-to-face acknowledgement that demands a moral response.

A classic line came from a 10-year-old boy, who said we differed from animals because 'We don't live in nature we live in technology'. This is a prevalent and enduring belief – we are technological beings and animals are not – despite the fact that there are a considerable number of publications about tool use by animals (Griffin 2001). Another student said: 'We can speak. We can do a lot more like ride a bike. We can talk to each other. They can too but we can't understand them.' This child recognized rather profoundly that humans and animals have their own forms of communication, even if they do not understand each other.

To clarify, I do not use the word 'bodies' in a biologically determined way; rather, I conceive of bodies as the place where beings reside. I am fully cognizant of the fact that many organisms are a composite of other beings. For example, humans house digestive bacteria, symbiotic mitochondria, and co-evolved viruses under and on our skin. As composite wholes, every animal body has different relational possibilities with the world.

> The point is that a world and a body are opened up for redistribution, disorganization, transformation, each is metamorphosed in the encounter, both becoming something other, something incapable of being determined in advance, and perhaps even in retrospect, but which nonetheless have perceptibly shifted and realigned.
>
> *(Grosz 1995: 200)*

The becoming-something-other in encounters with the world informed the younger children's stories. However, Myers (1998) in his nursery school research found that adults went out of their way to reinforce the species boundary, especially with respect to language. One nursery-school-aged boy named Joe refused adamantly to be persuaded by two adults that a class turtle could not talk. As Myers (1998) says, Joe had observed the turtle's active animate behaviour and believed the animal was capable of some version of intentional, meaningful language, no matter what the adults said. Joe showed an intriguing combination of ethological behaviour with his daily classroom observation of the turtle and stubborn resistance to authority, whereby he maintained that the turtle could communicate.

The second component integral to a kinship imaginary is the development of interspecies empathy. Cognitive ethologist Marc Bekoff studies social communication in numerous species and argues that we bond with animals through emotions, as the younger children demonstrated in their narratives. He gives us many examples of empathy between animals: for example, in a rather disturbing experiment they found that a 'hungry rhesus monkey would not take food if doing so subjected another monkey to an electric shock' (Bekoff 2007: 11). I am partial to Edith Stein, who studied with phenomenologist Edmund Husserl and developed a concept of sensual empathy (Stein 1989: 11) as a form of visceral, embodied knowing (somewhat comparable to Ken Shapiro's idea of kinaesthetic empathy). The children's empathetic skills need to be nurtured as valuable qualities that help increase our knowledge of ourselves, other species, and unknown encounters. It is important to note that empathetic feelings are not always pleasant ones, as the grade-five children were aware. One older child said he knew animals had feelings because 'if they get hurt they would moan'.

The trend of younger children telling more stories about animal friendship was consistent for all three animals, not just in terms of the emotional content, but also with regard to the actions depicted in the stories. On this basis, the norms of attachment, friendship, love, and care complement moral attention to responsibility, bonding, and sharing. Many ecological feminists discuss human connections with nature in non-hierarchical relationships of care and friendship; similarly, others acknowledge that friendship and civility are important initial steps for working with differences, and then we must go further into our understanding of difference and empathy across the species boundaries. What do children desire in terms of their relationships with animals?

Developmental psychologists know that moral growth depends on multiple social influences that guide children in analogous directions. To return to the children's experiences and to consider notions of environmentally sustainable citizenship, I rely on Rosi Braidotti as she revisits ethics of care and justice, telling us that 'care is the key to social accountability and responsible citizenship' (2006: 119). It must be troublesome for children in their moral development if they are thinking about their affinities with animals, while many adults and adult institutions are avoiding the topic or negatively reinforcing it as the children grow up. To counter the death-like grip of anthropocentric hegemony is a process that seems to need to be continually revisited, renewed, and re-imagined.

In a discussion on conservation education, David Orr (2001) states that an extraordinary level of empathy and a strong moral imagination are required to resolve divergent conservation problems. It could be that children are taught very early on to forget what they felt morally in relation to other animals. So adults may need to re-imagine, to recollect old emotions and resurrect feelings of empathy.

Children create their own culture by resisting, building on, and subverting values from the conventional adult world. We need to do much more child-centred research and fully explore the roles animals may or may not play in children's developmental and moral lives. Fortunately, there are exceptions; for instance, Joshua Russell's (forthcoming 2014) research on children's narratives about the death of companion animals. Still, by not doing more research with children themselves, we repeat the fundamental ethical failure: not knowing how to listen, attend with our senses, and be just and caring.

It is as good a time as any to exercise our ethical imaginations, to listen, and to radically re-imagine our environmental ethics to include interspecies relationships built on friendship, empathy, non-violence, and attentiveness to wild natural histories and relationships. I believe that we know ourselves as fully human, and feel our embodied existence, only insofar as we live in connection with, and have experiences of, other-than-humans.

Collective responsibility: education responds

Anthony Weston's (2004) call for deschooling environmental education is intriguing, as he says teachers must 'experience the human/other-than-human boundary as more permeable than our culture teaches us it is' (45). Indigenous knowledges have known and practised this respect for permeability for a very long time. In fact, Tewa scholar Gregory Cajete (2000) believes that Indigenous knowledges are similar to phenomenological thinking because they are both grounded in sensory, place-based lived experiences of the earth. Very recent scholarship in environmental education further enlarges North American Indigenous education research (Lowan-Trudeau 2012) and southern African Indigenous knowledges (Shava 2012), consequently deepening environmentally just, place-based pedagogical opportunities through insightful postcolonial lenses.

In response to ethical critiques, humane education has explicitly stated its obligation to Albert Schweitzer's initial formulation of a reverence for life and the importance of a circle of compassion (Weil 2004). Broadly defined, humane education commits to a life-affirming ethic and interconnectedness; it values clarification and democratic principles (Selby 1995: 49). Pedagogical research in humane education is demonstrated by Jan Oakley's (2009) excellent review of the educational and ethical dilemmas associated with school dissection practices, and Brazilian scholars Castellano, de Luca, and Sorrentino's (2011) insightful comparisons between anti-oppressive and humane educations. At a more expansive political level than humane education, there is a lively body of research that exists at the intersection of critical environmental education and human-animal studies (Oakley *et al.* 2010; Warkentin 2009; Watson 2006), and that shares many of the same ethical commitments as humane education, but posits some bold imagining and enacting of 'nature' as co-constructor in environmental education research.

Indeed, one of the most compelling voices to emerge at the intersection of critical pedagogy and animal studies is that of Swedish scholar Helena Pedersen. In her 2010 book, *Animals in Schools: Processes and Strategies in Human-Animal Education*, Pedersen deftly decodes the power of education's hidden curriculum to influence youth's understanding of the constructed limits to human-animal relationships. Employing participant observation in classrooms, interviews with teachers and students, and document analysis, her research uncovers how animal exploitation is produced, normalized, and reproduced in Swedish educational institutions. One can only hope that this small but international groundswell of research and interest in animal studies and education will surge onwards, causing waves to interrupt the extermination of wildness from all our lives. The intensity of affect for animals and its long-term biopolitical implications are alive and flourishing in young children's stories of animals. What do we really want to teach children about our animal others?

This chapter builds upon the idea that successful conservation and public education is often more limited by people's values and ethics (and how these are translated into policies and actions) than by current biological knowledge. The younger children's narrative ethics gesture towards a more radical, interdisciplinary understanding of animal subjectivity that could nourish biological conservation, inclusive environmental and animal ethics, and more politically focused humane education. The dream of a heterogeneous form of environmental citizenship requires that we keep and consider all the pieces and contradictions, particularly as they emerge from children's culture, because we never know what they might tell us, and where they might fit, as we keep imagining and re-imagining the possibilities for human-animal studies.

Education plays a critical translation and implementation role between children's research and liberatory pedagogies for children and animals. So as to honour children's narrative ethics and ideas about kinship imaginaries, I believe there is a need for forms of feral education[2] – teaching and learning that respects children's voices about animals across the wild/domestic spheres and that grants space for human-animal relationships to thrive. Feral education has the potential to disrupt the explicit curriculum, which is often in cahoots with the animal industrial complex, and also to let other animal voices be heard. As one child said in an interview, 'the world wouldn't be alive if there were no animals'. A kinship imaginary made up of children's stories of wild friendships, fears, and freedoms enlivens our day-to-day life, and helps ensure that a diversity of life will persist in our imaginations, in the material world, and in our joint dreams of the future.

Notes

1 Portions of this paper have been adapted (from Fawcett 2002: 125–39) with permission from the *Canadian Journal of Environmental Education*.

2 In earlier work (Fawcett 2009), I focused on feral sociality, but now I believe we already have the foundations for that amongst ourselves; it is a question of whether we nourish it or not. Feral sociality has all the promise of a wilder, wider co-mingling of bodily beings and ecologies embedded in particular times and spaces.

References

Acampora, Ralph R. (1999) 'Bodily Being and Animal World: Toward a Somatology of Cross-species Community', in H. Peter Steeves (ed.), *Animal Others: On Ethics, Ontology, and Animal Life*, Albany: SUNY Press, pp. 117–31.

Arquette, M. (1999) 'The Animals', in Haudenosaunee Environmental Task Force (eds), *Words That Come before All Else: Environmental Philosophies of the Haudenosaunee*, Ontario: Native North American Travelling College, pp. 82–101.

Barad, K. (2007) *Meeting the Universe Halfway*, Durham, NC: Duke University Press.

Bekoff, M. (2007) *The Emotional Lives of Animals*, Novato, CA: New World Library.

Berger, J. (1980) 'Why Look at Animals?', in *About Looking*, New York: Pantheon Books, pp. 1–26.
Bergman, C. (1990) *Wild Echoes*, Anchorage, AK: Alaska Northwest Books.
Bisgould, L. (2011) *Animals and the Law*, Toronto: Irwin Law Inc.
Braidotti, R. (2006) *Transpositions*, Cambridge, UK: Polity Press.
Cajete, G. (2000) 'Philosophy of Native Science', in *Native Science*, Santa Fe, NM: Clear Light Publishers, pp. 45–56.
Castellano, M., de Luca, A.Q., and Sorrentino, M. (2011) 'The Interface of Environmental and Humane Education as an Emerging and Relevant Dialogue: A Point of View from Brazil', *Canadian Journal of Environmental Education* 16: 93–105.
Cheney, J. and Weston, A. (1999) 'Environmental Ethics as Environmental Etiquette: Towards an Ethics-based Epistemology', *Environmental Ethics* 21: 115–34.
Code, L. (1987) *Epistemic Responsibility*, Hanover, NH: University Press of New England.
Damon, W. (1983) *Social and Personality Development*, New York: W.W. Norton.
Ehrenfeld, D. (2000) 'War and Peace and Conservation Biology', *Conservation Biology* 14.1: 105–12.
Evernden, N. (1992) *The Social Creation of Nature*, Baltimore, MD: Johns Hopkins University Press.
Fawcett, L. (2002) 'Children's Wild Animal Stories: Questioning Inter-species Bonds', *Canadian Journal of Environmental Education* 7.2: 125–39.
—— (2009) 'Feral Sociality and (Un)natural Histories: On Nomadic Ethics and Embodied Learning', in M. McKenzie, P. Hart, H. Bai, and B. Jickling (eds), *Fields of Green: Re-storying Culture, Environment and Education*, Cresskill, NJ: Hampton Press, pp. 227–37.
Fenton, B. (2001) *Just Bats*, New York: Facts on File.
Fudge, E. (2002) *Animal*, London: Reaktion Books.
Greene, M. (1995) *Releasing the Imagination: Essays on Education, the Arts, and Social Change*, San Francisco, CA: Jossey-Bass Publishers.
Griffin, D. (2001) *Animal Minds*, Chicago, IL: University of Chicago Press.
Grosz, E. (1995) *Space, Time and Perversion*, New York: Routledge.
Haraway, D. (2003) *The Companion Species Manifesto*, Chicago, IL: Prickly Paradigm Press.
Jonas, H. (1982) *The Phenomenon of Life*, Chicago, IL: University of Chicago Press.
Kahn, R. and Humes, B. (2009) 'Marching out from Ultima Thule: Critical Counter-stories of Emancipatory Educators Working at the Intersection of Human Rights, Animal Rights, and Planetary Sustainability', *Canadian Journal of Environmental Education* 14: 179–95.
Keller, E. Fox. (1983) *A Feeling for the Organism*, New York: W.H. Freeman.
Kellert, S. (1997) *The Value of Life*, Washington, DC: Island Press.
King, T. (2003) *The Truth about Stories*, Toronto: House of Anansi Press.
Levinas, E. (1999) *Alterity and Transcendence*, M. Smith (trans.), New York: Columbia University.
Livingston, J. (2007) *The John A. Livingston Reader*, Toronto: McClelland and Stewart.
Louv, R. (2008) *Last Child in the Woods: Saving Our Children from Nature-deficit Disorder*, Chapel Hill, NC: Algonquin Books.
Lowan-Trudeau, G. (2012) 'Moving Margins in Environmental Education', in R. Stevenson, M. Brody, J. Dillon, and A. Wals (eds), *International Handbook On Environmental Education Research*, New York: Routledge/AERA, pp. 404–8.
McHugh, S. (2011) *Animal Stories*, Minneapolis: University of Minnesota Press.
Merleau-Ponty, M. (1963) *Le visible et le l'invisible*, Paris: Gallimard; translated as *The Visible and the Invisible* (1968), Alphonso Lingis, trans., Evanston, IL: Northwestern University Press.
Midgley, M. (1978) *Beast and Man*, New York: New American Library.
Myers, O.E. (1998) *Children and Animals*, Boulder, CO: Westview Press.
Nabham, G. and Trimble, S. (1994) *The Geography of Childhood*, Boston, MA: Beacon Press.
Noske, B. (1989) *Humans and Other Animals*, London: Pluto Press.
Oakley, J. (2009). 'Under the Knife: Animal Dissection as a Contested School Science Activity', *Journal for Activist Science & Technology Education* 1.2: 59–67.
Oakley, J., Watson, G., Russell, C., Cutter-McKenzie, A., Fawcett, L., Kuhl, G., Russell, J., van der Waal, M., and Warkentin, T. (2010) 'Animal Encounters in Environmental Education Research: Responding to the "Question of the Animal"', *Canadian Journal of Environmental Education* 15: 86–102.
Orr, D. (2001) 'A Literature of Redemption', *Conservation Biology* 15.2: 305–7.
Pedersen, H. (2010) *Animals in Schools: Processes and Strategies in Human-animal Education*, West Lafayette, IN: Purdue University Press.

Plumwood, V. (2002) *Environmental Cultures*, London: Routledge.
Russell, C., Cameron, E., Socha, T., and McNinch, H. (forthcoming 2014) 'Fatties Cause Global Warming', *Canadian Journal of Environmental Education* 18.
Russell, J. (forthcoming 2014) 'Animal Narrativity', in J. Castricano and L. Corman (eds), *Animal Subjects II*, Waterloo: Wilfred Laurier University Press.
Salmon, E. (2013) 'Kincentric Ecology: Indigenous Perceptions of the Human-Nature Relationship', *Ecological Applications* 10.5: 1327–32.
Selby, D. (1995) *Earthkind: A Teacher's Handbook on Humane Education*, Staffordshire: Trentham Books.
Shapiro, K. J. (1990) 'Understanding Dogs through Kinesthetic Empathy, Social Construction, and History', *Anthrozoös* 3.3: 184–95.
Shava, S. (2012) 'Moving Margins in Environmental Education', in R. Stevenson, M. Brody, J. Dillon, and A. Wals (eds), *International Handbook on Environmental Education Research*, New York: Routledge/AERA, pp. 384–93.
Shepard, P. (1997) *The Others: How Animals Made Us Human*, Washington, DC: Island Press.
Stein, E. (1989) *On the Problem of Empathy*, W. Stein (trans.), Washington, DC: ICS Publishers.
Timmerman, N. and Osterlag, J. (2011) 'Too Many Monkeys Jumping in Their Heads', *Canadian Journal of Environmental Education* 16: 59–75.
Toadvine, T. (2003) 'The Primacy of Desire and Its Ecological Consequences', in *Eco-Phenomenology: Back to the Earth Itself*, Albany: State University of New York Press, pp. 139–53.
Van de Castle, R. (1983) 'Animal Figures in Fantasy and Dreams' in A.H. Katcher and A.M. Beck (eds), *New Perspectives on Our Lives with Companion Animals*, Philadelphia: University of Pennsylvania Press, pp. 148–73.
Warkentin, T. (2009) 'Whale Agency: Affordances and Acts of Resistance in Captive Environments', in S. McFarland and R. Hediger (eds), *Animals and Agency: An Interdisciplinary Exploration*, Leiden: Brill, pp. 23–43.
Watson, G. (2006) 'Wild Becomings: How the Everyday Experience of Common Wild Animals at Summer Camp Acts as an Entrance to the More-than-Human World', *Canadian Journal of Environmental Education* 11: 127–42.
Weil, Z. (2004) *The Power and Promise of Humane Education*, Gabriola, BC: New Society Publishers.
Weston, A. (2004) 'What If Teaching Went Wild?', *Canadian Journal of Environmental Education* 9: 31–46.
Williams, R. (1980) 'Ideas of Nature', in *Problems in Materialism and Culture*, New York: Verso, pp. 68–85.
Wilson, E.O. (1992) *The Diversity of Life*, New York: W.W. Norton & Co.

20

Mourning crows

Grief and extinction in a shared world[1]

Thom van Dooren

I remember most of all the Ho'okena bird, how after it lost its mate it cried out for weeks [. . .] a terribly high-pitched sound, like an inconsolable moaning. [. . .] The Ho'okena bird is so obviously looking for company, but there is none to be found – nowhere.

Glenn Klinger, quoted in M.J. Walters, *Seeking the Sacred Raven: Politics and Extinction on a Hawaiian Island*

Death, mourning, and that collective mode of dying called 'extinction' are painfully drawn together in this short quote. The bird in question, now long dead itself, was a member of that rarest of corvid species, the Hawaiian crow (*Corvus hawaiiensis*). At the time that biologist Glenn Klinger spoke these words, there were only three of these birds left in the wild. A couple of years later, in 2002, the last sighting of a free-living Hawaiian crow was made (Walters 2006: 245). Since this time, the only surviving participants of the species have lived in captivity, subjects of a long-running breeding and conservation plan (USFWS 2009).

This chapter explores the plight of the Hawaiian crow as a species, but it does so through a very particular lens: namely, mourning. Drawing on a broad range of material concerned with crow behaviour and ecology, my interest is in learning more about how these birds mourn for the deaths of others of their kind. Alongside this discussion, this chapter also draws on a philosophical literature in an effort to explore what it might mean for us to mourn for crows in a time of extinctions. Taken together, these two acts of mourning point to the possibility of our learning to mourn *with* crows for some of the many losses of life and diversity that take place within our shared world.

But this chapter is not just *about* mourning. In addition, it aims itself to *be* an act of mourning: to tell stories about the dead and dying that draw them into relationship with the living. In doing so, this chapter attempts to work across and break down the human exceptionalism that, as we will see, has so often dominated philosophical thought about death and our relationships with other animals and the broader environment. It is in part this exceptionalism that holds us distant, intellectually and emotionally, from our more-than-human world. Mourning offers us a way into an alternative space, one of acknowledgement and respect for the dead that undoes any pretence towards exceptionalism by drawing us into an awareness of the multispecies continuities and connectivities that make life possible for everyone inside our shared world.

The crow that was not a crow

If you had travelled into the dense volcanic forests of Hawaii's Big Island a hundred years ago, you may well have been lucky enough to catch sight of a Hawaiian crow. In fact, you may not even have had to look very hard. Deeply inquisitive by nature, Hawaiian crows seem to have frequently greeted early naturalists who made their way into the island's forests (Walters 2006). According to one of these naturalists, Henry W. Henshaw, in his 1902 *Birds of the Hawaiian Islands:*

> The bird, instead of being wary and shy, seems to have not the slightest fear of man, and when it espies an intruder in the woods is more likely than not to fly to meet him and greet his presence with a few loud caws. He will even follow the stranger's steps through the woods, taking short flights from tree to tree, the better to observe him and gain an idea of his character and purpose.
>
> *(quoted in Walters 2006: 63)*

As is perhaps implied by the vivid image that Henshaw's words paint, the forest was central to the life of these crows. Although sometimes venturing beyond its borders, Hawaiian crows lived primarily among the trees, relying on them for the invertebrates and forest fruits that made up the bulk of their diet (Banko, Ball, and Banko 2002). They even made use of the forest flowers, eating some whole, while probing and piercing others in search of nectar. As the island's largest forest bird, and a largely frugivorous one at that, the species is thought to have probably played an important role as a seed disperser, 'potentially influencing the composition and function of dry- and wet-forest ecosystems' (Banko *et al.* 2002).

Perhaps, in most of these habitat and dietary preferences, these birds do not really sound like 'crows' at all. The broad crow family (family *Corvidae*, often referred to as 'corvids') is composed of many different kinds of birds, including jays, magpies, ravens, and crows. But it is these last two types of predominantly black birds – crows and ravens, along with jackdaws and rooks, sometimes collectively called the 'true crows' (genus *Corvus*) – that most people think of when they hear the word 'crow'. While there are many species of 'true crow' around the world, the ones that many of us know best – those that make their homes among us, living in cities and rural areas – are in many ways very different kinds of birds to those found in Hawaii: we might think here of species like the American crow (*Corvus brachyrhynchos*), the Australian raven (*Corvus coronoides*), the house crow (*Corvus splendens*) in India and other parts of South Asia, not to mention perhaps the most successful member of the genus *Corvus*, the common raven (*Corvus corax*), a species that can now be found over fully half of the Earth's landed surface (Marzluff and Angell 2005: 47). All of these well-known species are omnivorous and opportunistic, generalists of the most blatant kind. They are willing and able to live in a wide range of habitats and situations, exploiting a similarly wide range of food sources. Much of this diet – at least the bit most visible to people – is now often composed of scavenged waste, whether carcasses collected along roadsides or rubbish pulled from bins or dumps. It was with these kinds of crow species in mind that biologist John Marzluff noted, 'If crows can be thought of as specialists in any way, they are specialists on people' (2005: 32).

But this urban scavenging lifestyle has often earned crows little fondness in people's hearts. This situation was, in fact, part of the motivation behind conservationists' decision to refer to the crow by its local Hawaiian name, *'Alalā*, thus emphasizing its considerable differences from many more well-known corvids and undoubtedly helping in efforts to raise funds and public concern for the future of the species.[2] As a fruit and forest specialist, the Hawaiian crow is already very different to a lot of other crows. But, importantly, it is also unlike many of these other species in terms of its response to human habitation. Where many other corvids have thrived in company with humans, the Hawaiian crow has instead been driven to the very edge of extinction.

The key problem for the Hawaiian crow, as with so many other island birds, has been rapid and ongoing environmental transformation. Hawaiian birds have had to survive through two waves of significantly different human occupation: first with Polynesian peoples, 1,500 to 2,000 years ago, and then with the arrival of Europeans and other peoples from the late eighteenth century onwards. In each case, many species have been lost or suffered severe impacts. Today, Hawaii has the dubious honour of being home to more endangered species per square mile than any other place on Earth (Restani and Marzluff 2002; Steadman 1995). While Hawaii is undoubtedly a particularly bad case, small islands all over the Pacific, and indeed around the world, are in a similar position. As Marzluff notes: 'In little over a thousand years we have extinguished more than half of all the bird species that occupied the lush islands of the tropical Pacific' (2005: 256).

The environmental change that has all but wiped out the Hawaiian crow has taken a variety of forms. At the most obvious level, the loss of large areas of forest has decreased the possible range of the species, while also reducing the availability of some food plants. These transformations have been incredibly widespread. As the US Fish and Wildlife Service's recovery plan for the species notes: 'There is no existing forest within the historical range of the 'Alalā that has not been substantially altered from its pre-European condition, much less from its condition prior to the colonization of the islands' (USFWS 2009: 1–10). In addition, the introduction of a range of animals to Hawaii since European arrival has produced new predators for crows, while also increasing vulnerability to existing predators. Newly arrived species like rats, mongooses, and cats attack crows and their eggs, while cattle, pigs, and other grazing animals have now also thinned out the understory in surviving forested areas, making crows more vulnerable to predation by the *'Io* (Hawaiian hawk, *Buteo solitarius*) – a species that is itself listed as endangered. In addition, humans have played a role as direct predators of crows, with farmers in the past even taking advantage of crows' curiosity, imitating their calls to attract and shoot them (Marzluff and Angell 2005: 259; Walters 2006: 62). Alongside all of these threats, introduced diseases – in particular, toxoplasmosis, avian malaria, and avian pox – have likely taken a huge toll on crows and a range of other birds, either killing them outright or significantly weakening them for other predators.

Today, the Hawaiian crow is extinct in the wild. Reduced to approximately 100 birds in captivity, it is widely thought to be the most critically endangered corvid on the planet (Banko *et al.* 2002: 25).

Despite the significant differences between the Hawiian crow and many of its corvid cousins – differences, for example, in terms of habitat and diet – the Hawaiian crow is most definitely crow-like in other important ways. In particular, Hawaiian crows likely share with other corvids a high degree of intelligence and a capacity for deeply social and emotional lives. Corvids are clearly among the most intelligent birds and perhaps animals more generally. So intelligent, in fact, that Nathan Emery has proposed – in the hominid-centric language typical of much human thought on intelligence – that corvids might reasonably be thought about as 'feathered apes' (Emery 2004), a view shared by many other biologists.[3] This is the intelligence that Henry Ward Beecher surely had in mind when he made the now oft-quoted statement that '[i]f men had wings and bore black feathers, few of them would be clever enough to be crows' (quoted in Marzluff and Angell 2005: 80). Whether it be jungle crows (Corvus Macrorhynchos) in Japan who have learned to use traffic lights and moving cars to open tough walnuts (2005: 240) or New Caledonian crows, another forest-dwelling island corvid, with their unique local cultures of tool construction and layers of metatool use (Hunt 1996; Taylor *et al.* 2007), corvids have, again and again, shown themselves to be highly intelligent.

Over the years a range of experiments concerned with the intellectual, social, and emotional complexity of corvid lives have pointed to highly sophisticated abilities previously thought absent in the bird world and perhaps restricted to humans, primates, dolphins, and a select few other animal species. In terms of cooperation and the coordination of behaviour, the consolation of mates after conflict, self-recognition, and the attribution of mental states to others, corvids seem remarkably good at understanding and interacting with conspecifics.[4] This doesn't mean that relationships are always perfectly amicable inside a murder of crows, or even for a mated pair, but it undoubtedly means that crows lead cognitively and emotionally rich lives (at least in the terms that matter enough to humans for us to measure them).

Throughout this chapter I have drawn on a large ethological literature on a range of corvid species, primarily studies concerned with true crows. In doing so, I hope to reinforce the link between the Hawaiian crow and the broader corvid family. While I understand conservationists' decision to refer to the species as 'Alalā – and perhaps to downplay this familial connection – I will instead use the name 'Hawaiian crow' throughout. My hope is that by the end of this chapter, having been drawn a little further into the world of this remarkable avian family, readers will consider the name 'crow' in the positive light that it deserves (if they don't already). In addition, I have drawn on this broad corvid literature for the simple reason that we just don't know that much about Hawaiian crows specifically, and they are now far too endangered to become the subject of these kinds of experiments. There is, however, every reason to believe that these crows share the cognitive and emotional attributes found among the other members of their genus – attributes that, it should be noted, have been shown to be possessed by both those generalist species with whom most of us live and a number of other island corvids with more specific dietary and habitat requirements (for example, the New Caledonian crow). And so, in drawing on this general corvid literature, my aim is to also say something as concrete and realistic as is now possible about the experiential world inhabited by Hawaii's crows, while at the same time exploring what we might learn about death and grief in a time of extinctions from thinking and mourning with corvids.

Death and human exceptionalism

Before turning to the crows themselves, however, there are some more general points on philosophy, death, and the nonhuman that deserve mention. Although taking a variety of different forms throughout the history of Western thought, the role of animals in thinking about death – as in so many other contexts – has almost always been as a foil for thinking 'the human'. Indeed, it seems fair to say that whenever animals have been mentioned in the same breath as death in Western philosophy, at least until very recently (see Plumwood 2002), it has inevitably been to distinguish something unique about human knowledge or experience, something that sets us apart from the rest of the animal kingdom.

A central strand of thought here has been the longstanding and widespread assumption that nonhuman animals do not 'know death'. Voltaire tells us that '[t]he human race is the only one that knows it must die', while, according to Schopenhauer, '[u]nlike man, animals, so to speak, live without knowing death' (quoted in Enright 1983: iv). In the twentieth century, this idea found perhaps its most ardent and eloquent support in the work of Martin Heidegger, according to whom the animal cannot 'die'. While, like all living things, animals will inevitably come to an end or 'perish', for Heidegger humans are unique in our relationship with that ending, in our ability to be consciously oriented towards our deaths, in his terms, to 'die' (1996).[5] This distinction is utterly central to Heidegger's larger philosophical project (Derrida 1993). The notion that the animal cannot die interacts in his work in a complex manner with a range of other ideas about how animals differ from humans – the animal has no language, has no 'hand', is 'poor in world' – each of these ideas informing and reinforcing each other in a way that ultimately yields a picture of humans as thoroughly and *essentially* different to the rest of the animal kingdom (Buchanan 2008: 45).

Knowledge of death here joins a long list of other 'lacks', other characteristics or attributes that are thought to ground an essential difference between humanity and animality: be it the possession of language, mirror self-recognition, rationality, moral agency, or any number of other characteristics (Calarco 2008: 75; Haraway 1989). In this context, death has become another of the 'propers' that Dominique Lestel has described: an essential and unique characteristic *proper* to the human that does not just make us different in the way that all animal species are different to each other, but somehow sets us outside of the sphere of animality (2011). In Val Plumwood's terms, these ideas about death form another important site for the development of philosophical modes of 'human exceptionalism'; that is, 'the idea that humankind is radically different and apart from the rest of nature and from other animals' (2007).

It might be argued, however, that knowledge of death does particularly important boundary-work in this kind of philosophical thought as the divide that it marks is often taken to be utterly central to the determination of whether or not a life is *meaningful*. Jeff Malpas articulates this kind of view when he argues that life in the absence of an understanding of one's own mortality would be 'devoid of interest, devoid of meaning' (1998: 120). As Jonathan Strauss has noted, this general notion is found in the work of numerous thinkers, including Paul Tillich, Herman Feifel, Gillian Rose, and others. In Feifel's words: 'the notion of the uniqueness and individuality of each one of us gathers full meaning only in realizing that we are finite' (Strauss 2000: 93).

But, as with all of the other capacities that animals have been said to 'lack', it is far from clear that knowledge of one's death is up to the task of dividing the animal kingdom up so neatly or finally. If we take seriously *specific* nonhumans, and the current scientific literature about them, examples abound of animals interacting with the dead in ways that, at the very least, must draw us to question these ideas: from foxes burying conspecifics, and gorillas caught up in obvious displays of profound grief (Bekoff 2007: 63–5), to the long periods of interaction with the bones of the dead that so often occur in elephant communities, sometimes covering them with leaves or branches, and at other times slowly and silently touching them with their trunks and feet (Poole 1996: 153–5). Reading longer accounts of these behaviours, it is often hard to believe that these animals do not have some notion of death, some concept that the other is no longer with us in the same way, and will not be again. What else might it mean to an elephant or a fox to bury or cover the body of another, or to return to their bones, again and again?

Ultimately, however, my intention in this chapter is not to directly refute these ideas about who 'knows death'. In the final analysis, this is an issue that cannot be resolved with certainty. In fact, as Derrida has noted, it is far from clear that we humans 'know death' or, for that matter, even know what it would mean to 'know death' (1993). Instead, my goal in this chapter is to shift the focus from knowledge of death to the experience of grief at its occurrence and the possibilities that this grief may open up for crows, humans, and others. Instead of reproducing a human exceptionalism that separates us off from the rest of the world in yet another way – as traditional philosophies of death have tended to do – my goal in taking up this focus is to explore some of the many ways in which death entangles us all in a more-than-human world.[6]

Keeping company with crows: the evolution of grief

This chapter opened with a Hawaiian crow crying out – seemingly in grief – at the death of its partner, a 'high-pitched [. . .] inconsolable moaning' (Walters 2006: 241). There are, however, relatively few accounts in the scientific literature of crows grieving, and those that do exist tend to be anecdotal in nature.[7] Perhaps most famously, in his popular book on animal behaviour, *King Solomon's Ring*, ethologist Konrad Lorenz tells the story of a mourning jackdaw. The bird was the last remaining member of a group of jackdaws kept by Lorenz that mysteriously disappeared from his aviary; the other birds likely either escaped or were killed by a predator. Lorenz recounts that:

> Her whole song was suffused with the emotion which obsessed her, with the sole desire of bringing back her lost ones by means of the 'Kiaw' call, 'Kiaw' and again 'Kiaw'. [. . .] Other sounds were scarcely audible in this song of woe. 'Come back, oh, come back!'
>
> *(Lorenz [1949] 2002: 163–4)*

In a recent interview, biologist John Marzluff also recounted witnessing behaviour among American crows that looked like grief: 'I've seen birds grieving, I think it would be fair to say. I've seen a bird that was perched above another bird that was dying – I don't know if they were related or mated, or what – but certainly that bird was very attentive and watching that other bird on the ground as it was dying.'[8]

These kinds of observations, now made in relation to a range of animals – elephants, some primates, crows, and others – are given further weight by recent comparative work in neurology that highlights important similarities in neural circuits and neurotransmitters between birds and mammals. It seems that the neural bases that enable grief in many mammals, including ourselves, are also found in these remarkable birds (Marzluff and Angell 2012). Over the past few decades, this kind of data has driven a growing acceptance that some nonhuman animals also experience grief at the deaths of others (Archer 1999; Bekoff 2007).

While theories differ slightly, at the core of the story that many evolutionary psychologists and ethologists now tell about grief is the notion that it is intimately entangled with the evolution of close social relationships, which are themselves desirable for the many fitness advantages that partner/group living can confer over individual existence (Archer 1999). I am partial to the influential version of this account first offered by Colin Murray Parkes, who saw feelings of love for others as essential to the maintenance of close relationships, and consequently understood grief as the 'cost of commitment': the cost of this evolved the ability to relate and *be* meaningfully with others (Archer 1999: 60). While there are obviously differences in the forms that this grief will take within and between species of animals, as John Archer has noted, there is now 'abundant evidence that reactions essentially similar to those shown by humans occur in social animals which [who] have lost or been separated from a social companion' (1999: 55).

Crows are highly intelligent and social birds, and are perhaps stereotypical of the species that we might expect to have evolved a capacity to grieve. Support for the existence of grief in crows is perhaps also offered by their expression of related emotional states that are connected to their close social relationships. As with grief, it has been argued that empathy likely evolved in animals as a result of selection pressures in social environments (de Waal 2008: 282). Recent studies on both rooks (*Corvus frugilegus*) and common ravens have revealed behaviours that point strongly to a highly developed capacity for empathy (Fraser and Bugnyar 2010a; Seed, Clayton, and Emery 2007). These studies found clear patterns of 'post conflict affiliation' among their subjects. On the basis of their findings with common ravens, Fraser and Bugnyar have argued that this behaviour likely plays an important conciliatory role, comforting and supporting a distressed partner. As these authors note, consolation requires a 'cognitively demanding degree of empathy' in which a bystander must 'first recognize that the victim is distressed and then act appropriately to alleviate that distress' (Fraser and Bugnyar 2010a: 1).

The link between grief and empathy is not at all straightforward. But, in the absence of more documentation of grief, these kinds of empathic responses seem to point to the kinds of emotional and social entanglements – the *kinds* of shared lives, the *being at stake in each other's company* – that would likely give rise to grief at the severing of a social bond.

In thinking in these ways about emotions like grief and empathy, we do productive work in undermining human exceptionalism by drawing our own responses to death into an evolutionary continuum. While Darwin's work should rightly have put human exceptionalism of this kind to rest long ago, as Plumwood has noted there is an important sense in which evolution has often been interpreted in a way that simply shifts the site of our exceptional nature. While *bodily* continuity with the animals is now readily acknowledged, '[t]he radical break or discontinuity that characterizes exceptionalist thinking has not been abandoned with modernity, but has been located elsewhere – in the human mind' (2007). While some emotions have been conceptually placed on the bodily side of this dualistic division – and there is a long history in the West of thinking about emotions as base bodily impulses (Despret 2004a: 37–8) – other more 'complex' emotional states, like grief, have instead usually been linked to 'developed' cognitive capacities, and as a result taken to be an exclusive possession of *Homo sapiens* (Bekoff 2006). In this context, paying attention to the *evolution* of grief goes some way towards unsettling this fallback exceptionalist position. While emotions like grief certainly take myriad forms among the many social mammals and birds, they are nonetheless shared in an important sense too, as is increasingly being shown in work on the neuroevolution of empathy and other emotions (Decety 2011).

In addition, paying attention to mourning crows enables us to understand a little better the experiential world that Hawaii's crows inhabit, at least for now. In doing so, we gain a greater sense of who these creatures might be, but also of what is lost in their disappearance. Far more than 'biodiversity' in any narrow sense, mourning crows remind us that whole modes of life, whole ways of living and dying in company with others, are disappearing here. Part of this loss will inevitably also be *ways of mourning*. Perhaps in the end what must be mourned at this time, alongside so many other things, is the diminishment of mourning itself: the loss of the rich and varied expressions of grief that have evolved on this planet over millions of years. As species disappear, or as their socialities become dislocated and fractured by violence and disturbance, their ways of being meaningfully together in death, as in life, are undermined and lost.

Mourning as relearning a shared world

But I suspect that corvids have more still to teach us about death and mourning. In exploring this possibility we might start with a funeral. While travelling through the Colorado mountains, ethologist Marc Bekoff witnessed a gathering of magpies in which four of these birds were standing around a fifth, likely killed by a car:

> One approached the corpse, gently pecked at it [. . .] and stepped back. Another magpie did the same thing. Next, one of the magpies flew off, brought back some grass, and laid it by the corpse. Another magpie did the same. Then, all four magpies stood vigil for a few seconds and one by one flew off.
>
> *(Bekoff 2007: 1)*

It is far from certain what these interactions might have meant for those birds on that day, or how widely similar practices might exist among other species – although since publication of this account, Bekoff has been sent numerous reports of similar behaviour among other corvids.[9] To my knowledge, no such funerals have been observed among Hawaiian crows, and perhaps we will never know for certain in what ways they marked the deaths of so many of their kind in recent decades – although the experience of the Ho'okena bird referred to above may give us an important indication.

Marzluff has also frequently encountered large gatherings of crows and ravens at sites of death. On several occasions he has even orchestrated these gatherings by placing a dead American crow – one found that way – back in the environment:

> In all those cases – I've done it several times – their response was the same. The birds come in, they see the dead bird, they immediately fly down and start scolding. They will land around that bird and make a lot of noise and scold. And then, being gregarious animals, they'll probably start preening and doing lots of other things, and then eventually they fly off. Personally, I think that that's what everybody sees when they say they've seen a funeral. Basically what's going on there is that the birds are learning about a very dangerous situation. [. . .] They're learning this is a dangerous place, or there is a dangerous predator, or some situation here that we need to know about and avoid in the future.

Death functions as a powerful stimulus to learning here. Marzluff's observations also indicate that the lessons to be learnt from death are very quickly taken up by crows. In fact, American crows have been known to avoid places where one of their kind has been killed for over two years – according to him, sometimes changing whole flight paths to avoid flying over such a place.

Clearly crows learn about danger from death, but this fact in no way undermines the possibility that they may also experience grief at such times. In fact, if death does provide an important opportunity for learning, this outcome would perhaps only be enhanced by a strong emotional response, be it fear or grief. And so this possible evolutionary *function* for crow gatherings at sites of death does not, of course, mean that this is also the *motivation* of individual birds in attending.[10]

In pointing to this potential learning opportunity, however, Marzluff and the crows that he knows remind us that there is more to mourning than the 'simple' expression of grief. In addition, as many psychologists and philosophers have insisted in relation to human grief, processes of individual and collective mourning do important work in allowing us to learn from and move through experiences of loss. This idea has been expressed in a range of different ways, but I am particularly drawn to philosopher and counsellor Thomas Attig's understanding of grieving as a process of 'relearning the world'. For Attig:

> As we grieve, we appropriate new understandings of the world and ourselves within it. We also become different in the light of the loss as we assume a new orientation to the world. As we relearn, we adjust emotional and other psychological responses and postures. We transform habits, motivations, and behaviors. [. . .] Some of what we took for granted in ourselves or in our life patterns is no longer viable or sustainable. Relearning the world thus requires that we make changes.
>
> *(Attig 1996: 107–8)*

In short, one of the core components of the way in which Attig understands grieving is as a more or less conscious process of learning and transformation to accommodate a changed reality.

What grief points to here is a particular kind of *shared* world or *shared* life. This is a way of being with others that, as far as we know, is unique to some mammals and birds: a particular sociality rooted in our being *emotionally* at stake in each other's lives. This possibility, this way of being with others, is a complex 'biosocial' achievement, requiring the coming together of evolutionary and personal histories as well as emotional and cognitive competencies, to produce embodied subjects who are unavoidably emotionally entangled with one another.[11] It is only inside these particular biosocial configurations that the passing of another out of the world can be experienced and felt as a genuine loss. But loss is not experienced in the face of all change or even death. It is not enough for two such beings to have lived alongside each other, in proximity to one another; rather, they must also in some way have become *at stake in each other*, bound up with what *matters* to each other. In other words, they must in some sense, more or less consciously, have come to inhabit a meaningfully *shared world*.[12]

Grief then, in Vinciane Despret's terms, is a very particular process of 'learn[ing] to be affected' (Despret 2004b: 131), in which the borders between self, world, and other are profoundly problematized (Despret 2004a: 209). This does not mean, however, that there is some sort of 'default state' in which we are unaffected by the world, to which we must later add an emotional life. There is no default, originary position; there is only becoming-together inside rich histories of inheritance, development, and relationship. In this context, learning *not* to be affected is equally a state that is produced, achieved through the cultivation of some relationships, some histories and understandings, and not others. As anthropologist Matei Candea makes clear, drawing on the work of Barbara Smuts, 'ignoring' one another is neither a simple nor an originary mode of being with others for social animals attuned intellectually and emotionally to their complex surrounds (2010).[13]

It is with this understanding of grief in mind – as a complex biosocial achievement – that I would like to explore 'our' response to the deaths of species like the Hawaiian crow. What does it mean that, in this time of incredible loss, there is so little public (and perhaps also private) mourning for extinctions? Why do the last expressions of so many species leave this world unnoticed and unmourned – except perhaps by the few conservationists on whose watch and sometimes in whose hands they pass away? (The others of their own kind being already gone, and so unable to mourn even if they once did.)

At the core of the answer that I would like to propose to these questions is our inability to really *get* – to comprehend at any meaningful level – the multiple connections and dependencies between ourselves and these disappearing others: a failure to appreciate all of the ways in which we are at stake in each other, all of the ways in which we share a world. This failure is, at least in part, rooted in the human exceptionalism that this chapter has explored. As Plumwood noted repeatedly throughout her long career, this kind of anthropocentric engagement with the world has important negative consequences for both humans and the many other living things with whom we share this planet. As she put it in an important posthumously published paper:

> When we hyperseparate ourselves from nature and reduce it conceptually, we not only lose the ability to empathize and to see the non-human sphere in ethical terms, but also get a false sense of our own character and location that includes an illusory sense of agency and autonomy. So human-centred conceptual frameworks are a direct hazard to non-humans, but are also an indirect prudential hazard to Self, to humans, especially in a situation where we press limits.
>
> *(Plumwood 2009: 117).*

The current anthropogenic 'mass extinction event' is clearly one of those situations in which we are ever more dangerously pressing up against the limits of resilience of various ecosystems.

In Plumwood's account, human exceptionalism is positioned as doubly problematic. In the first instance, it is implicated in the erasure of the significance of nonhuman others, in our inability to empathize with their suffering and mourn their deaths and ultimate extinction at our hands. Embedded in dominant cultural narratives, ideas like these are taken up and lived in ways that 'rearticulate' us as beings at stake in each other's lives in various ways. The affective separation of human exceptionalism holds the more-than-human world at arm's length: human exceptionalism plays a central role here in the active process of our learning *not* to be affected by nonhuman others (Despret 2004a: 140).

At the same time, however, Plumwood is attentive to the way in which human exceptionalism grounds a dangerous illusion in which the loss of nonhuman others is understood to never quite touch human lives and possibilities. No meaningfully shared world can emerge inside this conceptual space, and so the potential impacts of the loss of the Earth's diversity on our own prospects for sustainable and meaningful life are never quite grasped. As a result, we seem to have missed the real need for change – the need to relearn the world and our place in it – that death and grieving so often announce. As Marzluff's crows remind us, it can be very dangerous not to pay attention and make changes to behaviours in this context. But if the death of a single crow signals 'here lies danger' – a danger significant enough to avoid a place for years, to alter flight ways and daily foraging routes – then what must the death of a whole species of crows, alongside a host of others at this time, communicate to any sentient and attentive observer? How could these extinctions not announce *our* need to find new flight ways, new modes of living in a fragile and changing world?

Storied-mourning in a time of extinctions

My hope is that this chapter about grieving crows might itself function as a narrative form of mourning. As Ricoeur notes, 'the work of narrative constitutes an essential element of the work of mourning' (2007: 8). But this is so not just in the sense that stories help us to move on, to bear, or even to accept irreparable loss. In addition, stories play an important role in *communicating* this loss more widely, while at the same time helping to tease out the various ways in which loss matters, sometimes drawing distant listeners into a sense of felt connection and so affective involvement in a loss. A key part of this process is the 'fleshing out' of the dead that stories enable; the chance to capture and communicate a fuller notion of who has died and why they mattered. In Judith Butler's terms, 'to put together some remnants of a life, to publicly display and avow the loss' (2009: 39). In doing so, mourning might be an act of bearing witness to the deaths of so many individuals and species at this time in the Earth's history.

But, as they travel, stories also breathe new life into the dead, keeping them moving and enabling them to 'haunt' our lives and future possibilities. In this sense, storied-mourning does not attempt to recover and move on from a loss – to put the dead to rest – but rather, as Derrida has suggested, offers us the possibility of mourning as a deliberate act of *sustained* remembrance that requires us to interrogate how it is that we might 'live *with* ghosts' (Brault and Naas 2001; Derrida 1994: xviii; Ricciardi 2003). This is the kind of mourning that asks us – that perhaps demands of us, individually and collectively – that we face up to the dead and to our role in the coming into being of a world of escalating suffering, loss, and extinction.

While there is potentially a kind of respect and acknowledgement in this refusal to put the dead to rest, there is also an important sense in which the dead are 'put to work', a kind of 'use' of the dead that Derrida has frequently cautioned against as an unethical (but to some extent also unavoidable) facet of mourning (2001). And yet, as Derrida also acknowledges, 'we know better than ever today that the dead must be able to work. And to cause to work, perhaps more than ever' (1994: 120). The work to be done here is, first and foremost, a task of 'getting it' that these deaths, of individuals and of species, *matter* – that the world as we know it is changing and that new behaviours are necessary if life in its diversity is to go on. In this context, learning to mourn extinctions might also be essential to our and many other species' long-term survival.

It is not yet clear whether crows will make their way back into Hawaii's forests. The conservationists with whom I have spoken are hopeful, but also realistic about the many challenges that the species still faces: in particular, the need to restore forests and find ways to better protect birds from disease and predation. While I sincerely hope that this story has as happy an ending as is possible, this will not change the fact that countless birds have died and grieved, and that many generations of their kind will now be required to live in captivity for the species to have any hope of a future at all. Meanwhile, I cannot help but think of the literally hundreds of other species of Pacific birds who have already disappeared: just one part of our global impact on the diversity of forms of life over the past few human generations.[14]

It is in this context, inside deep histories of co-evolved affective bodies, that we are invited to mourn not just *for* crows, but *with* them. Hawaii's crows remind us that *if* we manage to find our way into a space of grief at this time, we will be just one species mourning among many, just one of the many forms of life on this planet that are experiencing this time of incredible loss through a lens of sadness and grief. In this context, mourning with crows is about more than any single species, or any number of individual species, but must instead be a process of relearning our place in a *shared world*: the evolutionary continuities and the ecological connectivities that make our lives possible at all. And so, learning to mourn might offer us a way into a fuller understanding of our living planet, of what it means and why it matters. As Attig has simply put it, albeit in a different context: 'In choosing to grieve actively, we choose life' (1996: 61). This has perhaps never been more true than it is now.

Acknowledgements

The first draft of this chapter was written while I was a visiting scholar in the History of Consciousness Program at the University of California at Santa Cruz in late 2010. Donna Haraway generously hosted this visit and provided thoughtful feedback on key ideas as the paper was drafted. I would also like to thank the many other people at UCSC who spoke with me about these and related issues during my visit, in particular Jennifer Reardon, Karen Barad, and Jake Metcalf. A draft of this paper was presented at The History, Philosophy and Future of Ethology: An International Collaborative Workshop (Macquarie University, February 2011). I would like to thank the participants at this gathering for their helpful feedback. Deborah Rose, Matt Chrulew, and Michelle Bastian have all also offered thoughtful advice on this paper. Finally, I would like to thank John Marzluff, Alan Lieberman, Richard Switzer, and Marc Bekoff, all of whom agreed to be interviewed and/or answer questions by email about their work on crows and mourning.

Notes

1 This chapter has been adapted from the author's forthcoming book, *Flight Ways: Life and Loss at the Edge of Extinction* (New York: Columbia University Press).

2 All quotations from Alan Lieberman are taken from an interview conducted by the author on 29 November 2010. Lieberman was at the time the Director of Regional Conservation Programs at the Institute for Conservation Research, San Diego Zoo, and has had a long-term involvement in Hawaiian bird conservation, including that of the Hawaiian crow.

3 See, for example, Emery and Clayton 2004; Heinrich and Bugnyar 2007; Marzluff and Angell 2005: 40; Seed, Emery, and Clayton 2009.

4 Several of these studies are discussed in more detail below. Also see Bugnyar 2011; Bugnyar and Heinrich 2006; Fraser and Bugnyar 2010a, 2010b; Pika and Bugnyar 2011.

5 I will primarily refer to Heidegger's *Dasein* as 'the human' in this chapter. While this description is clearly overly simplistic, in the context of the current discussion about Heidegger's distinction between the human and the animal, it is *Dasein*'s being as a human being that is its most salient feature.

6 I have used the terms 'grief' and 'mourning' interchangeably in this chapter, in opposition to more conventional usage which often reserves the latter term exclusively for humans. I suspect that this conventional usage stems from the fact that 'grief' is often used to refer to responses to loss in general, whereas mourning is specifically a response to the loss brought about by death (Attig 1996: 9). If animals are unable to understand death, however, then they are unable to experience this specific kind of loss and so unable to mourn properly – they are limited to grieving, as they would for any other lost attachment. As this chapter makes clear, I am not confident that things are this simple.

7 On anecdote as a respectable part of ethological study, see Bekoff 2007; Crist 1999. In addition, it must be kept in mind that there are often significant ethical problems with the construction of formal experiments to test for something like grief.

8 Unless otherwise noted, quotations from Marzluff refer to an interview conducted by the author on 13 November 2010 and subsequent email exchanges. Marzluff is a Professor of Biology at the University of Washington and an expert on corvid behaviour and conservation. He is also a former member of the USFWS 'Alalā Recovery Team.

9 Personal communication with Marc Bekoff, 30 November 2011.

10 On the difference between 'function' and 'motivation', see de Waal 2008.

11 The 'biological' and the 'social' are all mixed up here in ways that undermine the coherence of any concrete distinction between them. Sociality in all its multiple forms is rooted in specific biological capacities – in this case, capacities that we might label 'emotional' or 'cognitive'. In a related but distinct vein, plants and various micro-organisms are also engaged in ongoing 'social' relationships of their own kind – exchanging signs and meanings, communicating in ways that we often underestimate (Hall 2011). In this sense, sociality is perhaps a common feature of all life and should not be restricted to those organisms that possess similar enough modes of interaction to the human to be immediately recognizable as such (Hird 2009). In other words, our being social creatures, as well as the specific forms that this sociality takes, is in important ways a feature of our biological makeup. At the same time, however, biology has itself evolved within the context of very material processes of intergenerational life in the company of others. Sociable life produced the conditions for the evolution of various social capabilities, which in turn deepened and enhanced those social relationships. There is no sociality outside of its specific biological possibilities; nor is there any biological form that has not been shaped by its own particular social milieu.

12 It is with this understanding in mind that I would like to suggest, in contrast to some of Judith Butler's recent work on the topic (2004, 2009), that mourning is less about the 'recognition' of a valuable or 'grievable' life and more about the simple embodied reality of our being more or less affected by others, more or less constituted by their presence, more or less emotionally and intellectually bound up in their fate. It is this differential entanglement of affect that gives rise to the varied degrees of grief that accompany death and loss; we are simply more attached to and invested in the lives of some people, some animals, some environments, some jobs, and belonging, than we are in others. Not to mourn for the passing of some, or to mourn less for some than others, does not *necessarily* indicate a failure to recognize a life as 'grievable' or 'worth living', as Butler suggests – although I accept that in some limited cases it may mean precisely this. In most cases, however, the kind of threshold that Butler's terms imply – valuable or not, grievable or not – cannot do justice to the full spectrum of emotional responses that loss elicits, or fails to elicit, in the countless lives of all of those variously connected and entangled others that are left behind. In contrast, I would suggest that a 'failure' to grieve might more often result from an inability to 'get' (at various experiential levels) how one's own life and world are shared with these dying others.

13 Thanks to Michelle Bastian for pointing out this connection to Candea's wonderful work on meerkats.

14 For a detailed overview of avian extinctions in the Pacific region, see Steadman 2006.

References

Archer, J. (1999) *The Nature of Grief: The Evolution and Psychology of Reactions to Loss*, London and New York: Routledge.

Attig, T. (1996) *How We Grieve: Relearning the World*, New York: Oxford University Press.

Banko, P.C., Ball, D.L., and Banko, W.E. (2002) 'Hawaiian Crow (*Corvus hawaiiensis*)', in A. Poole (ed.), *The Birds of North America Online*, Ithaca, NY: Cornell Lab of Ornithology, available online at http://bna.birds.cornell.edu/bna, accessed 15 July 2011.

Bekoff, M. (2006) 'Animal Passions and Beastly Virtues: Cognitive Ethology as the Unifying Science for Understanding the Subjective, Emotional, Empathic, and Moral Lives of Animals', *Human Ecology Forum* 13(1): 39–59.

—— (2007) *The Emotional Lives of Animals*, Novato, CA: New World Library.

Brault, P.A. and Naas, M. (2001) 'To Reckon with the Dead: Jacques Derrida's Politics of Mourning', in J. Derrida, *The Work of Mourning*, P.A. Brault and M. Naas (eds), Chicago, IL; London: University of Chicago Press, pp. 1–30.

Buchanan, B. (2008) *Onto-Ethologies: The Animal Environments of Uexküll, Heidegger, Merleau-Ponty, and Deleuze*, Albany, NY: SUNY Press.

Bugnyar, T. (2011) 'Knower-Guesser Differentiation in Ravens: Others' Viewpoints Matter', *Proceedings of the Royal Society* 22: 634–40.

Bugnyar, T. and Heinrich, B. (2006) 'Pilfering Ravens: *Corvus corax* Adjust Their Behaviour to Social Context and Identity of Competitors', *Animal Cognition* 9: 36–76.

Butler, J. (2004) *Precarious Life: The Powers of Mourning and Violence*, London and New York: Verso.

—— (2009) *Frames of War: When is Life Grievable?*, London and New York: Verso.

Calarco, M. (2008) *Zoographies: The Question of the Animal from Heidegger to Derrida*, New York: Columbia University Press.

Candea, M. (2010) ' "I fell in love with Carlos the Meerkat": Engagement and Detachment in Human-Animal Relations', *American Ethnologist* 37(2): 241–58.

Crist, E. (1999) *Images of Animals: Anthropomorphism and Animal Mind*, Philadelphia, PA: Temple University Press.

de Waal, F.B.M. (2008) 'Putting the Altruism Back into Altruism: The Evolution of Empathy', *Annual Review of Psychology* 59: 279–300.

Decety, J. (2011) 'The Neuroevolution of Empathy', *Annals of the New York Academy of Sciences* 1231: 35–45.

Derrida, J. (1993) *Aporias*, T. Dutoit (trans.), Stanford, CA: Stanford University Press.

—— (1994) *Specters of Marx: The State of the Debt, the Work of Mourning, and the New International*, Peggy Kamuf (trans.), New York: Routledge.

—— (2001) *The Work of Mourning*, P.A. Brault (trans.), P.A. Brault and M. Naas (eds), Chicago, IL; London: University of Chicago Press.

Despret, V. (2004a) *Our Emotional Makeup: Ethnopsychology and Selfhood*, New York: Other Press.
—— (2004b) 'The Body We Care For: Figures of Anthropo-zoo-genesis', *Body and Society* 10(2/3): 111–34.
Emery, N.J. (2004) 'Are Corvids "Feathered Apes"?: Cognitive Evolution in Crows, Jays, Rooks and Jackdaws', in S. Watanabe (ed.), *Comparative Analysis of Minds*, Tokyo: Keio University Press, pp. 181–213.
Emery, N. and Clayton, N. (2004) 'The Mentality of Crows: Convergent Evolution of Intelligence in Corvids and Apes', *Science* 306: 1903–7.
Enright, D.J. (1983) *The Oxford Book of Death*, Oxford and New York: Oxford University Press.
Fraser, O.N. and Bugnyar, T. (2010a) 'Do Ravens Show Consolation? Responses to Distressed Others', *PLoS ONE* 5(5): 1–8.
—— (2010b) 'The Quality of Social Relationships in Ravens', *Animal Behaviour* 79: 927–33.
Hall, M. (2011) *Plants as Persons: A Philosophical Botany*, Albany, NY: State University of New York Press.
Haraway, D. (1989) *Primate Visions: Gender, Race, and Nature in the World of Modern Science*, New York and London: Routledge.
Heidegger, M. (1996) *Being and Time*, J. Stambaugh (trans.), Albany, NY: State University of New York Press.
Heinrich, B. and Bugnyar, T. (2007) 'Just How Smart Are Ravens?', *Scientific American* 296 (18 March): 64–71.
Hird, M.J. (2009) *The Origins of Sociable Life: Evolution after Science Studies*, New York: Palgrave Macmillan.
Hunt, G.R. (1996) 'Manufacture and Use of Hook-tools by New Caledonian Crows', *Nature* 379 (18 January): 249–51.
Lestel, D. (2011) 'The Philosophical Stakes of Ethology for the 21st Century', presented at The History, Philosophy and Future of Ethology, 19 February, Macquarie University, Australia.
Lorenz, K. ([1949] 2002) *King Solomon's Ring: New Light on Animal Ways*, London and New York: Routledge.
Malpas, J. (1998) 'Death and the Unity of a Life', in J. Malpas and R.C. Solomon (eds), *Death and Philosophy*, London and New York: Routledge, pp. 108–20.
Marzluff, J.M. and Angell, T. (2005) *In the Company of Crows and Ravens*, New Haven, CT; London: Yale University Press.
—— (2012) *Gifts of the Crow: How Perception, Emotion and Thought Allow Smart Birds to Behave Like Humans*, New York: Free Press.
Pika, S. and Bugnyar, T. (2011) 'The Use of Referential Gestures in Ravens (*Corvus corax*) in the Wild', *Nature Communications* 2, pp. 135.
Plumwood, V. (2002) *Environmental Culture: The Ecological Crisis of Reason*, London and New York: Routledge.
—— (2007) 'Human Exceptionalism and the Limitations of Animals: A Review of Raimond Gaita's *The Philosopher's Dog*', *Australian Humanities Review* 42, available online at www.australianhumanitiesreview.org/archive/Issue-August-2007/EcoHumanities/Plumwood.html, accessed 15 July 2013.
—— (2009) 'Nature in the Active Voice', *Australian Humanities Review* 46: 113–29.
Poole, J. (1996) *Coming of Age with Elephants*, London: Hodder and Stoughton.
Restani, M. and Marzluff, J.M. (2002) 'Funding Extinction? Biological Needs and Political Realities in the Allocation of Resources to Endangered Species Recovery', *BioScience* 52(2): 169–77.
Ricciardi, A. (2003) *The Ends of Mourning: Psychoanalysis, Literature, Film*, Stanford, CA: Stanford University Press.
Ricoeur, P. (2007) 'On Stories and Mourning', in P. Gratton and J.P. Manoussakis (eds), *Traversing the Imaginary: Richard Kearney and the Postmodern Challenge*, Evaston, IL: Northwestern University Press, pp. 5–13.
Seed, A.M., Clayton, N.S., and Emery, N.J. (2007) 'Postconflict Third-Party Affiliation in Rooks, *Corvus frugilegus*', *Current Biology* 17: 152–8.
Seed, A., Emery, N., and Clayton, N. (2009) 'Intelligence in Corvids and Apes: A Case of Convergent Evolution?', *Ethology* 115: 401–20.
Steadman, D.W. (1995) 'Prehistoric Extinctions of Pacific Island Birds: Biodiversity Meets Zooarchaeology', *Science* 267 (5201): 1123–31.

—— (2006) *Extinction and Biogeography of Tropical Pacific Birds*, Chicago, IL: University of Chicago Press.
Strauss, J. (2000) 'After Death', *Diacritics* 30(3): 90–104.
Taylor, A.H., Hunt, G.R., Holzhaider, J.C., and Gray, R.D. (2007) 'Spontaneous Metatool Use by New Caledonian Crows', *Current Biology* 17: 1504–7.
US Fish and Wildlife Service (2009) *Revised Recovery Plan for the 'Alalā* (Corvus hawaiiensis), Portland, OR: US Fish and Wildlife Service.
Walters, M.J. (2006) *Seeking the Sacred Raven: Politics and Extinction on a Hawaiian Island*, Washington, DC: Island Press.

21

Dead, dead, dead, dead, dead

Steve Baker

> Art exists [. . .] to make the stone *stony*.
>
> Viktor Shklovsky

One of the numerous short sections in Roland Barthes' book on photography, *Camera Lucida*, written shortly before his death, is called 'Flat death'. It reads as follows: 'With the Photograph, we enter into *flat Death*. One day, leaving one of my classes, someone said to me with disdain: "You talk about Death very flatly." – As if the horror of Death were not precisely its platitude!' (Barthes 1982: 92). I quote this at the outset, because my subject here is *flattening*, and *deadness*, and their near-opposites, *thickening*, and *liveliness* – as they apply to animal objects and to writing that addresses those objects, especially in the context of contemporary art.

Just as not all animal remains that are put on display are seen as specimens, not all dead animals are seen as equally dead. This is bluntly acknowledged in Rachel Poliquin's marvellous book on taxidermy, *The Breathless Zoo*, in which she writes, for example: 'When most people think of taxidermy, they think of hunting trophies. In a sense, trophies epitomize everything that is disliked about taxidermy in general. All taxidermy makes death overt, but just heads are decidedly deader' (Poliquin 2012: 151).

The remark is characteristic of Poliquin's approach to her subject. Although her book articulates the history of key episodes and themes in the development of taxidermy, running through it, like a refrain, is an emphasis on the viewer's presentness-of-experience rather than simply on the object's intended purpose within the museum or stately home or wherever. In the book's opening pages she insists that 'nothing can ever compare to the physical presence of the animal, admittedly dead and stuffed, but a physical presence nonetheless' (Poliquin 2012: 4). And she asks: 'Animal or object? Animal and object? This is the irresolvable tension that defines all taxidermy' (5). Later in the book she expands on her early assertion that all taxidermy 'is a disorienting, unknowable thing' (7) by writing as follows: 'Animals are not fixed entities fully explained by the hierarchies of natural order, or by recent cultural or political discourse, but rather are provocative forces, both ruthlessly physical and semantically ambiguous' (140).

Poliquin does touch on the use of taxidermy by contemporary artists, but her principal subject is its place in museums and private collections, where these ambiguous animal objects are indeed thought of primarily as 'specimens on display', as the subtitle of a 2012 conference put it.[1] At first glance, the idea of the specimen – 'an item or part typical of a group or whole', as Webster's has it – seems far removed from the characteristics of the contemporary taxidermic art object. The museum specimen is at least potentially informative; the art object is an altogether more awkward, less forthcoming thing. (Novelist Anna Burns describes one of her troubling and troublesome fictional characters as 'an example. Of something. Of Not Making It Easy' [Burns 2008: 11] – and it would be hard to better those words as a description of the animal object in the art of recent decades.) There is something wilful, deliberate, and specific about each use of a dead animal body by a contemporary artist, almost regardless of the artists' individual motivations, and it would be easy to take the view that there is therefore something 'livelier' about art's animal bodies than about the animal bodies of other contemporary institutions. But Poliquin's rhetoric, and her insistence on the animal object in the museum as more than purposeful, makes it productively difficult to hold to any very sharp distinction between the singular art object and the representative museum specimen, or indeed to hold to any very sharp distinction between the deadness of the object and the liveliness of that same object. This can only be a good thing, and my specific focus in these pages on ideas of deadness is a means, at least rhetorically, of trying to draw contemporary art's dead animals a little closer to those of other institutions.

The collision, or juxtaposition, of animals and death, or of animal bodies and deadness, takes many forms in recent art and in the discourses that circle its presentational practices. Art's dead animals, in other words, have no single or simple or secure meaning. Their deadness figures, but it also flails around. Let me now give a specific and detailed example to lend some weight to this observation.

New Zealand-based artist Angela Singer's conviction is that 'people need to see animals in a new way' and that 'artists can provide the new visual language'. The artist's role is to 'shock the viewer into a new way of seeing and thinking about the animal'.[2] Quite a number of her reworkings of trophy heads or of complete taxidermic animal bodies over the past decade have been encrusted with jewels, sequins, porcelain flowers, and other such materials that may connote beauty and elicit sympathy, but in a complex and often troubling manner (see Figure 21.1). Some of these works she describes as her 'memorial works': 'The animal, having no grave site, no bodily burial, becomes its own memorial', she writes. The artist turns the already-dead animal (or animal remains, to be more precise) into an object, a different kind of object, and the object then 'works', as it were, on the animal's behalf. Animal advocates unsympathetic to contemporary art sometimes criticize artists for 'objectifying' animals – Singer herself reports being accused of turning 'gallery walls into open graves' – but her work offers one of the clearest examples of the unsettling power of the animal-as-object.

In relation to what she calls the 'flawed dead animal' of her recycled and 'botched' taxidermy, she has said: 'I don't see an animal separate from myself; there is permeability to the boundaries separating other species from us. [. . .] [I]t draws me closer because it's not beautiful, not sentimental, not what animal art is meant to be' (Singer, quoted in Aloi 2008: 13). As she insisted some years ago, art of this kind 'should be done strongly, and for me that means using animal bodies that retain the look of a living body because the animal body speaks to the viewer's human body. Lines of body communication are opened up. In our gut we know human and animal are interdependent.'

How to make works that address that conviction is an ongoing challenge. Among her fairly recent works, *Spartle* (Figure 21.2) is particularly compelling because its look – somehow arrived at in the coming-together of the recycled taxidermy hawk, modelling clay, and wax of which it is comprised – is utterly baffling. There is something terrible in this object's

Figure 21.1 Angela Singer, *Pelt*, 2009, vintage taxidermy rabbit and mixed media. Courtesy of the artist

floundering, flailing, foundering, failing-to-be-an-animal. And its power as an object is in that instability, that inserting of an instability into human expectations of the natural world. It's a useful reminder that artworks are objects, not ideas, and they have to work as objects. Whether and how they do so will shape what they can convey or communicate. These things are determined by the *form* of the artwork, by the thickness of its form, and in many cases by its resistance to ready interpretation.

Singer explains that with *Spartle*, 'because it's quite an abstract work', she had initially to consider spatial constraints because she knew that she wanted to exhibit it in a glass dome, which influenced how big it could be, and in particular how far the sculpted clay shapes coming out of the bird's head could extend.[3] With works of this kind, she explains, 'when I get the taxidermy pieces they're usually damaged', so early work on a piece will typically involve some restoration and repair, 'just because they've come to me in this terrible condition'. A major problem is the fact that 'old taxidermy is such an unforgiving material':

Figure 21.2 Angela Singer, *Spartle*, 2009, vintage taxidermy hawk and mixed media. Courtesy of the artist

It's incredibly thin, the skin, it has no give whatsoever, it's like dealing with tissue paper, and it requires incredibly delicate handling, but at the same time because when I'm taking them apart there's usually metal inside, or bone, then there's this *wrenching*, so it's a combination of trying to be delicate with the outer while still busting through, getting through the inner workings.

Working with equipment ranging from 'quite delicate dental tools all the way up to the saws and hammers', Singer is acutely aware of 'the problems in handling an animal's body, handling it with respect but still being able to make the artwork I want to make. Just because there's so much pulling apart and destroying in order to create.' Her initial discomfort with working with taxidermy in this manner was something she had to overcome 'because I wanted to use that as the medium to express what I wanted to say about animals and how we treat them'. She openly acknowledges that 'it's bizarre that I'm handling these animals so roughly' in order to comment through her work 'on how we treat animals so badly'.

Returning to the taxidermy used in *Spartle*, she observed: 'This one, the hawk, was beyond repair, it's absolutely gutted and dried out, and the person who'd had it, their cat or dog had ripped all the innards out.' Generally the idea for a piece 'mainly emerges from working with the materials: even pieces that look like they've been put together *just so*, they just tend to happen, it's kind of like I know what animal or bird I want to work with, and if the person who's given me the work has told me its back story, that's really what I'm working with, and I just see what happens'.

In the case of the hawk, she says, 'I really responded to this head and the opening up at the back and wanted to create something to fill that'. The resulting modelling clay forms that flail around the bird's head had no clearly predetermined identity or purpose: 'when I was making it I wasn't sure whether these were going into the shell to fill it up and to create some kind of new body, or whether it was coming out of it. I just really liked this idea of fluidity.' This fluidity or open-endedness extends to the interpretation of the work: 'I thought it would be interesting to make this quite abstract form so that people would bring their own interpretation or their own questions to it, and I'm not really giving them very much.'

The elements of abstraction and ambiguity here do not represent a move away from the conscious awareness of animals with particular lives and animal bodies with particular histories. Of the pieces such as *Spartle* that have been placed under glass domes, Singer suggests: 'I think it almost says that they're so damaged that they need protection, they are parts, they're no longer part of a whole, they need some form of reverence [. . .] and they need to be honoured.'

What comes through clearly enough in Singer's comments on *Spartle* is the difficult but necessary strategy of taking liberties with animal form while simultaneously maintaining some sense of integrity in the process. Something of the hawk and its history persists in the piece, however obscurely. Rachel Poliquin has a great phrase for this stubborn presence: she calls it 'the truthiness of the animal'.[4] As she argues: 'The material truth of animal form obscures or neutralizes any propaganda motivating the image' (2012: 93). Again, she's thinking here of museum taxidermy rather than contemporary artists' use of the medium, but the statement could just as easily be applied to Singer's work. In a recent interview Singer has reflected on a shift in her own perceptions of her practice:

> When I first started out working with old taxidermy I thought that perhaps I was trying to get through an animal rights message with the work, but I've actually changed my mind over time, in that instead of trying to deliver an overt message, what I want the viewer to do is to think about the work and come to what they think about what I'm trying to say themselves.
>
> *(2011)*

Here, it may be that the relevant distinction is not between the animal object as museum specimen and the animal object as contemporary artwork, but rather between the work of language and the work of material animal form. Both Poliquin and Singer seem to have a hunch that this may be the case. Poliquin writes: 'This strangely queasy sense of knowing that develops during encounters with taxidermy might be called visceral knowledge: a bodily knowing that occurs in contact with physical things, a knowing that blurs emotion with materiality and may even defy reason, logic, and explanatory language' (39). And elsewhere: 'You can sense an understanding, but you cannot always draw that understanding into concrete language' (194). And Singer reflects that 'if I deliver written information with the artwork, then I think that pushes people away' (Singer 2011).

What is called for in these circumstances, I suggest, is a language that is somehow closer to its objects, enlivened by its objects: its dead objects. This would be a form of language that is not averse to handling its objects – handling them with writing, in a manner no less fleshy and familiar, and no less careful and attentive, than other forms of handling museum exhibits and artworks. (Or, indeed, of handling living creatures.)

Yves Le Fur begins a short essay called 'Displaced Objects on Display' with the following statement: 'I never understand it when people say that objects in museum galleries or collections are dead because they no longer exist in their original context. I rather feel that every piece or fragment has its own sight and voice, history and future' (Le Fur 2006: 6). It's an interesting observation partly because it links sight and voice – look and language – suggesting their proximity, and partly because it seems at least to imply that it may be the writer's responsibility to articulate *the voice of the object*, rather than simply invoking their own voice yet again.

On the whole, the emerging field of animal studies has not been particularly good at this. It generally pays surprisingly little attention to writing, and to the effects of different kinds of writing, and to the manner in which style and method may be entangled. It too often assumes – especially when it has a contentious artwork in its sights – that matters of meaning and morality are self-evident, transparent affairs, so that the writer has no difficulty in positioning themselves in relation to their perceived responsibilities.

Personally, I prefer the persistent unknowing of the overarching, awkward, but perfectly framed question posed by Poliquin in her book on taxidermy: 'what is this animal-thing now?' (7) – the blunt, clunky materiality of that hyphenated term 'animal-thing' almost guaranteed to rile those whose knee-jerk response is to take offence at any perceived linguistic objectification of animals, without pausing to attend to the productive imaginative work that may be undertaken by the words chosen by a writer in a particular context, a specific instance.

This is not – as it may sound – a call for animal studies to distance itself from the imperatives of animal advocacy. (Throughout the full two decades that the term 'animal studies' has now been in use in the arts, humanities, and social sciences, its relation to animal advocacy and to animal activism has been discussed and contested, and that shows no sign of changing.) Mine is a call, instead, for something rather closer to what Clifford Geertz called 'thick description' – especially in terms of his emphasis on ethnographic description as 'microscopic' and as characteristically based on what he called 'exceedingly extended acquaintances with extremely small matters' (1973: 21).

This may be a matter, for example, of going into a particular environment – an artist's studio, a museum's taxidermy collection – in order to look, and listen, attentively, patiently, before trying, as Garry Marvin puts it, 'to bring something back from that immersion [. . .] without meaning being imposed from the outside' (2011a). It's primarily about the gathering of information, intelligence, 'voice', and trusting in these and reporting on them. In the cases I've been discussing, it generally also involves shaping the terms with which to communicate what's happening in the realm of the non-verbal. Philosopher David Wood (who's also an artist) has a good phrase for this: he calls it 'presentational privilege' (2011).

How does this differ from the manner in which many cultural theorists and some animal studies scholars typically operate? It holds back from the rush to judgement, whether that judgement be aesthetic, political, or ethical. As the poet John Berryman scathingly put it in a quite different context: 'To be a *critic*, ah, / how deeper and more scientific' (1983: 401).

This is not, of course, to suggest that even the most painstaking description can ever be shorn of interpretation. Few would now question Geertz's assertion that ethnographic descriptions are necessarily 'themselves interpretations' and that they must thus also be understood as 'fictions, in the sense that they are "something made, something fashioned" ' (1973: 15). But they can represent a slowing-down, an inhibition of the urge to impose meaning or judgement or systematic order. This is where language matters. It may not be exactly what Geertz meant by 'thick description', but Poliquin's determined questioning of the very nature of her objects of study and (I hope) my own insistence on taking very seriously the often mistrusted voices of contemporary animal artists both edge towards a form of writing that is thick *with* description.

A writing of this kind is certainly one of the options available to animal studies. It does not need to be followed up immediately with evaluation – praise or blame – because it presents animal-things (difficult art objects or perplexing museum specimens) that are properly and necessarily open to multiple readings, and it need not be the job of animal studies to close off those readings, to close them down. It is a holding-open (like art's holding open). It does not settle matters. Even in relation to what may be seen as ethically complex or compromised exhibits, this refusal of closure is not an abnegation of responsibility, but rather the deliberate decision to take a specific form of action that presents that complexity without diminishing or diluting it, flattening it or deadening it. 'Coherence', Geertz reminds his readers, 'cannot be the major test of validity for a cultural description' (1973: 17).

Tom Tyler writes about this well in his fine book *CIFERAE*, which engages with the manner in which animals have figured in philosophical thought, where all too often, he notes, there is 'an implicit leveling at work':

> In deploying the notion of animality, or the animal, an inconceivable variety of living beings is flattened into a more manageable philosophical form. This compression of diversity, this indifference to difference, is described by David Wood as a 'deadening shorthand'.
>
> *(Tyler 2012: 41–2)*

These philosophical animals, Tyler observes, 'are often retained as mere ciphers, place-fillers who sustain an argument but remain faceless and interchangeable'. Or, employed as 'didactic exemplars', they end up as 'stereotypes, dulling fresh thinking by habitual repetition' (4).

It may be contentious to say so, but there is no lack of 'cipherous' writing, cipherous criticism, within animal studies: writing that doesn't feel the need carefully to look or to describe and instead just points in the general direction of, or gestures towards. It's evident, for example, in some of the writing about contemporary art that incorporates actual animals, living or dead. This is usually a sincere and ethically engaged but aesthetically flat writing that can sometimes amount to little more than a checklist of the ethical shortcomings of all-too-familiar examples – invariably including the dog allegedly left to starve to death as part of a gallery installation – and often accompanied by the assertion that such work cannot be 'art'. In making this observation, my point is not to defend or to condone these artworks. It is that this standardization of examples, which are discussed briefly but repeatedly (and by no means only in blogs, listservs, and online petitions), serves to dull and to level out both their differences *and* their heartlessness, their cruelty, their disengagement. Thick, detailed, attentive description of any one of them – or at least of any verifiable one of them – could have rendered that vivid, without the writer ever needing to voice their own strident condemnation of the work.

Turning back now to Rachel Poliquin, there is more to be said about the vivid, irrepressible visibility of the animal bodies that she characterizes as examples of 'bad' taxidermy:

> Bad taxidermy makes the craft of preservation too blatantly visible to inspire an emotive spectacle. [. . .] The same can be said of contemporary animal documentaries. Viewers do not want to see filmmakers at work. They want to experience the fantasy of a private view into nature's most intimate moments, as if they were unedited and unchoreographed.
>
> *(2012: 82–3)*

Here, bad taxidermy in the museum has very much the same effect as deliberately botched taxidermy, such as Angela Singer's, in the contemporary art gallery. Consciously or not, both seem to correspond to Jean-François Lyotard's useful definition of the postmodern as 'that which denies itself the solace of good forms' (Lyotard 1984: 81). And the notion of inspiring what Poliquin calls 'an emotive spectacle', 'a private view into nature's most intimate moments', is exactly the kind of uncritical, idealized conception of the nonhuman that has been so aptly criticized by Matthew Brower in the field of mainstream wildlife photography, which establishes what he calls 'a separation of human and animal that positions animals on the far side of a nature-culture divide' (Brower 2011: xvii). Brower sees in the banishing of the human from this genre of imagery 'a redeployment of the Garden mythos which envisions nature as pure and human beings as fallen and corrupting. Wildlife photography posits a vanishing nature corrupted by human traces that the photographer must work to overlook' (84).

It takes, perhaps, a deliberate clumsiness, a kind of falling over one's own feet, to bring human presence, human implication, human manufacture back into play while still keeping the life or death of the animal as the subject of the thing-being-made, whether it's a taxidermic object or a photographic image. In my own roadkill imagery over the past few years this has been something I've tried falteringly to address, and sometimes it's been done simply by having the bike on which I encounter these bodies while cycling the country lanes of Norfolk find its way into the image (Figure 21.3). In this particular image, the disconcertingly dislocated body of the pheasant is juxtaposed with the awkward-to-read image of the bike's strong shadows. It's not for me to say whether this even begins to undercut what might otherwise be regarded as the merely decorative (or gratuitously gory) display of dead animal form in splendid visual isolation, but at least it marks an actual encounter with a specific creature's lost life on one particular sunny spring morning.

The mere conjunction of the dead animal and the shadowy presence of its photographer is no guarantee of the nature of their relationship, of course. There exists, for example, a late nineteenth-century photograph by Carl Akeley, the hunter, scientific collector, and aspiring artist who would become assistant curator of the American Museum of Natural History, showing a dead Somali wild ass (Figure 21.4). It also includes the strikingly symmetrical shape of Akeley's own shadow, and, as Nigel Rothfels has noted, Akeley's memoir suggests that 'he wanted explicitly to implicate himself in this particular death' because he had come to doubt the legitimacy of certain kinds of collecting by means of hunting (Rothfels 2012: 134–5). Iconographically, there is a striking but entirely accidental similarity between the juxtaposed animal and 'human' elements in Akeley's image and my own. In each of them,

Figure 21.3 Steve Baker, *Roadside XXI*, 2012, photograph. Courtesy of the artist

the shadowy and rather ambiguous evidence of human presence may or may not be read as threatening to the adjacent animal bodies, and in both cases the shadow of the camera that links these shapes is interestingly absent.

The fact that the twenty-first-century cyclist behind the one camera is less likely to have been responsible for the animal's death than is the nineteenth-century collector behind the other camera does not diminish their shared interest in creating images of some kind of human implication in relation to the depicted animal bodies, and that of course raises legitimate questions about the nature of the photographers' interest in the animals themselves. Of Akeley's various photographs of animals killed during hunts, Rothfels writes: 'It is not that the animals don't matter in these works – the animals and their deaths are absolutely the central moral, emotional, and intellectual *matter*. But these works have never been about simply displaying death; they are attempts to make sense of death' (136).

Whether that assessment could also apply to my own roadkill work is an open question. It has never felt to me like an attempt 'to make sense' of animals' deaths – that would seem presumptuous, and too much like an attempt to settle things, to fix things. It is, undoubtedly, an attempt to *present* what certain killed animals looked like at a particular moment and that necessarily involves a shaping of things, a 'presentational privilege', which is where implication comes most clearly into play. The animal in the photograph (just as much as the taxidermic animal) could thus be said to be a *made* dead thing; not a thing made dead, but a dead thing, made. If it's to happen at all, the commemoration of a particular death – the attempt to keep that death 'visible' – will be more than the simple consequence of the camera happening to be

Figure 21.4 Carl Akeley, photograph of a Somali wild ass, 1896. © The Field Museum, #CSZ6011

pointed at the animal's remains. It has to be done from scratch, hacked out from scratch, *made* from scratch, in some weird blinkered process of lively reinvention: making the stone stony.

This is evidenced, I think, not so much in my own work as in this next example, in which encounters with roadkill also figure. 'I sometimes bump into things and make mistakes. I see a flat world.' These words are spoken by the dead dog Lily – who bumps into things because she has only one remaining eye – in artist Kathy High's *Lily Does Derrida*, which she subtitles 'A Dog's Video Essay' (High 2010–12).[5] In both its words and its imagery, it is a tale of flattened animals, thick language, and the collision of different deaths and different degrees of deadness. Loping around the house and garden, peering at the camera with her one remaining eye (Figure 21.5), and making the occasional disparaging remark about the two cats with whom she and High shared a home before her death, Lily bluntly announces, near the start of the video, 'I'm dead'. (She speaks in the voice of a human male, to dislocate things just a little further.) As she pads around the house, she muses on her condition: 'I've been reading a lot these days, and Jacques Derrida has caught my interest. Too bad he died, but then – I'm dead too, by the way.' And she amuses herself with *his* condition: ' "The animal," he says, "which is at unease with itself." We are not so *uneasy*. What is it about human animals which is so *uneasy* with us animal types?'

Soon after this, the first section of the video's roadkill footage appears on the screen (Figure 21.6). A small mammal lies in the road, dead, it seems, its mouth full of maggots,

Figure 21.5 Kathy High, video still from *Lily Does Derrida*, 2010–12. Courtesy of the artist

though an arrow momentarily appears on screen to point to its still-twitching back leg, as a large fly feasts on its face. And scrolling across the screen, as if to caption the spectacle, come Derrida's words, simultaneously voiced in dead dog Lily's male American accent: 'Derrida asks, "Does the animal dream?" Another way of asking, "Does the animal think?" "Does the animal produce representations?" ' And a string of others: ' "Does it die?" ', ' "Does it invent?" ', and so on (Figure 21.7). But far from the words framing how the animal imagery is seen, that blistering image of the roadkill and the insects spills over unstoppably into Derrida's famous text (2008), as does Lily's voice, now lurking in its pages for future readers.

The dead animal, at least in art's spaces, won't stay dead, stay down, stay still. It's almost the opposite of the Ted Hughes poem, 'View of a Pig', which begins: 'It was too dead. Just so much / A poundage of lard and pork. / Its last dignity had entirely gone' (Hughes 1997). And that question of dignity, which in my view High's video handles with such confidence and wit, is also important to Rachel Poliquin, who repeatedly, and I think unusually, insists on pointing back to the individual lives from which museum specimens derive. She writes, for example: 'the animal-thing on display was once a sentient creature: at once lifelike yet dead, both a human-made *representation* of a species and a *presentation* of a particular animal's skin' (Poliquin 2012: 41). And again, elsewhere: 'I would argue that these beasts deserve more respect because despite the death, despite the taxidermy, they remain pieces of the creaturely existence that all animals share' (222).

This view is not, of course, universally shared, even among artists. Polly Morgan, in a newspaper item that chose to herald her as 'the artworld's most famous taxidermist', was quoted as

Figure 21.6 Kathy High, video still from *Lily Does Derrida*, 2010–12. Courtesy of the artist

Figure 21.7 Kathy High, video still from *Lily Does Derrida*, 2010–12. Courtesy of the artist

Figure 21.8 Angela Singer, *Unexplained Recovery*, detail, 2010, vintage taxidermy stag and mixed media. Courtesy of the artist and the Weisman Heyman collection, New York

saying: 'Some people don't like you messing with dead animals but respecting the dead body of an animal is silly and hysterical' (quoted in Jones 2012: 11). Equally, as Garry Marvin has shown, the language of 'dignity and respect' is also explicitly employed by some hunters, with complete sincerity, in characterizing their own approach to the killing of the hunted creature and to its subsequent display (Marvin 2011b: 208). In both its making and its meanings, then, the taxidermic animal – 'this animal-thing' – is almost the epitome of the kind of unstable object that is so highly regarded in contemporary art.[6] In significant part, this is intimately tied to the thing's having-been-a-living-being (or, at least, to some of its materials having recognizably been part of such a being). The stubborn liveliness *and life* of the dead animal, the animal-thing in the museum, is not easily suppressed. As Poliquin puts it: 'this static thing in a very real sense is an animal still: the eyes may be glass, but the animal stares back' (50).

With that comment very much in mind, here is one final example of Angela Singer's work (Figure 21.8). The artist has written as follows about this piece and the series from which it derives:

> *Dead-Eyed* is a series of recycled taxidermy works with eyes that I have altered to make the dead animal look deader than dead. Viewers are so used to looking at taxidermied dead animals that they don't see the death anymore. I'm exploring ways to put more dead in the dead. *Unexplained Recovery* has a blind eye based on post-mortem clouding. It's not an accurate representation of a dead deer eye; it's more what someone might imagine a dead eye to look like.

And she speculates: 'When a viewer looks at a taxidermy glass eye they see what they expect to see – something that looks like an eye. I wonder if the viewer will feel uncomfortable to see instead a dead eye. Will they have a more emotional response? Will the viewer find the work creepy despite the animal body being covered in beautiful jewels? How will they feel about looking at the animal and the animal unable to return their stare?'[7]

Half-way through *The Breathless Zoo*, Poliquin summarizes her book's project by explaining: 'I am interested in the poetics of animal order' (2012: 115). Singer's playing with what viewers 'expect to see' is just one more example of contemporary art's handling of that acute, enlivening poetics.

Notes

1 The conference Activating Stilled Lives: The Aesthetics and Politics of Specimens on Display, organized by Mechthild Fend and Petra Lange-Berndt, was held at University College London in May 2012. A version this chapter was presented there as an invited paper.
2 Unless otherwise indicated, all quotations from Angela Singer in this paragraph and the next are drawn from her occasional correspondence with the author since 2001.
3 Here, and in the paragraphs that follow, all quotations from Angela Singer relating to the discussion of *Spartle* are drawn from the author's interview with the artist, Wellington, New Zealand, April 2010. These paragraphs on *Spartle* also appear as part of a more substantial assessment of Singer's work in Chapter 6 of *Artist | Animal* (Baker 2013).
4 Sadly, this resonant phrase was dropped at some point in the editing of Poliquin's manuscript.
5 All quotations from Lily's voiceover have been transcribed by the author.
6 For an interesting discussion of contemporary art's distancing itself from the idea of the 'stable object', see Boetzkes 2010: 20–4.
7 Angela Singer, email correspondence with the author, April 2012.

References

Aloi, G. (2008) 'Angela Singer: Animal Rights and Wrongs', *Antennae: The Journal of Nature in Visual Culture* 7: 10–17.

Baker, S. (2013) *Artist | Animal*, Minneapolis and London: University of Minnesota Press.

Barthes, R. (1982) *Camera Lucida: Reflections on Photography*, R. Howard (trans.), London: Jonathan Cape.

Berryman, J. (1983) 'Olympus', in Geoffrey Moore (ed.), *The Penguin Book of American Verse*, revised edition, London: Penguin, pp. 400–401.

Boetzkes, A. (2010) *The Ethics of Earth Art*, Minneapolis: University of Minnesota Press.
Brower, M. (2011) *Developing Animals: Wildlife and Early American Photography*, Minneapolis: University of Minnesota Press.
Burns, A. (2008) *Little Constructions*, New York: Harper Perennial.
Derrida, J. (2008) *The Animal That Therefore I Am*, D. Wills (trans.), New York: Fordham University Press.
Geertz, C. (1973) 'Thick Description: Towards an Interpretive Theory of Culture', in *The Interpretation of Culture: Selected Essays*, New York: Basic Books, pp. 3–30.
High, K. (2010–12) *Lily Does Derrida: A Dog's Video Essay*, DVD, 30 minutes.
Hughes, T. (1997) 'View of a Pig', in Paul Muldoon (ed.), *The Faber Book of Beasts*, London: Faber, pp. 275–276.
Jones, A. (2012) 'Silicon Pigs? Stags Stuffed with Bats? It's Taxidermy, But Not as We Know It . . .', *Independent*, 19 May, 'Independent Radar' section.
Le Fur, Y. (2006) 'Displaced Objects on Display', in J. Onians (ed.), *Compression vs. Expression: Containing and Explaining the World's Art*, New Haven, CT: Yale University Press, pp. 6–10.
Lyotard, J. (1984) *The Postmodern Condition: A Report on Knowledge*, G. Bennington and B. Massumi (trans.), Manchester: Manchester University Press.
Marvin, G. (2011a) 'Beastly Anthropology: Working in Human-Animal Worlds', inaugural professorial lecture, Roehampton University, London, 10 May.
—— (2011b) 'Enlivened through Memory: Hunters and Hunting Trophies', in S.J.M.M. Alberti (ed.), *The Afterlives of Animals: A Museum Menagerie*, Charlottesville: University of Virginia Press, pp. 202–17.
Poliquin, R. (2012) *The Breathless Zoo: Taxidermy and the Cultures of Longing*, University Park: Pennsylvania State University Press.
Rothfels, N. (2012) 'Trophies and Taxidermy', in J.B. Landes, P. Young Lee, and P. Youngquist (eds), *Gorgeous Beasts: Animal Bodies in Historical Perspective*, University Park: Pennsylvania State University Press, pp. 117–36.
Singer, A. (2011) 'Angela Singer: Artist Interview', YouTube video, uploaded by TheRealArtRoadshow on 26 May 2011, accessed 29 May 2012.
Tyler, T. (2012) *CIFERAE: A Bestiary in Five Fingers*, Minneapolis: University of Minnesota Press.
Wood, D. (2011) 'Animal Architects: The Bangladeshi Sand Crab', keynote presentation at the conference The Animal Gaze Returned, London Metropolitan University, London, October.

Index

For Product Safety Concerns and Information please contact our EU
representative GPSR@taylorandfrancis.com
Taylor & Francis Verlag GmbH, Kaufingerstraße 24, 80331 München, Germany

www.ingramcontent.com/pod-product-compliance
Ingram Content Group UK Ltd.
Pitfield, Milton Keynes, MK11 3LW, UK
UKHW050051200926
475702UK00019B/2778

9780367866716